T0320642

PHYSICS IN A
TECHNOLOGICAL
WORLD

PHYSICS IN A TECHNOLOGICAL WORLD

XIX General Assembly
International Union of
Pure and Applied Physics

**ANTHONY P. FRENCH
EDITOR**

L.C. Catalog Card No. 88-82612
ISBN 978-0-88318-591-9

Contents

Preface

The XIX General Assembly of the International Union of Pure and Applied Physics (IUPAP) was held at the National Academy of Sciences, Washington, D.C., during the period September 28–October 2, 1987. This was the first General Assembly to be held in the United States since the XIV Assembly (also at the National Academy) in 1972.

The 1987 Assembly was made the occasion of a special symposium under the joint sponsorship of The American Physical Society, the American Institute of Physics, and The National Academy of Sciences, and conducted as a joint activity of IUPAP and the Corporate Associates of AIP. A National Organizing Committee was formed under the chairmanship of D. Allan Bromley, President (1984–87) of IUPAP, with representation from APS, AIP, and the National Research Council. Financial support was provided in the form of grants from the National Science Foundation and the Department of Energy, plus contributions from the AIP Corporate Associates.

The symposium itself, with the theme "Physics in a Technological World," occupied the period September 30–October 2. In addition to the formal sessions, there was a reception and dinner in the evening of October 1. Tours were also arranged to the Goddard Space Flight Center and the National Bureau of Standards.

The various articles in this volume represent almost all of the presentations made at the symposium. There is no need to comment on them in this Preface; they provide their own testimony to the richness and vitality of physics today, and to the immense variety of ways in which physics and technology are interacting, with benefits flowing in both directions. The thanks of the organizers go to all the speakers who reported on their work at the many frontiers of physics and technology. It is interesting to note that, although the majority of the nineteen persons who gave papers now live and work in the United States, they represent twelve different countries of origin, as befits an international gathering of this kind.

The production of this volume is the work of the Books Division of the American Institute of Physics, under the management of Rita G. Lerner, aided by Michael Hennelly. The format of the book has been designed to make it a companion volume to *Physics 50 Years Later*, which is a collection of the papers presented at the XIV General Assembly of IUPAP—and which commemorated the 50th aniversary of the foundation of IUPAP itself. It is hoped that the present collection will be of both current and lasting value to anyone interested in the state of physics, both pure and applied, in 1987.

A. P. French
Massachusetts Institute of Technology
June 1988

IUPAP & AIP Corporate Associates

OPENING SESSION: Welcome and Introductory Remarks

(PRESIDING: Larkin Kerwin, President Designate of IUPAP)

WELCOME BY THE PRESIDENT OF THE NATIONAL ACADEMY OF SCIENCES (FRANK PRESS)

On behalf of the National Academy of Sciences and the United States Liaison Committee for IUPAP, chaired by Melvin Gottlieb, I, some time ago, invited IUPAP to hold its 19th General Assembly in the United States. That has now come to pass, and it is a pleasure to welcome you all as our guests. In particular, in a number of categories, I should like to welcome the IUPAP leadership: President Allan Bromley; Secretary-General Jan Nilsson; Larkin Kerwin, the President-Elect, and some 45 delegations from all over the world. A special welcome to the representatives of The American Physical Society and the American Institute of Physics which did all the work in arranging this convocation. It is an arrangement that worked very well between the Academy and the APS and the AIP—all of this under the supervision of the National Organizing Committee chaired by Allan Bromley. If the program and the activities associated with this General Assembly are successful, as I know they will be, I think the credit goes to the groups that I have mentioned.

It is also appropriate—and a sign of our times, you might say—that the AIP Corporate Associates are involved. From our own country's point of view it brings together U.S. corporate research leaders, and also the industrial physicists doing important research, with the academic researchers in physics. It is also, I think, a sign of the times in most countries represented here today—a sign of increasing relationships and liaisons between the industrial communities and the academic communities. The pace of scientific discoveries is so rapid, and the shortening of time between discovery and commercialization is a new phenomenon of our times; so this relationship, this connection is, I think, particularly appropriate.

I urge you all to enjoy the facilities of the Academy and this building, which is an important place in Washington, our nation's capital, for the scientists of this country. Please feel free to look at our art exhibits and, of course, the Einstein statue.

I hope you will take this opportunity to learn more about the National Academy of Sciences and the National Research Council, which is a remarkable organization. And, if you want to know more about what we do, please ask. Your American colleagues here in the audience will answer your questions or send you to me or my colleagues in this building so that you can learn more about what we do.

We are, of course, concerned about the health of our own institutions of science in the United States. But equally we are concerned with the global aspects of science in terms of freedom of communication, access of scientists to each other across boundaries, and the potential contributions of the world scientific community to global issues of concern to peoples everywhere. You have seen the recent volumes entitled *Physics Through the 1990s* outside. But we produce similar volumes for many different fields. If you want to learn in a one-hour reading session about the opportunities and the excitement in 30 or 40 different fields of science, please get a copy of our briefings; this is across all the sciences—biological, chemical, applied sciences, and certainly physics.

Your theme, "Science in a Technological World," is appropriate and important. Again, it is a sign of our times; and to see an organization like IUPAP involved with current issues in science, in technology, is to me a very healthy sign about the vitality of the physics community across the world.

I will never forget a conversation I had as a young professor with Murray Gell-Mann who was also a young professor at Cal Tech. We were in different departments, and most of you know Murray, and so will not be surprised at the way the conversation went. He said: "Frank, you're in the wrong field. There's only one field of science, and that's physics; because if you're a physicist, you can do anything—biology, chemistry—because it all goes back to first principles, and that's what physics is all about." You are in a great field, and I thank you for being here.

INTRODUCTORY REMARKS BY THE PRESIDENT OF
THE AMERICAN PHYSICAL SOCIETY (VAL L. FITCH)

On behalf of the 40 000 members of The American Physical Society I want to extend to all of you the warmest possible welcome. We are very pleased to be cosponsors of this General Assembly of IUPAP and this meeting of the Corporate Associates of the American Institute of Physics.

I want to take this opportunity to tell you something about The American Physical Society. It is the principal professional organization of physicists in the United States. The Society has two main functions. One is to organize meetings (about 20 per year) in which recent developments in physics are discussed. The other principal function is publication—publication of *The Physical Review*, *Physical Review Letters*, and *Reviews of Modern Physics*, and also *Bulletin of*

The APS in which abstracts of papers at the meetings are presented. These activities account for most of the budget of the Society which is about $12 million per year.

The Society through panels and committees, however, engages in many other activities which are highly significant. There is a Panel on Public Affairs which concerns itself with those important public issues that have a substantial physics content. Over the past 12 years this panel has initiated a number of studies that have provided an important input in the formulation of public policy. I mention, as examples, a study on reactor safety, which came out in 1975; then, after the famous Three Mile Island accident, a study of the radionuclide release from severe accidents at nuclear power plants.

Most recently, just published (July 1987) in *Reviews of Modern Physics*, is a study of Directed Energy Weapons. The work of the panel that produced this report followed the pattern established by the previous studies. A group of 16 recognized experts with unimpeachable qualifications was assembled, under the co-chairmanship of Professor Nicolaas Bloembergen of Harvard University and Dr. Kumar Patel of the AT&T Bell Laboratories. With the full cooperation of the SDI Office in Washington they were briefed, over a period of 18 months, on the details of the relevant programs, and then wrote a report which, in due time, was declassified. The intent has been to provide a kind of *vade mecum* on the subject of the science and technology of directed-energy weapons. The report is intended for people with technical training—physicists, engineers, scientists of all kinds—in order that they may have the facts with which to make their own judgments and also to prepare them for discussion of the issues with their fellow citizens.

As with all other studies that the APS has sponsored, the report, before release, was reviewed by an independent committee of experts. In this case the committee was composed of six individuals who had achieved high distinction for their work in relevant areas: particle accelerators, particle beams, and lasers. Also, considerable experience in directing large technical enterprises was present in the review committee. Among the six members were four ex-presidents of the Society.

The objective of the study was a technical report that was as free of subjective opinion as humanly possible, in the tradition of the best of physics. It was an enormous effort and I believe that the panel has succeeded brilliantly. The report was funded by the Carnegie Corporation and the MacArthur Foundation.

In addition to these special studies the APS has a staff member whose major effort is directed towards education, as well as a committee dedicated to improving physics education in general. There is also a committee whose special concerns are women in physics. Getting more women into the field is a special problem. With so few women going into physics, at least in this country, one is losing almost half the talent pool. There is a parallel committee on Minorities in Physics—a similarly under-represented portion of the population.

And, since this is a meeting of the International Union of Pure and Applied

Physics, I especially want to mention the work of the Subcommittee on International Scientific Affairs. Most recently this committee has proposed that a workshop on nuclear and particle physics be held which would provide a forum for physicists in the Eastern Mediterranean region. The study of physics, the doing of physics, is an activity that has no political boundaries, and this proposed workshop is intended to exploit that fact—to bring people together who otherwise would not have the opportunity. Of course, in this regard, I am very sensitive to the importance of international meetings. We are all indebted, enormously so, to IUPAP for its effectiveness in promoting international communication in physics.

Another important committee is one devoted to International Freedom of Scientists. Its charge is to be responsible for monitoring concerns regarding human rights for scientists throughout the world.

Last November (1986) the Council of The American Physical Society adopted a statement which has elements that I believe to be the essence of IUPAP. Among other things the statement said:

- We believe in the principle that science belongs to all humanity and transcends national boundaries. Science should serve, and has served in the past, as a bridge for mutual understanding and peace in a divided world.
- We believe that good science requires open communication. Committed to international scientific exchanges, we also affirm our traditional strong concerns for human rights and for freedom of expression in the conduct of science.

These words point to the abiding common interest of The American Physical Society and the International Union of Pure and Applied Physics, and also, I know, of the AIP and the Corporate Associates. It is one of the principal reasons for the great enthusiasm we have in being a cosponsor of this very interesting meeting.

INTRODUCTORY REMARKS BY THE CHAIRMAN OF THE GOVERNING BOARD OF THE AMERICAN INSTITUTE OF PHYSICS (HANS FRAUENFELDER)

For the American Institute of Physics I welcome all of you here to the Symposium on "Physics in a Technological World." Let me start by revealing one of the best-kept secrets that even most physicists are not aware of: The American Institute of Physics (AIP) and The American Physical Society (APS) are not identical! APS is the professional society of physicists; its members are individual physicists. AIP has no individual members; it consists of ten member societies, among them APS. One of its main functions is service for the ten member societies; another is support for physics education. In any case, I hope that you keep these organizations—both serving physics, physicists, and society—apart.

In looking at the program and the titles of the lectures of the IUPAP symposium, the impressions that I gained on reading the recently published Brinkman Report are reinforced. The Brinkman Report, prepared under the auspices of the National Academy of Sciences, covers most subfields of physics and surveys also some border areas. Most of us work in a small part of physics and are often tempted to assume that, whilst our own field is still healthy and vigorous, with lots of interesting problems, most of the other fields are slowly dying. I can vouch for this situation. I have talked to condensed-matter physicists who feel that nothing much has happened in elementary-particle physics in the last ten or twenty years, and I have talked to elementary-particle physicists who felt that beyond the work leading up to, say, the transistor, not much has happened in condensed-matter physics! Of course I am exaggerating a little, but it really is difficult both to work in a particular field and to keep up with what is happening elsewhere. But the Brinkman Report, like the earlier Bromley Report, reveals an enormous vitality in every branch of physics—even branches that had been given up as dead have come to life again and produced exciting results. Three aspects are particularly striking. First: fundamental discoveries are being made at a surprising rate. Second: The interactions with all neighboring fields, from mathematics to the life sciences, are becoming stronger and operate in both directions. Third: the impact of physics on technology continues to grow, and does so in every field. These three aspects are clearly evident at this meeting.

There are, however, also some clouds in the sky; I will mention only one. As physics becomes deeper and more sophisticated, the gap between us and even the educated layman is becoming larger. It is more and more difficult even for the educated layman to understand what is happening in physics and to experience the beauty of our science. Even journals designed for this purpose, such as *Scientific American*, cannot bridge this gap. We, therefore, must do everything we can to explain physics and its wonders to a broad audience, and we must also strive to attract more young students to the challenges of our field. If we fail, physics will become even more, for most people, a black box of often evil magic, rather than a source of new knowledge and understanding. The lectures at this symposium will, I hope, help to build bridges to other communities and explain some of the recent progress to a wide audience.

I should like to thank all those who have worked hard to make the meeting a reality, in particular Allan Bromley, Bill Havens, Bill Koch, and Don Shapero. I also thank the National Science Foundation, the Department of Energy, and the Corporate Associates of the American Institute of Physics for financial support.

Finally, as you have no doubt noted, an unprecedented event has happened. The first three speakers have all used less than their allotted time. We donate this time to Allan Bromley for his survey of physics!

FIGURE 1 Opening session of the Symposium, September 30, 1987 (left to right): Val Fitch, Frank Press, Hans Frauenfelder, and Larkin Kerwin. *Cecelia M. Brescia, AIP*

FIGURE 2 Outside the National Academy of Sciences during a break in the sessions. *Cecelia M. Brescia, AIP*

FIGURE 3 Allan Bromley in conversation. *Cecelia M. Brescia, AIP*

The State of Physics—A Tour
d'Horizon 1987/ D. ALLAN BROMLEY

1. INTRODUCTION

It has become traditional, in these IUPAP General Assemblies, for the retiring president to undertake an overview of the status of our science and of the notable developments that have occurred in the three years since the last General Assembly. There have been few, if indeed any, three-year periods that have presented more of a challenge, in such an overview, than these past three.

Physics, in terms of intrinsic promise of qualitatively deeper understanding, in terms of far-reaching new surprises, and in terms of equally far-reaching applications—throughout science and technology and into a great many facets of modern society—has never been stronger or more exciting.

Let me begin by putting this into context. We forget what enormous progress has been made during the past century in our understanding of nature. Figure 1 illustrates this with a few specific examples. It was only one hundred years ago that Henrich Hertz demonstrated that electromagnetic waves, crucial to much of modern civilization, were real and more than a mathematical figment of Maxwell's imagination. Seventy-five years ago, Rutherford's discovery of the nucleus and Onnes's discovery of superconductivity opened up vistas that are still very much alive and expanding; and fifty years ago Whittle's invention of the jet engine and Aitken's pioneering work on digital computers changed, irretrievably, the nature and scope of our lives.

The past three-year period—and this past year in particular—has been one of *supers* as listed in Fig. 2. I shall touch very briefly on each of these in this overview, and obviously my selection of topics will be a highly personal one.

D. ALLAN BROMLEY *began his life in Canada, but has spent most of his career at Yale University, working in the field of nuclear physics and serving as director of the A. W. Wright Nuclear Structure Laboratory. He has been extremely active in both national and international aspects of physics. This paper is his farewell address upon the completion of his term as President of IUPAP.*

PHYSICS

100 YEARS AGO

Hertz discovers electromagnetic waves.

75 YEARS AGO

Rutherford discovers the atomic nucleus

Onnes discovers super conductivity

50 YEARS AGO

Whittle invents the jet engine

Aitken begins work on a digital computer

Anderson discovers mesons in cosmic
radiation

Public television is introduced in the U. K.

FIGURE 1   Major developments in physics and in its applications in the past century.

2. CORNERSTONES OF PHYSICS

Two of the fundamental cornerstones of modern physics, of course, are relativity and quantum mechanics. Although generally taken for granted, both remain under fundamental study as new experimental techniques become available. Figure 3 shows the results of a recent repeat of the Michelson–Morley experiment in which the second-order Doppler shift of radiation from aligned ^{9}Be ions was monitored as the ions were carried in the Earth's rotating frame. During this rotation, the direction of the magnetic field rotates relative to the Earth's velocity through the preferred frame, if indeed it exists, and any violation of Lorentz invariance would cause the frequency of the Doppler shift to depend upon the direction and thus to vary periodically relative to the hyperfine frequency of unpolarized atoms. Such measurements now constrain the preferred frame parameter $[(c_0/c)^2 - 1]$ to be less than 3×10^{-22}, a result that is a factor of 10^6 smaller than that from the earlier classic Hughes–Drever experiments. It is perhaps appropriate, on this hundredth anniversary of the Michel-

A YEAR OF SUPERS

SUPERCONDUCTIVITY *- high temperature*

SUPERNOVA *- 1987 A - Magellanic Cloud*

SUPERCONDUCTING SUPERCOLLIDER *- Presidential*

approval

SUPERCOMPUTERS *- keys to understanding*

SUPERSYMMETRY *- beyond the GUTS*

SUPERSTRINGS *- the way to unification?*

FIGURE 2 The year 1987 has been characterized by a remarkable number of supers. Six of them are listed here.

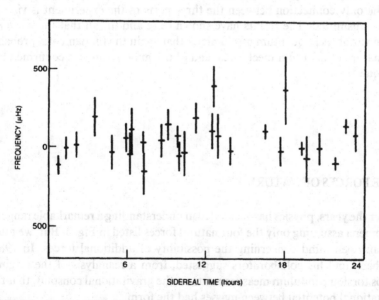

FIGURE 3 Recent results obtained at the National Bureau of Standards on the second-order Doppler shift of radiation from ^{9}Be ions in flight in the Earth's rotating frame. The frequency shift here is plotted in μHz as a function of sidereal time and, over the 24-hour period shown, the upper limit on the amplitudes of 12- and 24-hour variations is approximately 70 μHz. (Courtesy of J. D. Prestag, US National Bureau of Standards).

son–Morley experiment, to find it living on in new versions, confirming to ever-higher precision one of the deepest principles of physics.

As late as March 1947, however, it was Einstein who noted,

> *I cannot seriously believe in [the quantum theory] because it cannot be reconciled with the idea that physics should represent a reality in time and space, free from spooky actions at a distance.*

His problem dated back to the classic Einstein, Podolsky, and Rosen (EPR) paper of 1935 where the "spooky actions at a distance" are the acquisition of a definite value of a property by a system in one region by virtue of a measurement carried out in another of its regions. The EPR paper posed one of the classic *gedanken* experiments of physics, quite secure in the understanding that it would forever be safe from experimental examination, and an impressive literature has evolved from it on so-called *hidden variables* in quantum mechanics—variables introduced to reclaim the determinacy whose loss troubled Einstein. But, as illustrated in Fig. 4, using new technology and impressive ingenuity, Alain Aspect and his colleagues at Orsay have made the EPR experiment a reality. In essence, the experimental arrangement involves two detectors, *A* and *B*, and a source of particles, photons, or whatever, *C*. The source *C* is assumed to simultaneously emit information-carrying entities in the directions of *A* and *B*. Obviously, it is possible to prevent detection by interposing a shield between the source and a given detector, and it is also possible to increase the delay before a given detector fires by moving it farther from the source. It is important to note that the only connection between the three parts of the experiment is via the entity transmitted. The results have shown once and for all that there are *no* hidden variables in quantum mechanics—that again the fundamental probabilistic nature of quantum mechanics, and of the microcosm, are confirmed beyond question.

3. THE FORCES OF NATURE

Over the years, physics has succeeded in understanding a remarkable range of phenomena assuming only the four natural forces listed in Fig. 5. But, we must keep an open mind concerning the possibility of additional forces. In 1986, Fischbach and his collaborators suggested, from a reanalysis of the original Eötvös torsion pendulum measurements on the gravitational constant, that the gravitational potential between masses had the form

$$- G_0 \frac{m_1 m_2}{r}(1 + \alpha e^{-r/\lambda}) \, ,$$

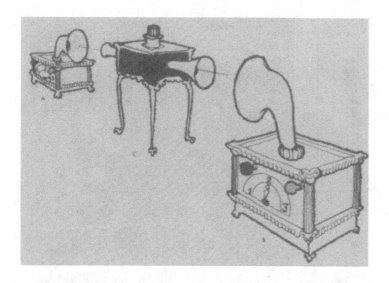

FIGURE 4   An experimental study of the Einstein–Podolsky–Rosen experiment. The upper panel is a very highly schematized and simplified version of the experiment, while the lower panel shows the actual experimental instrumentation in Alain Aspect's laboratory in Orsay, France. In the actual experiment, individual calcium atoms serve as the source and are excited by two lasers, visible here on either side of the photograph. The re-emitted photons then travel 6 m through evacuated tubes to polarization-sensitive detectors. (The upper figure is used with the permission of N. D. Mermin and the photograph was supplied by Alain Aspect.)

where G_0 is the value of the gravitational constant appropriate to large distances and α and λ are constants, the second of which is determined by the mass of the new field particle. Current evidence is confused. Geophysical and astrophysical data suggest that $100 \text{ m} \leqslant \lambda \leqslant 100 \text{ km}$, but recent studies by Thieberger at Brookhaven and by Adelberger at Seattle give conflicting results—respectively, positive and negative in the range $\leqslant 1 \text{ km}$.

One of the central thrusts of physics over the years has been the search for unification of the natural forces—for their understanding as different aspects of

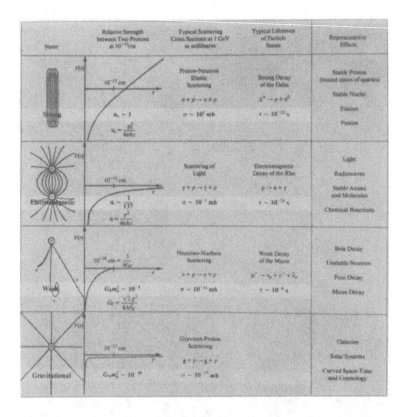

FIGURE 5 The four basic forces of nature are listed here together with an estimate of their relative strength as measured between two protons at one Fermi separation, typical cross sections for scattering induced by these forces at 1 GeV, typical lifetimes of particle states governed by these forces, and finally, several representative effects.

one superforce governing the behavior of all matter in our universe. This was Einstein's dream, and significant progress is being made toward realizing it. A hundred years ago, Maxwell succeeded in unifying electricity and magnetism; within the past decade Weinberg, Salam, and Glashow's suggested unification of electromagnetism and the weak nuclear force has received overwhelming experimental support. Current theories, as shown in Fig. 6, suggest that at sufficiently high energies (10^{17} GeV) and temperatures (10^{30} K) the electroweak and strong nuclear forces come together—as they are presumed to have done before 10^{-35} sec after the Big Bang. It is the hope that at still higher energies and temperatures a further unification with gravitation occurs, but this region, inside the Planck era in the original universe, is still shrouded in mystery.

As is obvious here we have come full circle from the study of the very large—cosmology—to that of the very small—elementary particle physics—in a spectacular demonstration of the unity of physics and of science. Let me then begin with a quick overview of recent progress in particle physics.

4. ELEMENTARY PARTICLE PHYSICS

Elementary particle physics, perhaps more than any other subfield, pushes the technological frontiers and to reach its frontiers requires major accelerators. Figure 7 shows the almost-completed Stanford Linear Collider—designed effectively as a Z^0 factory—to make possible detailed studies on the electroweak interaction. Such electron–positron colliders have the great advantage that quantum electrodynamics gives us a remarkably detailed understanding of the electromagnetic interaction and thus simplifies analysis of collision events free from the additional complexities that are present with hadronic probes. Figure 8 shows such a collision, as recorded at the PETRA accelerator complex in Hamburg, and illustrates the now-familiar jets which provide one of the most direct pieces of evidence for the physical existence of quarks and antiquarks produced in $e^+ - e^-$ collisions. If the particles arose directly from the annihilation they would be expected to follow widely diverse paths, and the fact that they are focused into the jets suggests that each develops from a single precursor—either a quark or an antiquark.

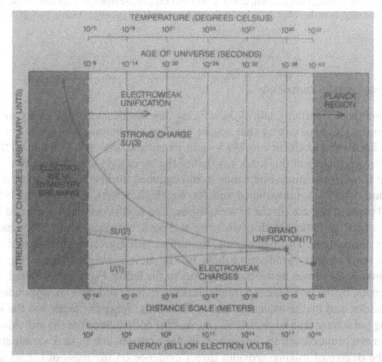

FIGURE 6   At sufficiently high temperatures (10^{30} K), or sufficiently small separation (10^{-33} m), or high enough energy (10^{17} GeV), the electroweak and strong forces are believed to unify. Gravity does not enter this unification until one enters the Planck region above temperatures, times, distances, and energies of 10^{32} K, 10^{-43} sec, 10^{-35} m, and 10^{19} GeV, respectively. (Courtesy of M. B. Green.)

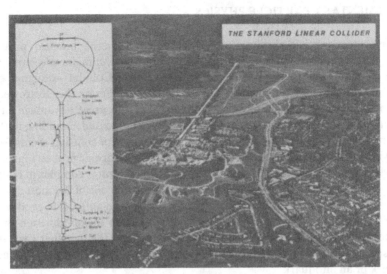

FIGURE 7   An aerial view of the Stanford Linear Collider with a schematic illustration of the collider layout on the left. The two-mile Stanford Linear Accelerator appears in the center of the photograph and the experimental hall constructed about the collider collision point is shown in the foreground slightly to the left of the axis of the accelerator itself. (Courtesy of Martin Perl, SLAC.)

A. Time Reversal Invariance

Again from Hamburg, one of the striking new discoveries of the past year is the amazingly large mixing that occurs between the bottom-flavored, B^0 meson—discovered at Cornell in 1983—and its antiparticle $\overline{B}^0$. Previously, the only known mixing of this kind was that of the more than ten times lighter K_0 meson from whose study Fitch and Cronin deduced the first clear-cut evidence for time reversal noninvariance and for which they were awarded the 1980 Nobel prize in physics. In the new experiment an upsilon particle, produced in the annihilation of the colliding electrons and positrons, decays into a B^0 meson and its antiparticle. Through mixing, the antiparticle converts into a second B^0 meson and the decay products of those two B^0 mesons are identified in Fig. 9. The mixing, as shown in the box diagrams to the left of Fig. 9, is mediated, in both cases, by the three $2e/3$ charged quarks u, c, and t and their corresponding antiquarks, and by the exchange of two W bosons. From the facts that the B^0 mixing parameter is as large as 20% and that the contributions of the individual quarks are proportional to the squares of their masses, the t quark dominates and the degree of mixing provides direct evidence on the mass of the so far unobserved t quark. Beyond this, these new data provide vital new information on quark mixing (specifically on the V_{td} element of the Kobayashi–Maskawa mixing matrix) and on time reversal invariance. Although not yet fully analyzed, these preliminary results are reproduced by the so-called standard model

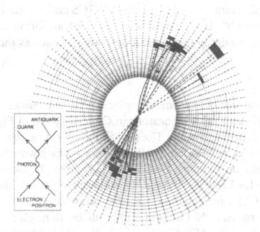

FIGURE 8   Typical jet structure resulting from high energy electron–positron collisions—as measured in the JADE detector of the PETRA accelerator at DESY in Hamburg. The collision region is surrounded by a 2.4-m cylindrical detector and the particles resulting from the collision are detected in the individual cells of this cylindrical array. Charged particles are shown here as solid lines and neutral particles as dashed ones. The inset schematic shows the interaction that is assumed to have taken place. (Courtesy of Chris Quigg, Fermilab.)

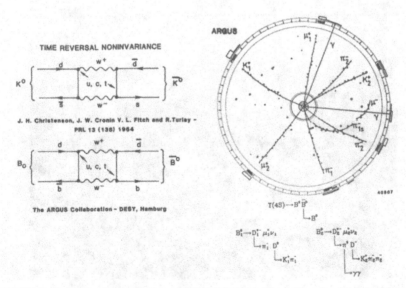

FIGURE 9   Recent results—again from the DESY facility in Hamburg—showing the mixing of the bottom-flavored B^0 meson and its antiparticle. On the right is shown the experimentally observed event detected in the ARGUS detector. On the left of this figure is shown the symmetry between the mixing of the strange flavored K_0 meson and of the bottom-flavored B^0. Until now, the mixing of the K_0 meson provided the only experimental access to time reversal noninvariance. (Courtesy of Roy Schwitters, Harvard.)

of elementary particles (see Fig. 17 below). The right panel of Fig. 9 shows a particularly simple event, recorded by the ARGUS collaboration, in which an upsilon decays into a B^0-$\overline{B^0}$ pair of which the latter transforms into a second B^0 and the unambiguous decay products of both are recorded as shown.

B. New Facilities

Some idea of the size of elementary particle physics facilities is provided by the aerial view of the CERN laboratory in Geneva shown in Fig. 10 with the Geneva airport in the foreground. The superimposed white ring shows the current LEP (Large Electron Positron collider) ring—which would also house the proposed large hadron collider (LHC) at 8 TeV currently under discussion at CERN. In Phase I, LEP will achieve a beam energy of 50 GeV and, in Phase II, 100 GeV. TRISTAN, the Japanese electron–positron collider, operates at 50 GeV and may be raised to 60 GeV but will be unable to reach the threshold for $W^\pm$ and Z^0 production. HERA, the new facility at Hamburg, will collide 30 GeV electrons with 800 GeV protons; BEPC, the Beijing electron–positron collider, will have a beam energy of 2.8 GeV; and UNK, a 3 to 5 TeV proton accelerator, is under construction at Serpukov in the Soviet Union.

FIGURE 10   An aerial view of the CERN facilities in Geneva, with the Geneva International Airport in the foreground. The large white ring superimposed on this aerial view is the tunnel location for the LEP (large electron/proton collider) and will also serve as the tunnel for the Large Hadron Collider now under discussion—if and when it is approved for construction. The smaller rings show the existing proton synchrotron and super proton synchrotron that have been in use for many years at CERN. (Courtesy of CERN.)

Detection instrumentation in this high-energy regime is also physically large and monetarily daunting. Figure 11 is a view of the UA1 detector used by Rubbia and his collaborators in the discovery of the $W^{\pm}$ and Z^0 bosons for which he and Van der Meer were awarded the 1984 Nobel prize in physics.

The accelerator at the Fermi National Laboratory, shown in Fig. 12, has recently been upgraded to permit study of proton–antiproton collisions in the energy range up to 2 TeV. In Fig. 13, I show two views of the so-called CDF (Collider Detector at Fermilab) detector system; the upper panel shows the Italian, Japanese, and US flags as a reminder of the truly international flavor of these activities.

FIGURE 11   A view of part of the UA1 detection system used by Rubbia in the discovery of the intermediate bosons at CERN. (Courtesy of CERN.)

FIGURE 12   An aerial view of the Fermi National Accelerator Laboratory (FNAL) at Batavia,
Illinois. The main ring of this fixed target proton synchrotron is 6.3 km in circumference. It has
recently been upgraded for operation as a proton/antiproton collider with a total collision energy of
2 TeV. (Courtesy of Leon Lederman, FNAL.)

Figure 14 shows one of the first events studied with this CDF at 1.8 TeV, the
highest energy studies carried out thus far, again showing the distinctive jet
structure signature of the underlying quarks and their scattering interactions.
The lower panel is a reconstruction of this event showing the energy deposited in
different calorimeter cells as a function of the rapidity. This plot emphasizes the
jet structure and indicates a center-of-mass energy of some 130 GeV for the
quark–quark collision.

I have already noted the international aspect of this activity; but it involves
collaboration on a much grander scale. Figure 15 reproduces the title page of a
recent CDF report and includes 229 participant scientists, ten US universities,
three US national laboratories, as well as two Japanese and two Italian groups.
There is a very important and often forgotten fact illustrated here. In contrast to
the traditional picture of discoveries coming from the dedicated and often lonely
efforts of the single investigator—as many still do—the fact is that in elementary
particle physics, and increasingly in other subfields, the voyage to the frontiers
where new discoveries are being made is forever denied to the individual, and
while individuals still can, and must, play leadership roles in major new discoveries, the work would be quite impossible, and the discoveries forever denied to
us, without the dedicated participation of literally hundreds of other able scientists—as illustrated here. Such cooperation on the frontiers of knowledge is both
new and important.

With the drive toward ever higher center-of-mass energies delivered to ever
smaller volumes in the hope of materializing new heavier particles, in particle
physics collider accelerators have come to dominate the international scene.
This is shown rather dramatically in Fig. 16 showing existing and planned
collider facilities.

FIGURE 13 Two views of the CDF (Colliding Detector at FNAL) system installed at the Tevatron Facility at FNAL. Note the Italian, Japanese, and US flags hanging from the rafters in the
upper photograph, reflecting the three-nation participation in development and use of this facility.
The lower photograph shows the assembly of the innermost detector system surrounding the collision regions. (Courtesy of Roy Schwitters, Harvard.)

PROTON–ANTIPROTON COLLISIONS
AT
FERMILAB
THE CDF COLLABORATION

(10 US universities, 3 US national laboratories, 2 Japanese & 2 Italian groups)

(229 scientists)

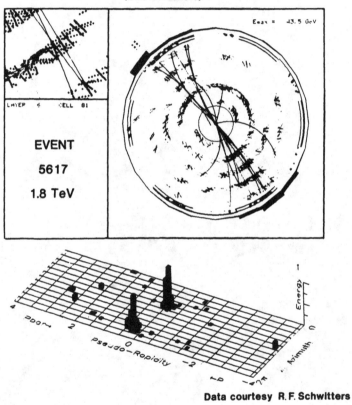

Data courtesy R. F. Schwitters

FIGURE 14   One of the first proton–antiproton collisions studied with the CDF facility at 1.8 TeV total energy. The upper panel shows, on the right, a reconstruction of the event from the firing of individual segments in the cylindrical detectors surrounding the collision point; the figure at the upper left simply expands a section of the detection system to illustrate the large amount of detail made available. The lower panel is an alternate presentation of the data from this event explicitly showing the energy involved in the quark event under study. (Courtesy of Roy Schwitters, Harvard.)

CDF Collaboration - June, 1987

F.Abe[o], D.Amidei[c], G.Apollinari[k], G.Ascoli[g], M.Atac[d], P.Auchincloss[n],
A.R.Baden[f], A. Barbaro-Galtieri[i], V.Barnes[l], E.Barsotti[d], F.Bedeschi[k],
S.Belforte[m], G.Bellettini[k], J.Bellinger[g], J.Bensinger[b], A.Beretvas[n], P.Berge[d],
S.Bertolucci[e], S.Bhadra[g], M.Binkley[d], R.Blair[a], C.Blocker[b], J.Bofill[d], A.W.Booth[d],
G.Brandenburg[f], A.Brenner[d], D.Brown[f], A.Byon[l], K.L.Byrum[e], M.Campbell[c], R.Carey[f],
W.Carithers[i], D.Carlsmith[c], J.T.Carroll[d], R.Cashmore[1], F.Cervelli[k], K.Chadwick[l],
T.Chapin[m], G.Chiarelli[m], W.Chinowsky[i], S.Cihangir[o], D.Cline[c], T.Collins[d], D.Connor[j],
M. Contreras[b], J.Cooper[d], M.Cordelli[e], M.Curatolo[e], C.Day[d], R.DelFabbro[k], M.Dell'Orso[k],
L.DeMortier[b], T.Devlin[n], D.DiBitonto[o], R.Diebold[a], F.Dittus[d], A.DiVirgilio[l], R.Downing[g],
G. Drake[d], T.Droege[d], M.Eaton[f], J.E.Elias[d], R.Ely[i], S.Errede[d], B.Esposito[e], A.Feldman[f],
B.Flaugher[n], E.Focardi[k], G.W.Foster[d], M.Franklin[f], J.Freeman[d], H.Frisch[c], Y.Fukui[h], I.Gaines[d],
A.Garfinkel[l], P.Giannetti[2], N.Giokaris[m], P.Giromini[e], L.Gladney[j], M.Gold[i], K.Goulianos[m],
J.Grimson[d], C.Grosso-Pilcher[c], C.Haber[i], S.Hahn[j], R.Handler[c], R.M.Harris[i], J.Hauser[c],
Y.Hayashide[o], T.Hessing[o], R.Hollebeek[j], L.Holloway[g], P.Hu[n], B.Hubbard[i], P.Hurst[g], J.Huth[d],
M.Ito[o], J.Jaske[c], H.Jensen[d], U.Joshi[n], R.W.Kadel[d], T.Kamon[o], S.Kanda[o], I.Karliner[g],
H.Kautzky[d], K.Kazlauskis[n], E.Kearns[f], R.Kephart[d], P.Kesten[b], H.Keutelian[g], Y.Kikuchi[o],
S.Kim[o], L.Kirsch[b], S.Kobayashi[3], K.Kondo[o], W.Krishuk[l], U.Kruse[g], S.Kuhlmann[l], A.Laasanen[l],
W.Li[c], T.Liss[c], N.Lockyer[j], F.Marchetto[o], R.Markeloff[c], L.A.Markosky[c], M.Masuzawa[o],
P.McIntyre[o], A.Menzione[k], T.Meyer[o], S.Mikamo[h], M.Miller[j], T.Mimashi[o], S.Miscetti[e],
M.Mishina[h], S.Miyashita[o], H.Miyata[o], N.Mondal[d], S.Mori[o], Y.Morita[o], A.Mukherjee[d],
A.Murakami[3], Y.Muraki[4], C.Nelson[o], C.Newman-Holmes[d], J.S.T.Ng[f], L.Nodulman[a],
J.O'Meara[d], G.Ott[c], T.Ozaki[o], S.Palanque[d], R.Paoletti[k], A.Para[d], J.Patrick[d], R.Perchonok[d],
T.J.Phillips[f], H.Piekarz[b], R.Plunkett[m], L.Pondrom[c], J.Proudfoot[a], G.Punzi[k], D.Quarrie[d],
K.Ragan[j], G.Redlinger[c], R.Rezmer[a], J.Rhoades[c], L.Ristori[k], T.Rohaly[j], A.Roodman[c],
H.Sanders[c], A.Sansoni[e], R.Sard[d], V.Scarpine[o], P.Schlabach[g], E.E.Schmidt[d], P.Schoessow[a],
M.Schub[j], R.Schwitters[f], A.Scribano[k], S.Segler[d], M.Sekiguchi[o], P.Sestini[k], M.Shapiro[f],
M.Sheaff[c], M.Shibata[o], M.Shochet[c], J.Siegrist[i], V.Simaitis[g], J.Simmons[c], P.Sinervo[j],
M.Sivertz[5], J.Skarha[c], D.A.Smith[g], R.Snider[c], L.Spencer[b], R.St.Denis[f], A.Stefanini[k],
Y.Takaiwa[o], K.Takikawa[o], S.Tarem[b], D.Theriot[d], J.Ting[c], A.Tollestrup[d], G.Tonelli[k],
W.Trischuk[f], Y.Tsay[c], K.Turner[d], F.Ukegawa[o], D.Underwood[a], C.vanIngen[d], R.VanBerg[j],
R.Vidal[d], R.G.Wagner[a], R.L.Wagner[d], J.Walsh[j], T.Watts[n], R.Webb[o], T.Westhusing[g],
S.White[m], V.White[d], A.Wicklund[a], H.H.Williams[j], T.Winch[c], R.Yamada[d],
T.Yamanouchi[d], A.Yamashita[o], K.Yasuoka[o], G.P.Yeh[d], J.Yoh[d], F.Zetti[k]

CDF Member Institutions

[a] Argonne National Laboratory-[b] Brandeis University-[c] University of Chicago
[d] Fermi National Accelerator Laboratory- [e] INFN, Frascati, Italy
[f] Harvard University-[g] University of Illinois-[h] KEK, Japan
[i] Lawrence Berkeley Laboratory-[j] University of Pennsylvania
[k] INFN, University of Pisa, Italy-[l] Purdue University
[m] Rockefeller University- [n] Rutgers University-[o] Texas A&M University
[p] University of Tsukuba, Japan-[q] University of Wisconsin

Visitors

[1] Oxford University, England- [2] INFN Trieste, Italy- [3] Saga University, Japan
[4] ICRR, Tokyo University, Japan- [5] Haverford College, Haverford, PA.

FIGURE 15   The title page from a recent preprint from the CDF collaboration listing the scientific and institutional participants in this activity.

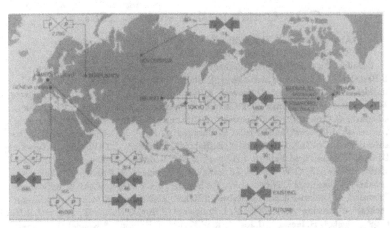

FIGURE 16   The distribution of colliding beam accelerator facilities worldwide, including both existing and planned accelerators. In each case, the facility is labeled according to the particles that they collide and the maximum total collision energy in GeV that will be available. (Courtesy of J. D. Jackson, California, Berkeley.)

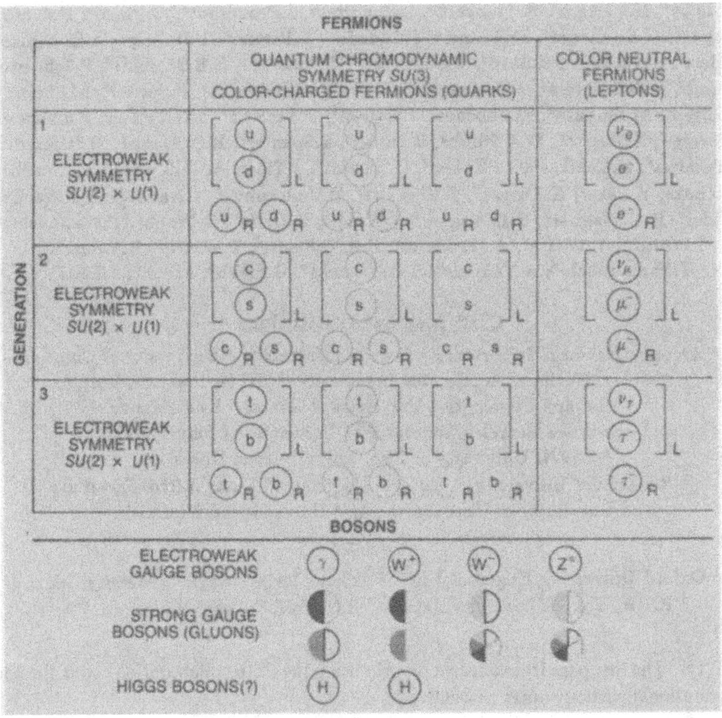

FERMIONS			
	QUANTUM CHROMODYNAMIC SYMMETRY SU(3) COLOR-CHARGED FERMIONS (QUARKS)		COLOR NEUTRAL FERMIONS (LEPTONS)

FIGURE 17   The standard model of elementary particle physics combines quantum chromodynamics and the electroweak theory. The fermions are grouped in the table to constitute three generations of particles displaying the underlying symmetry. (Courtesy of M. B. Green.)

C. The Standard Model

From all this work—both experimental and theoretical—has emerged what is now known as the standard model. This is illustrated in Fig. 17; in this model all matter is composed of basic fermions whose interactions are mediated by the exchange of so-called gauge particles. These basic fermions include the quarks, which carry one of three color charges, and the leptons, which carry no color charge. In this figure the color charges are represented as red (R), green (G), and blue (B), but, of course, these are only convenient labels. Quarks also carry electroweak flavor; six flavors are known—up, down, strange, charmed, bottom, and top. The leptons are not subject to the strong force and include the electron, the muon, and the tau together with their corresponding neutrinos. The forces involving all these particles are transmitted by gauge bosons, which in the electroweak case include the photon and the three massive bosons, W^+, W^-, and Z^0. There are eight strong gauge bosons or gluons carrying color and anticolor charges and there may also be a Higgs boson, which could be a possible source for the nonzero mass of all the other particles. The subscripts L and R indicate the handedness of the individual particles. For all except the neutrinos there is an antiparticle of opposite handedness, and the left-right asymmetry in this standard model, which is most notable in the absence of righthanded neutrinos, is a signature for the fact that the weak interaction distinguishes among particles on the basis of handedness, i.e., does not conserve parity.

The standard model also assumes the existence of four interactions—electromagnetic, strong, weak, and gravitational; and it assumes only three generations of quarks and leptons although the evidence against a fourth and more generations is far from complete as yet. The model has been remarkably successful in reproducing experimental observations and in suggesting new and challenging experiments. But open and puzzling questions remain.

Perhaps the most challenging of these concerns the origin of mass. In the simplest version of electroweak dynamics it is postulated that the spontaneous symmetry breaking in the electroweak interaction, leading to the observed great mass difference between the photon and the $W^\pm$ and Z^0 bosons (all of which in a symmetric theory would have zero mass), arises from an interaction with an electrically neutral field, the Higgs field, which, if it exists, permeates the universe and assumes a nonzero background value even in a perfect vacuum. This is no longer as distressing as it once might have been, given the rapidly increasing indications that the vacuum is a highly sophisticated medium capable of a wide variety of interesting fluctuation and other phenomena.

The interaction of the particles of nature with the Higgs field gives them an energy—and thus a mass—with respect to the vacuum. In this simplest of models the Higgs field would be represented by a single Higgs particle as suggested in Fig. 18. Unfortunately, existing theory does little to constrain the mass of the Higgs particle itself; if the mass were less than 50 GeV/c^2 it should be found at existing or planned electron–positron colliders and if between 50 and 200 GeV/c^2 it should be detectable at the Tevatron. For masses above 200 GeV/c^2,

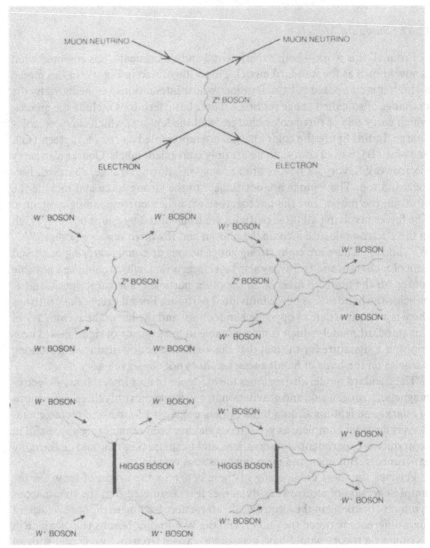

FIGURE 18 One possible source for the mass exhibited by many of the elementary particles appearing in a normalized electroweak theory is the postulated Higgs boson. By definition, a renormalized theory is one that can be applied to calculate experimentally observable quantities to any desired precision whereas a nonrenormalizable theory has no predictive power beyond a certain limit. Without the Higgs boson, the electroweak theory can successfully reproduce the scattering of neutrinos from electrons, as shown in the top panel. The theory fails, however, when it is used to study the interaction of W bosons with one another, as in the middle panel. Specifically, the theory predicts that at energies above about 1 TeV the probability of scattering one W boson off another exceeds unity—a clearly nonphysical result. In this case, the electroweak theory can be renormalized by introducing a Higgs boson as shown in the bottom panel. Predictions which are physically reasonable are obtained by effectively subtracting the graphs in the bottom panel from those in the middle. (Courtesy of M. J. G. Veltman.)

its lifetime for decay into two W or two Z particles will be so short—and its width correspondingly so large—that it may not make much sense to discuss it as a single particle. In any event, if the Higgs particle is as massive as 1 TeV/c^2 then electroweak theory predicts a whole series of quite new phenomena in that energy regime. This reflects a fundamental problem with electroweak theories. If they are simply extrapolated to energies above 1 TeV they characteristically predict probabilities for certain interactions that exceed unity. The theories as they stand must therefore be incomplete.

D. Supersymmetry

In the absence of any evidence for the existence of a Higgs particle, a totally different approach to the origin of mass that is under widespread study involves supersymmetry. In a supersymmetric world, as shown in Fig. 19, every parti-cle—including the Higgs boson, if it exists—would have a partner identical to it in every aspect except spin; to every ordinary fermion there would correspond a supersymmetric boson and to every ordinary boson a supersymmetric fermion. The supersymmetry field thus is different from all others in physics in that it treats fermions and bosons on an equal footing and can transform one into the other.

On the other hand, if the supersymmetric partners differed only in spin, and if they in fact exist, many would already have been discovered. Since none have, this leads naturally to the suggestion that, as in the case of the electroweak symmetry, the supersymmetry is also broken in such a way as to give all the supersymmetric partners masses that are beyond current experimental reach. This constitutes one of the arguments for new higher energy accelerators.

E. Superstrings

Despite all its successes, the standard model fails in one very glaring aspect. It is unable to incorporate gravitation with the other natural interactions.

In an effort to remedy this problem, Green and Schwarz, and now many other theorists, have turned in recent years to the development of so-called super-string theories. *Ab initio*, these theories postulate the validity of special relativi-ty, of quantum mechanics, of supersymmetry, and also insist that all the forces of nature be included.

Within the theory, which is cast in a 10-dimensional universe—six of which compactified in the earliest instants of the universe to leave our familiar three spatial dimensions and one temporal dimension—the elementary particles are thought of as one-dimensional curves (strings) rather than points, with charac-teristic strings of the Planck length $(\hbar G/c^3)^{1/2} = 1.6 \times 10^{-33}$ cm and character-istic masses of the Planck value $(\hbar c/G)^{1/2} = 1.2 \times 10^{19}$ GeV/c^2. That this has a certain logical appeal follows from Fig. 6, where it is shown that the Planck state is necessarily the one relevant to a unified theory containing gravitation.

Obviously string theories can involve two quite distinct topologies; the strings can be open-ended or closed into loops. Currently there are three known consis-

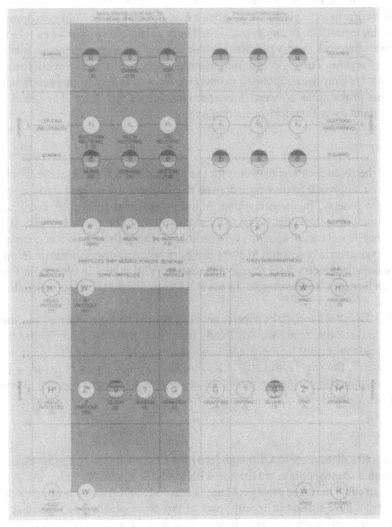

FIGURE 19 Supersymmetry requires that for every ordinary particle there exists a superpartner having similar properties—except for spin. Here the fermions that constitute matter are shown in the upper part of the figure whereas the bosons that mediate the forces of nature are shown in the lower part. The bracketed number, which appears below each of the particles, is its approximate mass—when known—in GeV. Fermion antiparticles are not shown in the figure but the antiparticles are shown in the case of the bosons. None of the superpartners have yet been observed experimentally. (Courtesy of H. E. Haber and G. L. Kane.)

tent string theories; Type I involves both open and closed strings while Type II and heterotic theories involve only closed loops. There is currently no way of choosing between them.

Figure 20 is a highly schematic representation of some string dynamics. Here an elementary string (a) can vibrate in different modes to represent two different particles (b) and (c); a single string (d) can divide (e) to represent a decay process or join (e) → (f) to represent a fusion process. An open string (g) can bend (h) and finally join ends to form a closed loop (i); two such loops can then join to represent a fusion process of (j) → (k) or separate, (k) → (j) to represent a decay.

In string theory, the time evolution of the universe is postulated to be that in Fig. 21. There is an aesthetically pleasing aspect to the group-theoretical de-

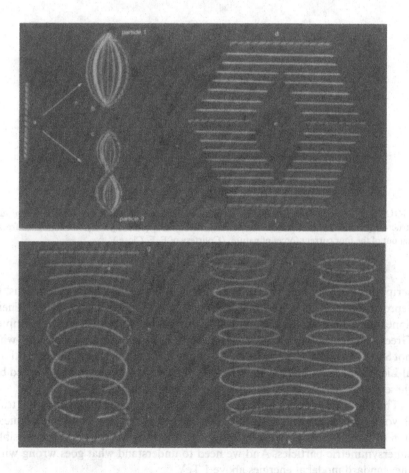

FIGURE 20   A schematic illustration of the interaction of elementary strings. (Courtesy of F. E. Close.)

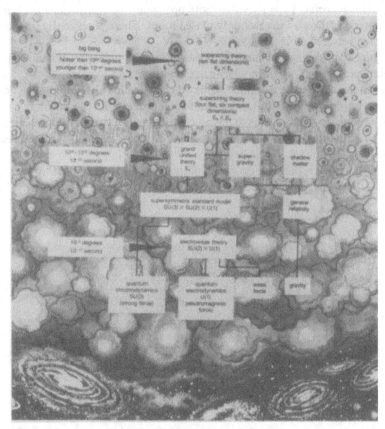

FIGURE 21   A very highly schematic view of the evolution of the universe—indicating the group structure governing at each stage in the condensation following the Big Bang and during the freezing out of the four current forces of nature. (Courtesy of F. E. Close.)

scriptions involved here in that the starting universe, in ten flat dimensions, is represented by $E_8 \times E_8$ and the daughter, grand-unified theory, by E_6. Their appearance here avoids the natural question that arose when, for example, Green and Schwarz originally suggested SO(32) as the governing group—why not SO(37) or SO(70)? E_8 and E_6 are two of the three most complex exceptional Lie groups and if nature, for whatever reason, has chosen to be governed by these groups, there simply isn't room for greater complexity!

Thus far, string theory remains completely impregnable to experimental test. Beyond that fundamental difficulty, we need to understand how compactification occurs, and why? We need to search for Higgs bosons and for possible supersymmetric particles. And we need to understand what goes wrong with the standard model at energies above 1 TeV.

Although we have still a long way to go, we have reason for the first time in history to be hopeful that we can integrate quantum gravity into our picture of the universe, that we can look forward to a unified understanding of the forces of nature and that, indeed, we *may* be on the trail of a completely unified theory that can incorporate not only the forces, but also all matter, into an overarching theory of physical reality. This remains a dream—but what a dramatic and marvelous one! And, moreover, it is a dream that would not have attracted serious attention even three years ago at our last General Assembly.

F. The Superconducting Supercollider

For all the questions noted above we need higher energies. It bears noting that, at 2 TeV, the Tevatron will permit detailed study of phenomena only up to about 0.3 TeV, and if we want to study phenomena in the range 1 to 5 TeV, we need accelerator energies $\geqslant 20$ TeV.

This year, President Reagan announced that his Administration would support the construction of the proposed Superconducting Supercollider Accelerator (SSC). As designed, this facility will have colliding 20 TeV beams of protons to yield 40 GeV in the collision center-of-mass. Neither the site nor the construction schedule has yet been established. To give some indication of the size of this proposed facility, Fig. 22 shows it superimposed on a Landsat photograph of Washington, D.C. Depending upon design details it will have a ring circumference of about 120 km, to be compared to the 27 km circumference of the LEP (or LHC) ring at CERN. Even on the most optimistic schedule, the SSC cannot be available until the latter part of the 1990s. In the meantime, however, there are continuing tests, to ever greater precision, of standard model predictions accessible to lower energy accelerators or, indeed, to studies requiring no accelerator at all. I shall mention only two of these latter.

G. Proton Decay

A characteristic of all the standard models is that they require that the proton decay. As shown in Fig. 23, there are now, internationally, seven major detector systems in operation, or under construction, in the search for this decay. All are installed underground to minimize cosmic-ray background. Three of the experiments use water Cerenkov detectors while the other four use detectors in which iron plates are interleaved with layers of position-sensitive particle detectors.

While the simplest of the unified theories, developed by Georgi and Glashow, [minimal SU(5)] predicts a proton half-life of less than 2.5×10^{31} years, the experimental data, thus far, set a lower limit of 1.7×10^{32} years; this would appear to rule out the minimal SU(5) theory with its predicted $p \rightarrow \pi^0 + e^+$ decay mode. There are, however, many more complex decay modes possible. Although candidate events exist for some of them there is, as yet, no general consensus that proton decay has ever been observed.

For the decay to occur, a field that couples hadrons and leptons is required; a proton lifetime as short as 10^{30} years would correspond to a field particle (the so-called lepto-quark) having mass $\approx 10^{15}$ GeV/c^2, i.e., heavier than an average

FIGURE 22   The proposed superconducting supercollider (SSC) ring is here superimposed on a
Landsat photograph of the Washington, DC area. For orientation, National Airport is visible just
south of the trifurcation of the Potomac River, and the Mall, running from the Lincoln Memorial to
the Capitol, is the horizontal structure just above the trifurcation. Six interaction regions have been
sketched here as has the injector complex.

bacterium! At the current limit of 1.7×10^{32} years, a human would have to live
some 2500 years before a single one of his or her body protons decays.

The search for proton decay continues; the standard models require that it
decay, and further study is of crucial importance.

H. Double Beta Decay

Over the past 40 years, a continuing effort has been made to detect another
exceedingly feeble decay process, that of double beta decay, traditionally con-

sidered a nuclear rather than a particle-physics activity. The possibility of the decay in a nucleus such as ^{136}Xe—which is under active current study—arises from the pairing phenomenon in nuclei which results in even–even nuclei being bound more strongly than odd–odd ones. This leads to the two dashed parabolae in the isobaric cross section of the nuclear valley of stability shown in Fig. 24.

In ordinary beta decay, lepton and momentum conservation require the emission of an electron antineutrino in the final state. It was recognized long ago that if the neutrino were, in fact, a Majorana particle, i.e., if the particle and antiparticle were identical, then the double beta decay process could occur without neutrino emission and with emission of two electrons with a fixed total energy.

If, on the other hand, the neutrino and its antiparticle are distinguishable, then two-antineutrino emission is required and the electron energies cover a spectrum up to a nuclear-mass-determined maximum.

This latter process has very recently been observed directly by Moe *et al.*, in a technological *tour de force*, in the decay of ^{82}Kr and has been found to have a half-life of about 10^{20} years.

The magnitude of this lifetime provides another critical test for the standard model, as well as providing information concerning a possible small but nonzero neutrino mass. Extraction of such information, however, depends critically on detailed knowledge of the nuclear wave functions involved, and only recently have these been explored to the level of precision where the nuclear system can truly begin functioning as a microscopic laboratory wherein some of the most fundamental questions in physics can be probed.

	SPONSORING INSTITUTIONS	LOCATION	DEPTH (EQUIVALENT METERS OF WATER)	DETECTOR MASS (METRIC TONS)	DETECTION METHOD
WATER CERENKOV DETECTORS	University of California at Irvine, University of Michigan, Brookhaven National Laboratory, Cleveland State University, University of Hawaii, California Institute of Technology, University College Warsaw	Morton Thiokol salt mine, Painesville, Ohio	600 1,600	8,000 TOTAL 3,300 FIDUCIAL	2,048 FIVE-INCH PHOTOMULTIPLIERS ON ONE-METER SURFACE GRID
	KEK, University of Tokyo, University of Tsukuba	Kamioka metal mine	825 2,400	3,000 TOTAL 1,000 FIDUCIAL	1,000 20-INCH PHOTOMULTIPLIERS ON ONE-METER SURFACE GRID
	Harvard University, Purdue University, University of Wisconsin	Silver King mine, Park City, Utah	525 1,500	700 TOTAL 420 FIDUCIAL	704 FIVE INCH PHOTOMULTIPLIERS ON ONE-METER LATTICE, MIRRORED WALLS
LAYERED TRACKING DETECTORS	Tata Institute, Osaka City University, University of Tokyo	Kolar gold fields, South India	2,500 7,600	140 TOTAL 100 FIDUCIAL	1,600 PROPORTIONAL GAS COUNTER TUBES
	CERN, Frascati Laboratory, University of Milan, University of Turin	Mont Blanc tunnel, French–Italian border	1,850 5,000	150 TOTAL 100 FIDUCIAL	47,000 LIMITED STREAMER TUBES
	Orsay, École Polytechnique, Saclay, Wuppertal University, Tufts University	Fréjus tunnel, French–Italian border	1,550 4,200	160 TOTAL	1,500 PLASTIC FLASH-TUBE PLANES; 200 GEIGER TUBES
	Argonne National Laboratory, University of Minnesota, University of Oxford, Rutherford Laboratory	Soudan iron mine, Soudan, Minn.	675 1,800	30 (PROTOTYPE)	HEXAGONAL DRIFT TUBES

FIGURE 23 A listing of the detection systems currently active in the search for proton decay. In this figure, listed for each detector are the sponsoring institutions, its location, its equivalent depth beneath the surface (in meters of water), the mass of the sensitive portion of the detector (in metric tons), and finally, the detection method.

NUCLEAR DOUBLE BETA DECAY

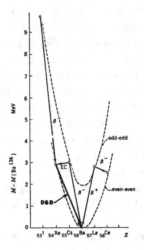

Zero Neutrino DBD $^{136}Xe \longrightarrow {}^{136}Ba + 2e^-$

Two Neutrino DBD $^{136}Xe \longrightarrow {}^{136}Ba + 2e^- + 2\bar{\nu}_e$

Michael K. Moe, University of California, Irvine

$$^{82}Kr \longrightarrow {}^{82}Sr + 2e^- + 2\bar{\nu}_e$$

Halflife $\cong 10^{20}$ years

Reference: Science News, **132** 148, September 5, 1987

FIGURE 24   A schematic illustration of the double beta decay process in the $A = 136$ isobaric system. Although zero, neutrino double beta decay has not yet been identified, very recently Michael Moe of the University of California at Irvine has detected two-neutrino double beta decay of ^{82}Kr with a half-life of approximately 10^{20} years.

5. NUCLEAR PHYSICS

These past three years have been exciting ones in nuclear physics as new instrumentation has opened up whole new domains to study. We tend to forget that, in the seventy-five years since Rutherford's discovery of the nucleus, our studies have been very much limited to cold, low-spin nuclei near their ground states, and specifically to phenomena reflecting shell-model effects in their surface regions and collective effects involving nonspherical equilibrium shapes.

The study of hot nuclei has advanced dramatically during this past year as a consequence of new studies with beams of heavy ions both at low Van de Graaff energies and at the highest energies available from the largest accelerators at CERN and Brookhaven. Exciting new frontiers also await new electron accelerators currently under construction in several centers.

A. Renaissance in Nuclear Spectroscopy

At the lower energies a renaissance in nuclear spectroscopy, comparable to that worked by the advent of the laser in atomic spectroscopy, reflects the development of large, new, high-resolution, high-multiplicity detection systems typified by the TESSA-III unit at the Daresbury Laboratories in the U.K. Using this facility, Twin and his associates have, in a single set of measurements, *doubled* the angular momentum that can be examined in discrete nuclear quantum states; as shown in Fig. 25, electromagnetic deexcitation of states having 60 units of angular momentum is clearly visible. These data are evidence for a new, relatively stable shell structure in highly deformed, hot nuclei having pronounced football shapes with major/minor axis ratios of 2:1.

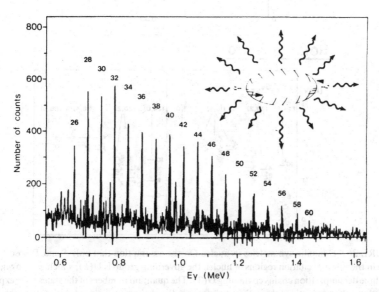

GAMMA RADIATION FROM SUPER DEFORMED DYSPROSIUM NUCLEI

Peter Twin et al. Daresbury Laboratory

FIGURE 25 A section of the gamma-ray spectrum from superdeformed ^{152}Dy nuclei produced at the Nuclear Structure Facility at Daresbury, England. These superdeformed nuclei have major axis ratios of 2:1 and the indicated peaks in this spectrum are those de-exciting states with the indicated number of units of angular momentum. (Courtesy of Peter Twin, Daresbury.)

Group Decomposition

$$\begin{aligned}
&\nearrow U(5) \supset O(5) \supset O(3) \supset O(2), &&\text{(I)}\\
U(6) &\to SU(3) \supset O(3) \supset O(2), &&\text{(II)}\\
&\searrow O(6) \supset O(5) \supset O(3) \supset O(2). &&\text{(III)}
\end{aligned}$$

Quantum Numbers

$$\begin{array}{ccccccccc}
U(6) & \supset & O(6) & \supset & O(5) & \supset & O(3) & \supset & O(2)\\
\downarrow & & \downarrow & & \downarrow & & \downarrow & & \downarrow\\
N & & \sigma & & \tau & (\nu_\Delta) & & L & & M_L
\end{array}$$

Mass Formulae

$$E^{(I)}(N, n_d, v, n_\Delta, L, M_L) = E_0^{(I)} + \epsilon n_d + \alpha n_d(n_d + 4) + \beta v(v + 3) + \gamma L(L + 1),$$

$$E^{(II)}(N, \lambda, \mu, K, L, M_L) = E_0^{(II)} + \kappa(\lambda^2 + \mu^2 + \lambda\mu + 3\lambda + 3\mu) + \kappa'L(L + 1),$$

$$E^{(III)}(N, \sigma, \tau, \nu_\Delta, L, M_L) = E_0^{(III)} + A\sigma(\sigma + 4) + B\tau(\tau + 3) + CL(L + 1).$$

Vibrator Spectra

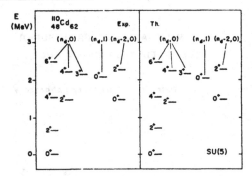

Rotor Spectra

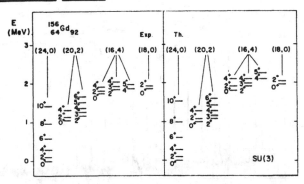

FIGURE 26   Dynamic symmetries in nuclei. On the assumption that only S and D bosons are active in the lower excitation regions of nuclei, the governing group is $U(6)$, which, as shown, has only three decomposition chains containing $O(3)$. The quantum numbers of the states to be expected are those characteristic of the individual groups in the chain, as indicated for the $O(6)$ case. Use of the corresponding Casimir operators leads to the mass formulae shown. The lower two panels of this figure compare the predictions of these formulae with the experimental situation for cases I and II. The agreement in case III is equally good. (Courtesy of F. Iachello, Yale.)

Detailed calculations applying Landau theory to these hot nuclei predict first- and second-order phase transitions between prolate, oblate and spherical shapes in the temperature range from 1 to 3 MeV, and early measurements on the giant dipole resonances of these hot nuclei appear to provide experimental signatures for at least one of these transitions thus far. Obviously this new area of study is expanding rapidly.

In the past, nuclear physicists have used a wide variety of models—shell, liquid drop, and collective—to characterize the behavior of the nuclear many-body system; but each characteristically had its own limited regions of applicability and there were large intervening regions where no model appeared adequate. All this has changed as a consequence of the recognition of underlying dynamic symmetries in nuclear structure.

B. Symmetries in Nuclei

All the old models were, in essence, geometric ones, wherein a postulated potential was inserted into a Schrödinger equation whose solutions were matched as closely as possible to experimental observations. In recent years, Iachello, Arima, and Talmi have developed a parallel algebraic approach where the basic entities are nucleon Cooper pairs (bosons), now having both zero and two units of angular momentum; the spin one configuration is unbound for well-understood reasons and the higher-spin bosons come into play only at relatively high excitation energies. Group theory then provides the guide for assembling these bosons to yield nuclear model wave functions and energy levels. As shown in Fig. 26, the governing symmetry for low excitations is $SU(6)$ and the only three-group decomposition chains which contain an $O(3)$ group—as required to yield good angular momentum quantum numbers—are those labeled, respectively, as $SU(5)$, $O(3)$, and $O(6)$ corresponding, respectively, to the familiar vibrator, symmetric rotor, and triaxial rotor, i.e., to the old standard models. The new symmetry-based approach is not limited to these special cases or to any particular regime and for the first time provides a relatively simple unified model for the structure of *all* nuclei. As illustrated in this figure, it is a simple matter to write down the Casimir operators for the groups involved and thus to obtain, in closed form, expressions for the model excitation spectra. The extent to which these reproduce experimental data is illustrated in the two examples of Fig. 26.

This renaissance in the theory of nuclear structure has paralleled that in experimental technology, and has provided a powerful new impetus to structural studies and understanding.

While the underlying symmetries apparent here result in families of related states in single nuclei it was quickly recognized that if even-mass nuclei could be understood as comprising these nucleon Cooper pairs—bosons—then odd-mass nuclei with a valence fermion outside a bosonic core would require a supersymmetric theory for their description. In this case, in contrast to that of Fig. 26, the mass formula relates the ground state masses of all nuclei in a region of the periodic table as well as many of their excited quantum states. This is

POSSIBLE SUPERSYMMETRY IN NUCLEI

A j=3/2 VALENCE NUCLEON COUPLED TO AN O(6) TRIAXIAL ROTOR CORE

SUPERGROUP DECOMPOSITION

$$U(6/4) \supset U^{(B)}(6) \otimes U^{(F)}(4) \supset O^{(B)}(6) \otimes SU^{(F)}(4) \supset \text{Spin}(6) \supset \text{Spin}(5) \supset \text{Spin}(3) \supset \text{Spin}(2)$$

$$\downarrow \qquad \downarrow \qquad \downarrow \qquad \downarrow \qquad \downarrow \qquad \downarrow \qquad \downarrow \qquad \downarrow$$

$$\mathcal{N} \qquad N \qquad M \qquad \Sigma \qquad \sigma_1 \sigma_2, \sigma_3 \quad \tau_1, \tau_2(\nu_\Delta) \quad J \qquad M_J$$

MASS FORMULA

$$E(\mathcal{N}, N, M, \Sigma, \sigma_1, \sigma_2, \sigma_3, \tau_1, \tau_2, \nu_\Delta, J, M_J) = \mathcal{E}_0 + \mathcal{E}_1 N + \mathcal{E}_2 N^2 + A\Sigma(\Sigma + 4)$$
$$+ A'[\sigma_1(\sigma_1 + 4) + \sigma_2(\sigma_2 + 2) + \sigma_3^2]$$
$$+ B[\tau_1(\tau_1 + 3) + \tau_2(\tau_2 + 1)] + CJ(J + 1)$$

SUPERMULTIPLETS

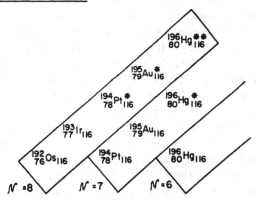

FIGURE 27   Possible dynamic supersymmetry in nuclei. In this case of a $j = 3/2$ valence nucleon coupled to an O(6) triaxial core, the supergroup decomposition and the mass formula equivalent to those of Fig. 26 are shown. The best current example of a dynamic supersymmetry in nature appears to be that which holds in the region centered on ^{193}Ir. (Courtesy of J. Cizewski, Yale.)

illustrated in Fig. 27 for the case of a $j = 3/2$ valence nucleon outside an O(6) core. The supersymmetric model now predicts families of quantum states in a whole range of neighboring nuclei as well as transition selection rules among them. Cizewski *et al.* have found that, in the region of ^{193}Ir, such a model reproduces all available data in remarkable fashion, and they and others are now searching for other regions that may display similar characteristics. If real, this would represent the first example of a supersymmetric situation observed in nature.

Within the past year, Iachello *et al.* have attempted to extend this algebraic, symmetry-based approach to the description of unbound nuclear scattering and reaction systems. Building upon the success of the interacting boson model for nuclear structure, Alhassid, Iachello, and Shao have developed a symmetry-based approach to the description of nuclear interactions. As in the structure case, this is an algebraic theory that is extendable to an arbitrary number of channels. In Fig. 28 it is applied to the $^{16}O + {}^{24}Mg$ interactions, and shows, as indicated, elastic scattering, inelastic scattering to the first excited state of ^{24}Mg, transfer of an alpha particle to the ground state of ^{28}Si, and transfer of an alpha particle to the first excited state of ^{28}Si. The striking thing here is that there has been no normalization of the theoretical curves relative to the experimental data and a single set of parameters has sufficed to give remarkably good representation of the data for all four of these reactions simultaneously. The hope is that this can be generalized to provide a unified treatment of nuclear interactions, but as yet very few experimental data are available to test this new approach. Although still at a very preliminary stage, early results such as those shown in Fig. 28 are very encouraging. If successful, this approach promises, for the first time, a unified description of all nuclear interactions.

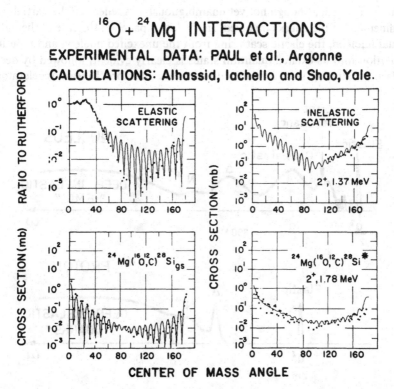

FIGURE 28   A possible unified description of nuclear interactions. (Courtesy of Y. Alhassid, Yale.)

C. Nuclear Studies with Electron Beams

I noted earlier that nuclear studies have been largely limited to the nuclear surface. This limitation is being removed as new, higher energy, higher duty-cycle, and higher beam-quality electron accelerators are becoming available. Recently, for example, Frois and his collaborators at Saclay have shown that a $3s$ proton, deep in the heart of nuclei in the lead region, possesses a classic shell-model wave function and thus that nucleons deep in the nuclear interior retain their surface characteristics.

The major challenge here lies in going to high enough energies to map out the transition from hadronic to quark matter as the nucleons effectively heat up and melt—deconfining their constituent quarks into a quark–gluon plasma. Figure 29 illustrates schematically the kind of spectra obtained in the scattering of high-energy electrons from a typical nucleus and then, at higher energies, from a nucleon. In the upper panel, one finds elastic and inelastic scattering from low-lying discrete nuclear states, and then the giant resonance excitations followed by quasielastic processes and by structure which represents the excitation of excited states of the participant nucleons. Finally, at even higher energy, the deep inelastic interactions begin. These are referred to as EMC phenomena (as a consequence of their discovery by the European Muon Collaboration) and have been interpreted, although not yet unambiguously, as reflecting the partial de-confinement of the quarks involved. In the lower panel, in the case of the individual nucleon, the elastic scattering from the unexcited nucleon and then its excitation into its excited quantum states is clearly evident, followed by deep inelastic scattering such as that studied at the Stanford Linear Accelerator,

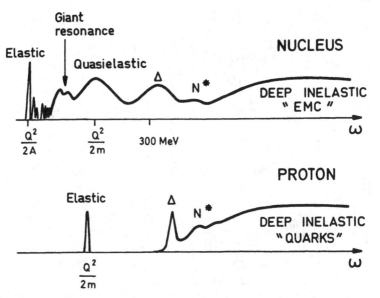

FIGURE 29   A schematic representation of the spectra from high-energy electron interactions with a typical nucleus and with a proton. (Courtesy of B. Frois, Saclay.)

where the electrons interact with the individual quarks in the proton. In both cases there appears clear evidence for the excitation of the individual nucleons to their excited quantum states—the Δ and so-called nucleon isobars N*—as well as for the eventual participation of individual quarks in the interaction.

Only now are adequate facilities (such as the CEBAF in Virginia, which can probe this entire range of phenomena) under construction. Although the transition from hadronic to quark matter is a complex one, and although experimental studies on this transition region will be difficult, the complexity promises a rich reward in new understanding of the fundamental structure of matter.

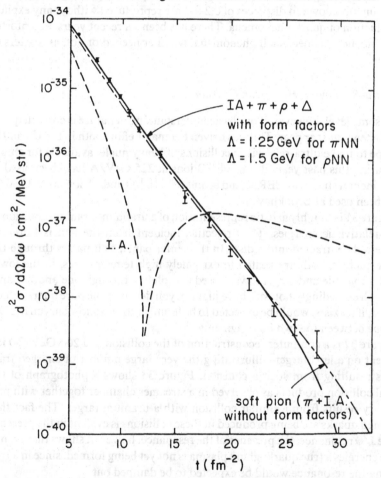

FIGURE 30 The cross section for electrodisintegration of the deuteron (as measured at Saclay) plotted against momentum transfer. The dashed curve labeled I.A. is from an impulse approximation calculation assuming the presence only of the neutron and proton in the deuteron. The dot-dash curve is that obtained by adding the presence of soft pions, in the simplest possible model, to the impulse approximation and, as illustrated, it provides a remarkably good reproduction of the experimental results. The solid curve is that obtained by adding, in addition, the effect of rho meson exchange, as well as the possible excitation of the nucleon to the delta resonances. (Courtesy of G. E. Brown, Stony Brook.)

A cautionary note is in order, however, as illustrated in Fig. 30. What is shown here is the cross section for electro-disintegration of the deuteron plotted effectively against momentum transfer. As is obvious, an impulse approximation calculation assuming the presence of only the nucleons fails to represent the data even qualitatively. Addition of soft pions in the simplest possible model results in a dramatic improvement; addition of effects that would be attributable to ρ meson exchange, or to nucleon excitation to the Δ isobar, change the results hardly at all since their contributions are of equal magnitude and opposite sign. What *is* surprising is that these data, which correspond to probing the deuteron wave function down to distances of 0.2 fm, are reproduced without any explicit introduction of quark phenomena. There has been, in recent years, an unfortunate tendency to inject such phenomena, whether needed or not, as matters of fashion.

D. Ultra-Relativistic Heavy-Ion Collisions

At some level, however, if our basic understanding of nature is correct, quark phenomena *must* appear. This has driven the major effort both in the US and in Europe to make ultra-relativistic collisions of heavy nuclei available for studying during this past year. Beams of ^{16}O ions at 225 GeV/A have been used in experimental studies at CERN, and beams of both ^{16}O and ^{32}S ions at 15 GeV/A have been used at Brookhaven.

Figure 31 is a highly artistic representation of a uranium–uranium collision at ultra-relativistic energies. The projectile, incident from the left, is Lorentz–Fitzgerald contracted into a disk. In the center panel, as it passes through the target nucleus, both are heated to extremely high temperatures; in the lower panel, projectile and target have passed completely through one another and, although exceedingly hot, still have high baryon density. The pure quark–gluon plasma, if it exists, would be expected to be found in the very low baryon density *firetube* between the two hot fragments.

Figure 32 is a computer reconstruction of the collision of a 200-GeV ^{16}O ion incident on a lead target—illustrating the very large number of charged fragments resulting from such a collision. Figure 33 shows a photograph of the actual collision products as observed in a streamer chamber together with preliminary results from a similar collision with a uranium target. The fact that charmed quarks are being produced in these collisions is evident in the presence of the J/ψ resonance; the presence of the resonance, however, shows that even at these energies a true quark–gluon plasma is not yet being formed, since in a true plasma the resonance would be expected to be damped out.

Given the large number of charged particles shown in Figs. 32 and 33, it becomes possible to use the Hanbury–Brown, Twiss interferometric technique, first evolved to measure the diameter of Sirius, to measure the size of the region in these heavy-ion collisions from which the fragments originate, i.e., to answer the question of whether only a part of the system is heated to very high temperatures. The results of a preliminary analysis are given in Fig. 34, suggesting that

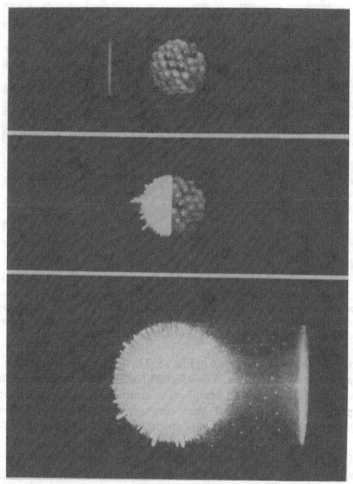

FIGURE 31 A highly schematic representation of an ultra-relativistic uranium projectile colliding with a uranium target. (Courtesy of David Shirley, California, Berkeley.)

indeed the hot source is much smaller than the gold target, which has a radius of
about 9 fm.

Contrary to some expectations that at these energies nuclei would be entirely
transparent to one another, the past year's experiments have demonstrated con-
clusively that at 15 GeV/A the projectile is essentially stopped in heavy targets
whereas at 200 GeV/A some transparency is evident. In order, however, to
hope to see a true quark–gluon plasma, a fifth state of matter, substantially
higher energies are required; only then will the firetube of Fig. 31 be adequately
separated from the hot target and projectile regions.

A facility that would produce such energies—200 GeV/A in the center-of-
mass system using colliding 100 GeV/A beams of projectile nuclei ranging up to
gold—is the proposed Relativistic Heavy Ion Collider (RHIC) at the Brookha-

FIGURE 32   A computer reconstruction of a streamer chamber photograph of the interaction of a 200 GeV per nucleon beam of ^{16}O ions with a gold target. This reconstruction illustrates clearly the high multiplicity of these high-energy, heavy-ion events and the difficulty involved in extracting physics from them. (Courtesy of the NA-35 group and W. Willis, CERN.)

ven National Laboratory as shown in Fig. 35. This facility, however, has not yet been approved for construction. A new facility has also been proposed at CERN which would yield 225 GeV/A beams up to uranium, but again it has not yet been approved for construction. It bears emphasis that in a fixed target mode a 200 GeV/A beam corresponds to only about 20 GeV/A in the center-of-mass system.

All of the present and proposed relativistic heavy ion facilities have as their goal the study of the equation-of-state of nuclear matter—of interest both in physics and in cosmology and astrophysics. A version of this phase diagram, plotted as temperature *versus* density, is shown in Fig. 36. Recent measurements at the Bevalac at the Lawrence Berkeley Laboratory have demonstrated that in collisions of heavy nuclei at roughly 2 GeV per nucleon, hydrodynamic effects lead to substantial shock-wave generation and thus to density increases several times normal. Because of the finite size of the nucleus, the deconfinement-chiral transitions are smeared over a finite range, whereas in infinite nuclear matter this would be a first-order phase transition. As indicated, the transition from the quark–gluon plasma into hadronic matter occurred at low baryon density but at high temperature in the early universe. Similarly, in supernova events, the assumption is that the transition is again traversed but now at relatively low temperature and at correspondingly higher density. The purpose of the proposed new heavy ion accelerators will be to traverse it in controlled fashion. This is entirely new physics with the possibility of probing entirely new kinds of matter and of behavior.

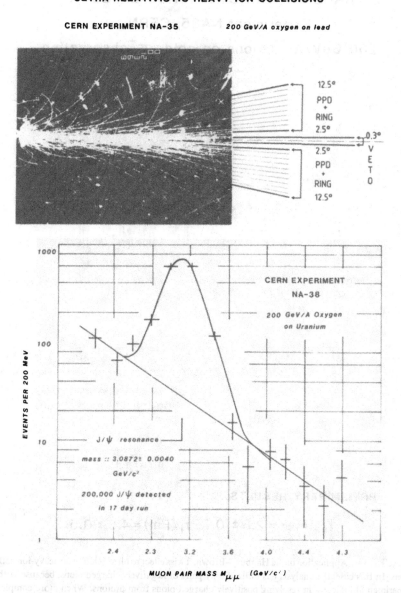

FIGURE 33 Ultra-relativistic heavy ion collisions with 200 GeV per nucleon oxygen beams. The upper panel shows a streamer chamber photograph of the interaction with a lead target and, in addition, shows the detector structure which follows the streamer chamber to detect those events having significant transverse energy components. The lower panel shows the unmistakable presence of the J/ψ resonance. (Courtesy of NA-35 and NA-38 groups and W. Willis, CERN.)

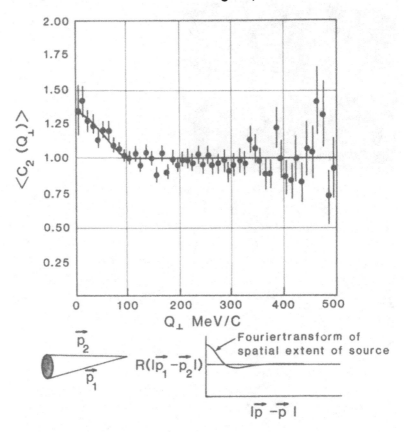

HANBURY – BROWN TWISS EFFECT

Experiment NA35, CERN

200 GeV/A ^{16}O ions on gold; π^- observation

PRELIMINARY RESULTS:

$$r_{\parallel} \text{ (Fm)} = 2.8 \pm 0.7; \quad r_\perp \text{ (Fm)} = 4.5 \pm 0.5$$

FIGURE 34   Application of the Hanbury–Brown, Twiss effect to ultra-relativistic heavy-ion collisions. In this case, the analysis has utilized only pairs of negatively charged pions, because of the
experimental difficulty in resolving positively charged pions from protons. When more complete
analyses are available, these positively charged particles will also be included in the analyses and will
provide much improved statistics. It is a consequence of the high multiplicity of these heavy-ion
events that enough pairs can be found in a single collision to allow the interferometric definition of
the size of the source volume. As shown here, the indicated source dimensions are substantially less
than those of a gold nucleus and therefore suggest that the oxygen projectile heats a substantially
smaller fraction of the gold to the high temperatures responsible for the subsequent particle emission. (Courtesy of W. Willis, CERN.)

FIGURE 35 An aerial view of the Brookhaven National Laboratory site showing the components of the proposed Relativistic Heavy Ion Collider (RHIC) project. At the lower right, the twin MP tandem accelerators produce pulsed beams of heavy ions, which are taken via the beam transport tunnel to the existing Alternating Gradient Synchrotron (AGS) for acceleration to 15 GeV per nucleon. Currently the Booster Synchrotron, shown to the left of the AGS, is under construction and, when completed, it will not only increase the proton intensity in the AGS and therefore its kaon output by a factor of four but also will expand the range of heavy ions available for injection into the main RHIC ring from sulfur to gold. After acceleration in the AGS ring, all beams will be transferred into the large RHIC ring at the upper part of the photograph, which remains from the earlier construction of the ISABELLE accelerator. This larger ring will yield 100 GeV per nucleon counter-rotating beams and thus up to 200 GeV per nucleon in the center of mass of the collision. The fact that so much of the construction and facilities remaining from the ISABELLE project can be incorporated into RHIC makes it a particularly cost-effective proposal. (Courtesy of N. Samios, Brookhaven.)

The pion condensate sketched in Fig. 36 has not yet been detected experimentally despite extensive searches; the pion wave functions would be expected to display coherence in such a condensate analogous to that characteristic of a laser. No evidence has yet been found, either, for the predicted gas and liquid phases at densities less than those normal to nuclei, although these, too, should be accessible in heavy ion collisions.

In contrast to particle physics, where the goal is that of delivering ever more energy to ever smaller volumes, here the goal is one of delivering ever more energy to relatively large volumes containing large numbers of nucleons—and quarks—so that the possibility of entirely new collective phenomena exists.

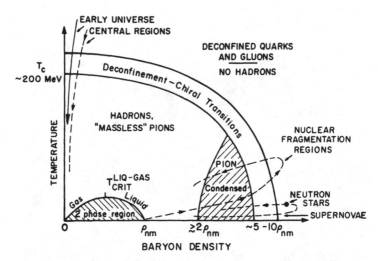

FIGURE 36   A phase diagram for nuclear matter, plotted here as temperature versus baryon density. The finite nuclear size results in the hadronic–quark matter transition being smeared out as shown. As yet no evidence exists for the predicted pion condensate at higher than normal nuclear density, nor for the liquid and gas phases predicted at substantially less than normal density. As shown, the hadronic–quark transition region was clearly traversed in the early universe and is traversed in gravitational collapse. The trajectory followed in a typical heavy-ion collision is indicated and is that which will be studied extensively at the new ultra-relativistic heavy-ion accelerator facilities. The estimate of 200 MeV for the zero baryon density transition temperature comes from current lattice gauge calculations.

E. The High Intensity Frontier

These activities involving both electron and heavy-ion projectiles mark the high-energy frontier of nuclear physics. The high-intensity frontier is quite different in its focus on very low cross section, very rare events that can only be studied with extremely intense projectile beams. Here, too, as shown in Fig. 37, it has been possible to test the standard model, in this case using the Los Alamos Meson Physics Facility, in the scattering of neutrinos by electrons, a notoriously low-cross-section process. These data represent a technological *tour de force* and it is encouraging that the standard model appears able to reproduce them in acceptable fashion.

F. Positrons from Heavy Nuclear Collisions

Before leaving the question of heavy-ion collisions, I must mention a much studied and still puzzling new phenomenon at the interface with atomic physics. Several years ago, Greenberg and his collaborators showed that the velocities characteristic of very heavy ions in collision are so much smaller than those of orbital electrons, and the ratio of dimensions similarly small, that during heavy-ion collisions transient electronic shell structures form about the collision partners, and X radiation and Auger electrons are observable into and out of these shells. Clearly in this way it becomes possible to prepare transient atomic systems with effective element numbers in the range up to 188 and above where quantum-electrodynamic and relativistic effects, rather than being high-order corrections, become dominant phenomena. In studies on collisions of nuclei in the uranium region, Greenberg and collaborators, as well as groups led by Kienle and by Kankeleit, several years ago, found the surprising results shown in Fig. 38. Superimposed on the now well-understood continuous positron spectra was a sharp positron line indicating the presence of a long-lived entity asso-

Neutrino-electron scattering produces recoil electrons in the forward direction. The experimental results show clearly this forward peak.

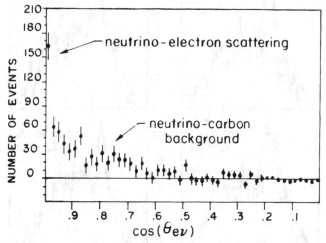

From the events in the forward peak, the experimenters extract the cross-section,

$$\sigma(\nu_e e^-, exp) = 9.8 \pm 2.5 \; E_\nu(GeV) \times 10^{-42} \; cm^2$$

This agrees well with the standard model (WSG) prediction, which includes the W–Z interference,

$$\sigma(\nu_e e^-, theory) = 9.2 \; E_\nu (GeV) \times 10^{-42} \; cm^2$$

Experiment 225-Los Alamos Meson Physics Facility

FIGURE 37 Preliminary results from a study of neutrino–electron scattering carried out at the Los Alamos Meson Physics Facility. The experimental cross section is well reproduced by the standard model prediction which included W–Z interference. (Courtesy of G. T. Garvey, Los Alamos.)

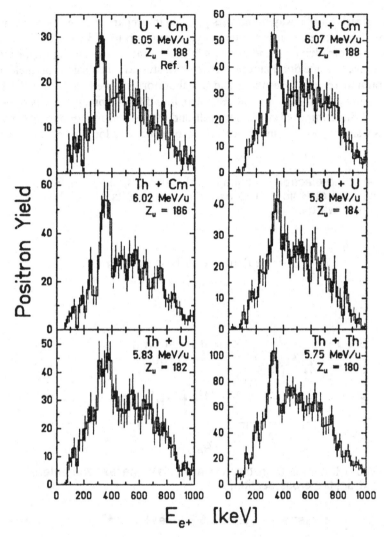

FIGURE 38 Positron spectra from selected heavy nuclear collisions in the regions of the respective Coulomb barriers. While the underlying continuous positron spectrum is now relatively well understood, the sharp peak in the vicinity of 350 keV is not at all understood. In some of these data there appears possibly statistically significant evidence for more than a single peak. In the sum spectra, in addition to the main peak, there now appear to be several additional higher energy peaks. (Courtesy of J. S. Greenberg, Yale.)

ciated with the collisions. More recently, in a very difficult experiment—inasmuch as the electron backgrounds are some 10^7 larger than those of positrons—Greenberg and associates showed that these positron lines were in time coincidence with electron lines of equal energy and that the peak obtained by summing the electron and positron lines was sharper than either by itself, suggesting cancellation of Doppler shifts in back-to-back emission from a source at rest in the collision center-of-mass.

These data have given rise to much theoretical activity and speculation, but as yet the origin of these monoenergetic leptons associated with heavy ion collisions remains a puzzle. The obvious suggestion that they arise in the decay of a new particle such as the much-sought axion appears to be ruled out by other experimental data, and resolution of the source of these leptons from super-atoms awaits further experimental work both on heavy-ion collisions and on direct electron–positron scattering at high resolution.

6. ATOMIC PHYSICS

In turning to somewhat more standard atomic physics I am reminded of I. I. Rabi's comment:

> *The twentieth century started in 1897 with Thomson's discovery of the electron.*

To a remarkable degree, twentieth-century technology has been based on our ability to manipulate electrons and atomic structure.

A. Trapped Atoms

One of the most interesting recent discoveries, originally suggested by Hänsch and Schawlow, and independently by Wineland and Dehmelt, is that the mechanical forces exerted by laser light can dramatically lower the temperature of a sample of atoms or ions, allowing very-high-resolution atomic spectroscopy and ultra-low-temperature atomic energy studies. Laser cooling slows the atoms to a virtual standstill from their initial velocity of some 1000 m/sec. Once cooled, slowly-moving ions and atoms can be trapped for indefinite periods in magnetic bottles and held for ultra-precise measurement. Figure 39 shows the intersection of a laser beam from the left and a beam of sodium atoms from the right. The slowed atoms, now virtually at rest at the end of the solenoid, fan out into the skirt about the laser beam. The trapping limitation is currently set by the levels of residual vacua attainable and thus the collision rate with residual atoms; periods of many hours appear attainable in the near future.

An alternative trapping technique involves use of the dipole mechanical forces exerted by laser beams. This technique was first suggested by Letokhov in 1968 and modified by Ashkin in 1978. Figure 40 shows a schematic view of the laser trapping system, and Fig. 41 shows a sample of about 200 000 sodium atoms trapped as what is now known as *optical molasses* at the intersection of six laser beams. The ball of atoms is some 7 mm in diameter and is made visible by the light that the atoms scatter out of the laser beams. The temperature of the ball is approximately 240 μK.

These cooling and trapping techniques are quite new, and application to a wide range of problems in atomic and other physics is only now beginning.

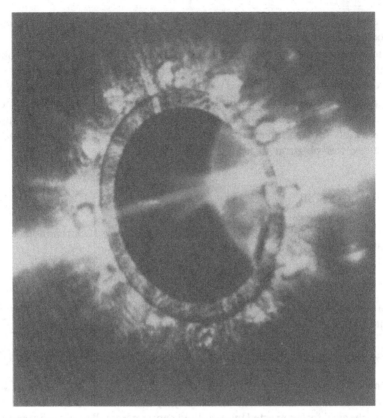

FIGURE 39   A beam of sodium atoms traversing a magnetic solenoid from the right-hand side meets a laser beam entering from the left-hand side of the figure. Having been brought to a halt by the radiation pressure of the laser light, the sodium atoms fan out into the conical skirt that appears in this photograph about the laser beam. (Courtesy of William D. Philips.)

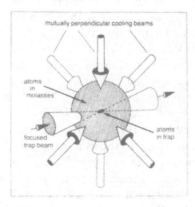

FIGURE 40   A highly schematic illustration of the multi-laser system used to produce optical molasses and to trap atomic species by radiation pressure exerted by the laser beams.

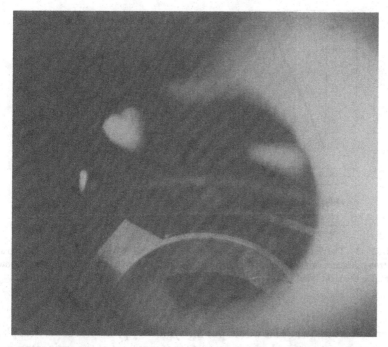

FIGURE 41 An illustration of optical molasses holding approximately 200 000 sodium atoms at a temperature of about 240 μK. The trapped sample, which appears as a small ball in the center of the photograph, is roughly 7 mm in diameter. (Courtesy of S. Chu, AT&T Bell Laboratories.)

B. Trapped Elementary Particles

Dehmelt and Gabrielse have used a strong magnetic field and an electrostatic quadrupole field in a Penning Trap configuration, and in 1984 held a single electron in the trap for ten months. Figure 42 shows a remarkable measurement made on that electron interacting with an electromagnetic wave. The single electron begins to respond strongly at the rest-mass cyclotron frequency, on the downward sweep, reaching an energy of 0.8 eV in this figure before jumping back to noise level. On the upsweep, in contrast, no resonance is observed, resulting in the striking hysteresis curve shown in the right panel.

A bistable hysteretic interaction involving the simplest of all electromagnetic systems—a lone electron in an electromagnetic field—might seem surprising and something to be expected only of a much more complex system; bistable optical interaction of radiation with matter is well known, for example, but there it depends upon nonlinear macroscopic properties of the medium involved. The single-electron hysteresis turns out to be a simple consequence of the relativistic mass increase of the electron with increasing velocity, and the effect is observable at kinetic energies below 10 eV where the mass increase is only a few parts per million. More recently, the Seattle group has succeeded in trapping a single proton, and their goal is to trap a single antiproton in order to measure its parameters with a precision comparable to that with which those of

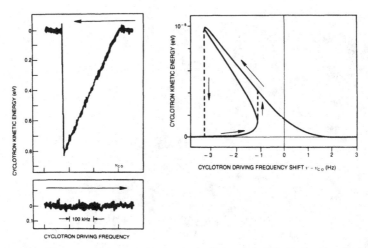

FIGURE 42   The response of a single electron in the University of Washington Penning Trap as a cyclotron resonance driving frequency is swept downward (upper figure) and then upward (lower figure). On the right-hand side is the hysteresis curve. (Courtesy of G. Gabrielse, Seattle.)

the proton are known. This would put the symmetry principles of particle physics to a stringent test, and a very accurate measurement of the antiproton mass would provide a precise test of the fundamental *CPT* theorem.

C. Ultra-precise Measurements

Atomic physics continues to set the standards for precision in physical measurements. This is illustrated, for example, in Fig. 43 showing the achievement of an improvement by a factor of more than 10^6 in attainable precision during the past twelve years. The top panel shows a conventional infrared absorption spectrum in the wavelength region of 10 microns; the next magnification below it, obtained with diode lasers, shows an expanded section of that spectrum. Although entirely new structure appears in this section of the spectrum, the resolution is limited by Doppler broadening that reflects the thermal motion of the molecules in the sample under study. In the next expansion of the spectrum, the Doppler broadening has been avoided by use of saturation spectroscopy; here the resolution is limited only by the frequency instability of the laser. Again, however, this expansion reveals previously hidden structure in the molecular spectrum. Finally, in the last magnification, the jitter in the laser frequency has been sharply reduced as a result of electronic control and the resolution in this part of the figure is typical of the maximum that can be obtained; the limitation on the resolution here follows directly from the Heisenberg uncertainty principle and is fundamentally imposed by the finite time during which the molecules are under observation. Here, again, a single sharp line in the previous magnification has been revealed to comprise a whole complex of spectral lines.

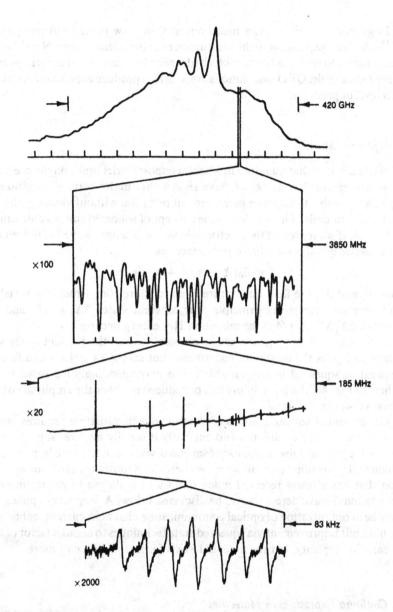

FIGURE 43 Molecular spectroscopy, as in the case of sulfur hexafluoride shown here, has increased in resolution by a factor of more than a million within the past twelve years. Shown here are spectra of sulfur hexafluoride with progressive increases in resolution, from top to bottom, by factors of 100, 20, and 2000, respectively. The resolution in the lowest panel is limited simply by the Heisenberg uncertainty principle and is fundamentally set by the finite time during which the molecules were under observation. (Courtesy of the National Research Council.)

In general, atomic physics measurements are now possible to one part in 10^{15}–10^{16}, corresponding to the measurement of the distance from New York to Los Angeles to within a fraction of a hydrogen atom, and, what is perhaps even more remarkable, QED calculations are able to reproduce experimental data to this level of precision.

D. Squeezed States

While it is true that quantum mechanics imposes strict limits on the precision of measurement, Slucker *et al.* have shown that under certain conditions it becomes possible to improve measurement precision without violating the uncertainty principle. This has led to the concept of squeezed states as illustrated in Fig. 44. If we represent the electric field vector of a light wave (or indeed any appropriate quantum oscillator parameter) as

$$E = E_0(X_1 \cos \omega t + X_2 \sin \omega t)$$

then X_1 and X_2, the field quadrature operators, are complementary variables and obey an uncertainty principle. For coherent states $\Delta X_1 = \Delta X_2$ and the product $\Delta X_1 \Delta X_2 = 1/4$ is the minimum Heisenberg uncertainty.

For a *squeezed state,* which has no classical analog, the product of the variances still gives the minimum uncertainty, but $\Delta X_1 \neq \Delta X_2$. Noise can thus be reduced or squeezed in one variable if it is correspondingly increased in the other. In Fig. 44, the uncertainty can be reduced in either the amplitude or the phase of an electric wave.

Generation of squeezed states of electromagnetic radiation requires phase-sensitive nonlinear techniques and currently it usually involves some form of four-wave mixing. Obviously, use of squeezed-wave techniques to improve precision will be justified only under rather heroic circumstances, but one application that has already received much discussion is its use to reduce noise in gravitational-wave detectors—to be discussed below. A long-term application may be in optimization of optical communication channels; current technology would limit improvement via squeezed state techniques to about a factor of two in carrying capacity, but the potential exists for factors of ten or more.

E. Coulomb Explosions in Molecules

Figure 45 illustrates a new and powerful technique for studying molecular systems as developed by Gemmel and his collaborators, making use of the so-called Coulomb explosion technique. When a molecular ion—in this case $C_2H_3{}^+$—traverses a thin target foil, the molecular bonds are broken and the component nuclei are driven apart by mutual Coulomb repulsion. When these nuclei are detected in a two-dimensionally sensitive detector, the striking pattern shown in the lower panel of the figure is obtained and, from measurements on it, the details of the original molecular wave function can be disentangled.

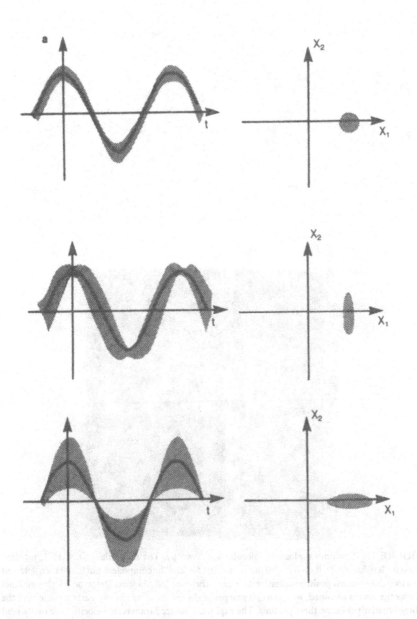

FIGURE 44 An illustration of squeezed states as an approach to reducing the attainable quantum noise limits. The time variation of the electric vector of an electromagnetic field is shown by the black sinusoidal curve in all three panels of this figure. The cross-hatched bands indicate the uncertainty for a coherent state in the upper panel and for states with reduced fluctuations in the amplitude in the central panel and in the phase in the lower panel. The corresponding X_1 and X_2 graphs on the right-hand side of the figure show the corresponding variances in these quadrature amplitudes. (Courtesy of C. M. Caves.)

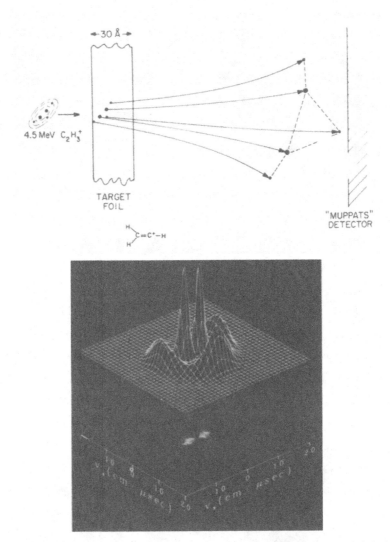

FIGURE 45   Coulomb explosion molecular spectroscopy. The molecular ion, $C_2H_3^+$, incident from the left-hand side, is dissociated in a thin target foil and the component nuclei are then detected in a two-dimensional position-sensitive detector—the so-called Muppats Detector. The lower panel shows the results obtained, with the two sharper peaks representing the two carbon nuclei and the three broader peaks the three protons. The explosion-induced transverse velocities are easily read from the lower figure. (Courtesy of D. Gemmel, Argonne.)

F. The X-Ray Laser

The search for an x-ray laser, because of its potential applications in holography of biological systems, lithography for integrated circuit production and as a tool for fundamental spectroscopy, has been in progress almost since the discovery of the laser itself. In 1985, Matthews and his Livermore group reported the first definite evidence for lasing in the soft x-ray wavelength range. In these studies, two very-high-power laser beams (5×10^{13} W/cm^2) vaporized a selenium foil, creating a plasma of neonlike ions which were subsequently excited by collisions with electrons in the plasma. Subsequent transitions between $J = 2$ and $J = 1$ states of the $2p^5 3p$ and $2p^5 3s$ levels produced amplified radiation at 20.63 and 20.96 nm as shown in Fig. 46. These two wavelengths were amplified by factors of about 700 as compared to the intensity of spontaneous emission lines. The use of neon-like ions follows a suggestion of Zharikin of the Moscow Institute of Spectroscopy.

Sukewer (of Princeton Plasma Physics Laboratory) and his associates have also reported amplification of soft x rays by factors of ~ 100; in their experiments, laser light was used to vaporize a carbon target and ionize the resulting atoms completely. While trapped magnetically, the resulting plasma cools radiatively as electrons repopulate the higher states, and the resulting hydrogen-like ions have population inversions between levels with principal quantum numbers $n = 3$ and $n = 2$ with output radiation at 18.2 nm.

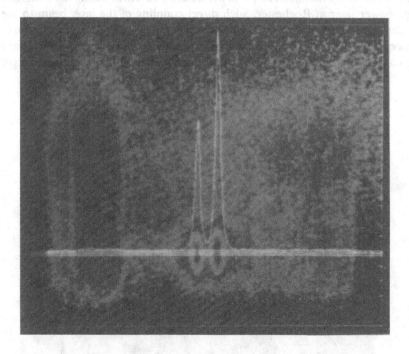

FIGURE 46   Evidence for soft-x-ray lasing obtained in a selenium plasma at Livermore. The lasing lines appear in this computer-enhanced photograph of data from a transmission grating spectrograph. In the plot, wavelength is on the horizontal axis and time on the vertical one. The superimposed white curve is a microphotodensitometer trace plotting intensity versus wavelength with the two amplified peaks appearing at 20.6 and 20.9 nm. (Courtesy of H. D. Shay, Livermore.)

7. PLASMA PHYSICS

A. High-Power Lasers

The development of high-power lasers, such as those used at Livermore, was, of course, driven by the thermonuclear fusion power program. Substantial progress has been reported in this area as well as in the magnetically confined plasma approach to fusion. Figure 47 is an overview of the NOVA laser system at Livermore—currently the world's most powerful such installation. A single, low-energy, nanosecond pulse of light is split into ten parallel paths with partially reflecting mirrors. Each of these ten pulses is then amplified by 400-foot-long chains of 16 glass laser amplifiers grouped in successively larger diameter stages. The resulting high-energy, 2.5-foot-diameter, infrared light pulses are steered toward targets using large movable mirrors. As these infrared laser beams approach the 16-foot-diameter vacuum vessel inside which the target is mounted, they pass through precision mosaic arrays of crystal plates that can be adjusted to convert the infrared beams to either green or ultraviolet light through frequency doubling or tripling.

Figure 48 is a streak camera photograph showing the compression and ignition of a glass microballoon target of deuterium and tritium using the OMEGA high-power laser at Rochester. Such direct coupling of the laser beam to the fusion target has now passed out of favor, as far as an approach to fusion is concerned, in favor of an indirect approach wherein the laser energy is delivered into a high-atomic-number cavity (*hohlraum*) surrounding the fusion target

FIGURE 47   An artist's sketch of the NOVA laser facility at Livermore. The technician in the foreground gives some sense of the scale involved. (Courtesy of Lawrence Livermore National Laboratory.)

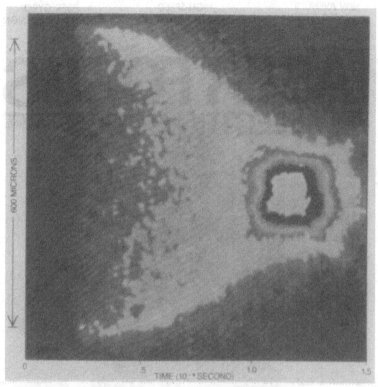

FIGURE 48   A streak camera photograph of the compression and ignition of a microballoon glass target containing deuterium and tritium gas. This photograph was taken at Rochester using the OMEGA laser facility. (Courtesy of R. McRory, Rochester.)

and the resulting soft x rays compress and ignite the target. Although in the USA much of this work remains classified for military reasons, the Japanese, even in 1984, were already publishing data on this approach, as shown in Fig. 49.

Currently, in the open literature, the highest temperatures and power densities have been reported by Tsahiris *et al.*, in Europe, using 300-psec pulses of 1.3-nm light from the ASTERIX III iodine laser into gold cavities from 250 to 1000 nm in dimension, to yield power densities of $\sim 3 \times 10^{13}$ W/cm^2. Very recently, however, this group has reported power densities of 4×10^{15} W/cm^2 corresponding to an effective temperature of 5×10^6 K.

In this area of high-power lasers, McCrory and his collaborators at Rochester have recently announced the development of a *chirped pulse* laser amplification system that is capable of power densities of 10^{18} W/cm^2. Chirped pulse amplification expands and compresses the duration of short laser pulses using optical fiber and diffraction grating techniques. This same Rochester group has also developed an electro-optical sampling device capable of measuring transient electronic signals with hundred femtosecond (10^{-13} sec) resolution, an improvement by a factor of ~ 100 on preexisting state-of-the-art devices.

NEW AVENUES   HIGH-SPEED   FIBER GYRO
FOR LASER FUSION   PHOTON DETECTORS   THEORY AND DESIGN

A PennWell Publication   February 1984

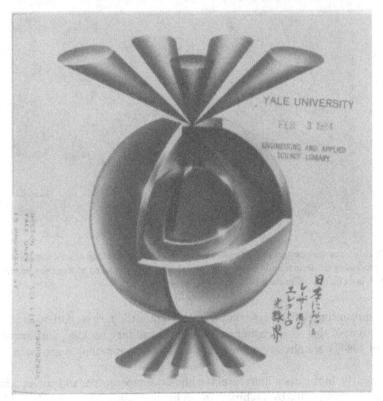

LASERS AND ELECTRO-OPTICS IN JAPAN

FIGURE 49   The cover of the February 1984 issue of the *Laser Focus* magazine, which featured laser and electrooptic activities in Japan. The cover photograph shows a fusion target surrounded by a heavy-metal *hohlraum*, which converts the energy of the incident radiation into soft-x-rays which in turn compress the target. Details of such studies are classified in the United States but are partially available in the open literature both in Japan and in Europe.

B. Free-Electron Lasers

Since its development by Madey at Stanford, the free-electron laser has been an attractive candidate for high-laser-power applications but, until recently, no such device was able to convert more than 5% of the electron beam power into microwave radiation. Recently a Livermore–Berkeley team led by Sessler has made a dramatic improvement in this efficiency. In all free-electron lasers, the output radiation of wavelength λ_s is produced as the electrons traverse a spatially periodic, transverse magnetic field or wiggler with "wavelength" λ_w. The resonant condition is that the optical wave travel one optical wavelength farther than a beam electron during the time the electron travels λ_w. This results in the wiggler equation

$$\lambda_s = \frac{\lambda_w}{2\gamma^2}\{1 + a_w^2/2\}$$

where γ is the electron's Lorentz factor and a_w is a signature parameter for the specific wiggler—a parameter proportional to the product of the wiggler wavelength and magnetic field. Clearly λ_s varies inversely with the square of the electron energy, so that as energy is extracted from the beam the resonance condition is destroyed. Sessler *et al.* avoided this problem by tapering the wiggler and changing a_w. The magnetic field intensity measured by a_w is constant for about one meter, then decreases to maintain the resonance condition as the electrons lose energy to the laser beam. As a consequence, the laser power grows along the wiggler length. The experimental results are shown in Fig. 50.

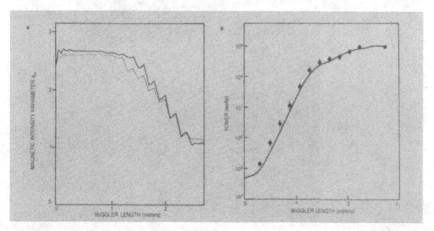

FIGURE 50   Results with the Berkeley–Livermore free electron laser. The figure on the left-hand side shows the field profile of the tapered wiggler and that on the right-hand side the corresponding power output. The magnetic field intensity measured by the parameter a_w is constant for about one meter, then decreases to maintain the resonance condition as the electrons lose energy to the laser beam. As a consequence, the laser power grows along the wiggler length. The experimental results are shown as the dark curve on the left and as the data points on the right, while the numerical evaluations are shown by the light curve on the left and by the solid curve on the right. The degree to which these reproduce the experimental data is gratifying. (Courtesy of A. Sessler, California, Berkeley.)

Decreasing the magnetic field strength along the length of the wiggler enabled the efficiency to be increased from below 5% to in excess of 40%, and resulted in an output power amplification in the wiggler of over four orders of magnitude, as shown on the right of the figure. Further improvements are almost certainly possible in this very rapidly moving field of high-power lasers and their applications.

C. The Status of Fusion Power

So where do we stand with respect to thermonuclear power? Figure 51 provides a summary overview. The crude measure of success in this field has traditionally been the Lawson criterion $nt = 10^{14}$ cm^{-3} sec. The contour curves on the figure indicate the gain achieved with deuterium–tritium fuel. The shaded region is that which will lead to ignition of that fuel. A variety of magnetic and inertial confinement systems are shown indicating the steady progress that is being made toward achieving ignition. None of the existing devices have satisfied all the conditions for ignition simultaneously, although several are close to that goal. But because the engineering of an economically viable fusion power system will be orders of magnitude more demanding than was the case for

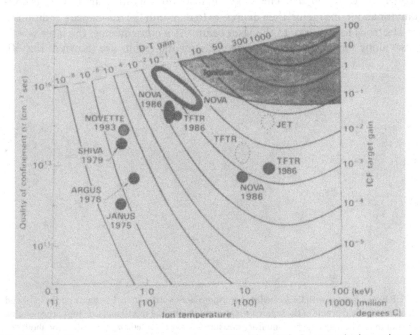

FIGURE 51 The search for thermonuclear fusion. Plotted here is the Lawson criterion against the achieved ionic temperature in the plasma. A variety of magnetic and inertial confinement systems are shown indicating the steady progress that is being made toward achieving ignition. It bears emphasis, however, that the engineering problem that remains in converting a laboratory demonstration into an economic power source is truly formidable. (Courtesy of Lawrence Livermore National Laboratory.)

FIGURE 52   A view inside the Princeton TFTR Tokomak. Some of the engineering complexity referred to in the text is evident in this photograph where a workman adjusts some of the complex magnetic windings required to control the Tokomak plasma. (Courtesy of H. Furth, Princeton.)

fission, economic fusion power lies at least several decades into the future. Figure 52, which shows part of the TFTR Tokomak at Princeton, illustrates the complexity of the magnetic systems involved.

8. CONDENSED MATTER PHYSICS

Few areas of modern science have been studied as intensively as the physics of semiconducting material; semiconductors are a hallmark of our era. But still, as indeed in all of physics, even our most thoroughly studied subfields continue to present us with dramatic surprises.

A. Quantized Hall Effect

Among these recent surprises was the integral quantized Hall effect for which von Klitzing received the 1985 Nobel prize in physics. Figure 53 illustrates this effect. The ratio of the voltage drop across a sample to the current flowing along it is called the Hall resistance. In the normal Hall effect, that resistance increases steadily and linearly as the strength of the magnetic field is increased. In the quantized Hall effect, in contrast, as shown in the upper panel of Fig. 53, the Hall resistance shows plateaus which coincide with the disappearance of the sample's electrical resistance. On the plateaus, the Hall resistance remains constant while the magnetic field strength is varied. At each of these plateaus, the Hall resistance is precisely given by h/ne^2, where n is integral, and thus provides a new approach to the determination of h and e as well as a very convenient calibration technique in precise measurements.

It has been possible to understand these phenomena in terms of the systematic lowering of the Fermi level in the sample relative to its Landau bands; localized states created by the presence of impurity atoms in the sample are crucial to this understanding since they act as electron reservoirs, so that over a range of magnetic fields the extended states in the Landau band are either completely empty or completely filled.

The fractional quantum Hall effect is more complex. For example, when the lowest Landau band is one-third filled, a plateau is observed experimentally where the Hall resistance equals $3h/e^2$. In order to understand this special stability of fractionally filled bands it has been necessary to include explicitly the electron–electron interactions and to use a wave function that depends simultaneously on the positions of all particles in the system.

Laughlin, in 1983, succeeded in constructing a wave function with the required stability when the fraction of filled states had values such as 1/3, 1/5, 2/3, 4/5, and 6/7—i.e., to the reciprocal of an odd integer or to one minus such a fraction. It is an interesting prediction of such a model that if one adds an extra electron to a system in which the Landau level is one-third full, the extra charge should appear in three distinct places in the sample and at each place precisely $\frac{1}{3}e$ should appear. These fractional charges, now called quasiparticles, behave very much like normal charged particles, and it is their behavior which is now believed to underlie the fractional Hall effect.

B. Aharonov–Bohm Effect

Another phenomenon that yields information on these same fundamental constants is the Aharonov–Bohm (AB) effect first discussed in 1959 but recently observed in a new guise by Webb and his colleagues at IBM. As originally proposed, the AB effect was an interference phenomenon expected for an electron propagating around a magnetic field in a field-free vacuum. The system used by Webb *et al.* is shown in Fig. 54. A gold ring, one micron in diameter (see top panel) provides two alternate paths for current flowing from left to right.

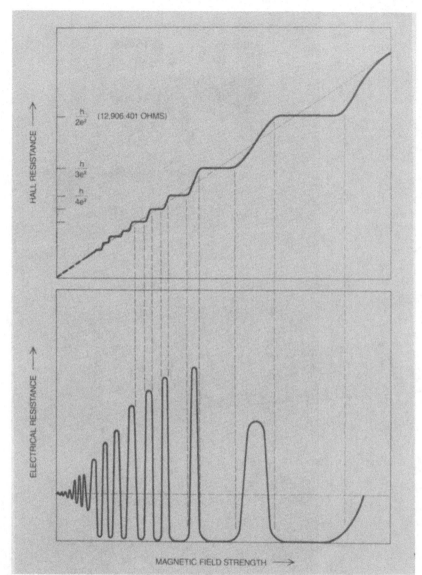

FIGURE 53 The quantized Hall effect. As shown in the upper panel, the Hall resistance shows plateaus which coincide with the disappearance of the sample's electrical resistance. On the plateaus, the Hall resistance remains constant while the magnetic field strength is varied. At each of these plateaus, the value of the Hall resistance is precisely equal to Planck's constant divided by an integer multiple of the square of the electron charge. (Courtesy of B. I. Halperin, Harvard.)

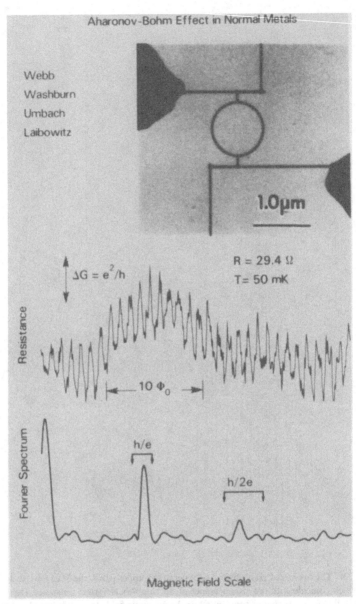

FIGURE 54   The Aharonov–Bohm Effect in a normal metal (Webb *et al.*, IBM). The gold ring shown in the top panel was fabricated using integrated circuit techniques and consists of a poly-crystalline gold ring with a diameter of 1 micron and a line width of 0.03 microns. The central figure shows the interference effect as the field is varied. The principal period of oscillation corresponds to a field change of 76 G, which in turn corresponds to a flux change of h/e through the ring. The Fourier transform of the data in the central panel (shown in the lower one) also shows a second harmonic peak corresponding to a field change of 38 G or to a flux period corresponding to $h/2e$. The enhancement in the Fourier transform near zero field corresponds to a longer scale aperiodic fluctuation of the magnetoresistance reflecting the magnetic flux that traverses the metal of the ring itself. (Courtesy of P. Chaudhari, IBM, Yorktown.)

The Aharonov–Bohm effect appears as a change in the magnetoresistance as a current is sent through the ring and the magnetic flux threading it is varied. The central figure shows the quantum interference effect clearly as the field is varied. The principal period of oscillation corresponds to a field change of 76 G, which in turn corresponds to a flux change of h/e through the ring. The Fourier transform of the data (shown in the lower panel) also shows a second-harmonic peak corresponding to a field change of 38 G or to a flux period corresponding to $h/2e$. Both of these peaks persist out to field intensities of 80 kG—in contrast to other measurements made with cylindrical samples where the second harmonic peak very rapidly damps out.

Recent observation of h/e AB oscillations in a gallium arsenide heterostructure, by Datta and his colleagues at Purdue, suggests the possibility of application in relatively large semiconductor devices where the electron wavelengths are much longer than in metals.

C. High-Temperature Superconductivity

Beyond question, however, the most dramatic development in condensed matter physics in this past year has been the discovery of high-temperature superconducting materials. The early story is truly an international one—and one that does not depend upon the availability of huge machines or large groups. The original focus on a ceramic appears to have come from work of Michel and Ravenau in the early 1980s at the University of Caen in France; in 1983 this was picked up by Müller of IBM (Zurich) while reading in a monastery garden in Sicily. Returning to Zurich he teamed up with Bednorz and found that a lanthanum-barium-copper ceramic gave evidence of superconductivity at 35 K, higher than any previously reported. Müller and Bednorz published their results in *Zeitschrift für Physik* in mid-1986 but stirred up relatively little interest except in the laboratories of Tanaka in Tokyo and of Chu in Houston. All three groups—Zurich, Tokyo, and Houston—reported their latest results on December 18, 1986 on the last day of The American Materials Society meeting in Boston; by the following day literally hundreds of laboratories, worldwide, were attempting to fabricate and study the new ceramics. It soon became clear that while almost anyone could enter the game, only those with access to sophisticated measuring equipment (e.g., for Meissner effect studies) could remain in it for long. By January 28, 1987, Chu and his colleagues had reported superconductivity at 93 K in an ytterbium-barium-copper-oxide ceramic.

Figure 55 plots the transition temperatures attained as a function of time. It bears noting that on the basis of simple extrapolation we would not have expected these new high-temperature materials before about 2190!

As a consequence of detailed x-ray and neutron diffraction studies on the ceramic samples the structure has been identified as a perovskite. It was also found that its resistivity was strongly anisotropic and that it depended critically upon the oxygen content, which in turn was very dependent upon the exact fabrication recipe. Figure 56 shows the structure as now understood.

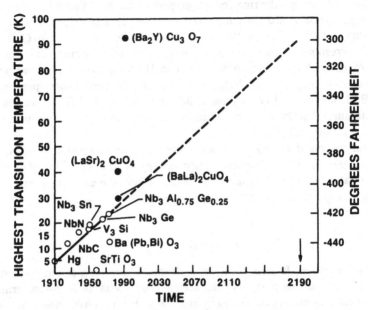

FIGURE 55   A plot of the highest transition temperature, in degrees Kelvin, for attainable super-conductors, as a function of time. Simple extrapolation of the progress over the past 75 years would have suggested that the present high-temperature material might not have been expected before the year 2190! (Courtesy of S. Buchsbaum, AT&T Bell Laboratories.)

More recently, there have been repeated reports of much higher transition temperatures: Stacy and Cohen of Berkeley reported a dramatic resistance drop between 280 and 292 K but were unable to repeat the measurement; Chu reported loss of all resistance at 225 K but then found that only 1% of the sample showed the Meissner effect; by replacing some of the ceramic's oxygen with fluorine, Ovshinsky produced a material showing bulk superconductivity at 155 K and filamentary Meissner effect at 260 K, but again these measurements appear to have been nonreproducible—presumably reflecting instability in the samples.

Obviously, apart from its intrinsic interest, work on these high-temperature superconductors is being driven by the prospect of important and widespread applications. But a number of critical and very challenging basic questions remain to be answered before substantial application appears feasible. Among them are: how to increase the critical current density; how to control the oxygen content; how to improve the stability over time; how to improve the formability of the ceramic; and how to raise the critical temperatures even higher—with the ultimate goal being a T_c comfortably above room temperature.

Chaudhari and his collaborators at IBM (Yorktown) have fabricated thin films of 90 K superconductor that carry currents in excess of 10^5 A/cm^2 at 77 K and up to 2×10^6 A/cm^2 at 4.2 K. Geballe and his collaborators at Stanford

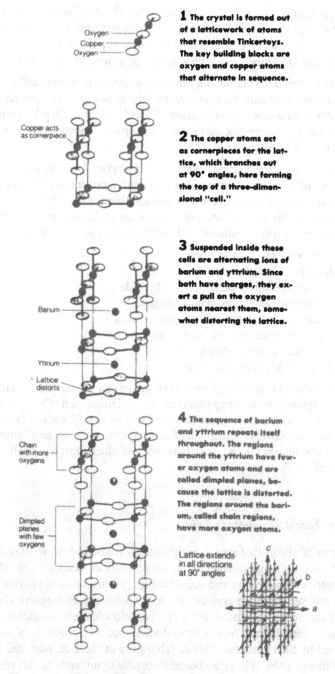

1 The crystal is formed out of a latticework of atoms that resemble Tinkertoys. The key building blocks are oxygen and copper atoms that alternate in sequence.

Oxygen
Copper
Oxygen

2 The copper atoms act as cornerpieces for the lattice, which branches out at 90° angles, here forming the top of a three-dimensional "cell."

Copper acts as cornerpiece

3 Suspended inside these cells are alternating ions of barium and yttrium. Since both have charges, they exert a pull on the oxygen atoms nearest them, somewhat distorting the lattice.

Barium

Yttrium

Lattice distorts

4 The sequence of barium and yttrium repeats itself throughout. The regions around the yttrium have fewer oxygen atoms and are called dimpled planes, because the lattice is distorted. The regions around the barium, called chain regions, have more oxygen atoms.

Chain with more oxygens

Dimpled planes with few oxygens

Lattice extends in all directions at 90° angles

c
b
a

FIGURE 56 The structure of the high-temperature superconducting ceramic fabricated from yttrium, barium, and copper oxide has been determined through intensive x-ray and neutron diffraction studies. The material is highly asymmetric as is the superconductivity in it and appears to depend critically upon the oxygen content of the ceramic. (Courtesy of G. Maranto, *Discover Magazine,* August 1987.)

have reported similar thin films carrying 7×10^5 A/cm^2 at 77 K and 5×10^7 A/cm^2 at 4.2 K. These values compare favorably with the 10^6 A/cm^2 capacity of most current metallic, e.g., niobium–titanium, superconductors.

Early last December, in a review article in *American Scientist,* I speculated:

> *I am confident that within the coming decade materials will be discovered that have transition temperatures into the superconducting state well above the temperature of liquid hydrogen. On a longer time scale, I am optimistic that room temperature superconducting material will be found. . .*

As usual, with speculations of this sort in physics I was far too conservative! And I would now move forward my speculation concerning the availability of room temperature superconductors to some time before 1990.

Turning briefly to the matter of application of the new superconductors, in some order of expected utilization I would have to list the following:

> Computer interconnections
> Josephson junction devices
> Semiconductor–superconductor hybrids
> Nuclear magnetic resonance imaging
> Underground electrical transmission cables
> Large electrical generators and motors
> Electrically powered automobiles
> Magnetically levitated trains.

Figure 57 shows a state-of-the-art NMR image of a human head—a new noninvasive diagnostic technology that is revolutionizing much of medicine. Higher-field superconducting magnets for use in this work will not only permit better resolution but will also permit finer focusing of the scans on particular elements to yield new insight into internal dynamics and biochemistry—the fundamental processes of life itself.

D. Icosahedral Symmetry in Solids

Figure 58 shows that even in as thoroughly studied a field as crystallography, where it has been stated categorically for generations that all the possible crystal symmetries were present and accounted for, nature can still provide surprises. What one sees here are crystals of a new aluminum–manganese alloy having clearly defined icosahedral symmetry (five-fold symmetry) which is supposedly strictly forbidden in two or three dimensions. The original observation was reported by Shechtman and his collaborators at the U.S. National Bureau of Standards in 1984. It has now become possible to understand this very surprising result as the result of a quasi-periodic lattice formed by the intersections of families of parallel planes of atoms whose characteristic spacings alternate between two fixed values in a Fibonacci sequence—the same sequence (ordered in a way neither periodic nor random) that appears repeatedly in nature, for

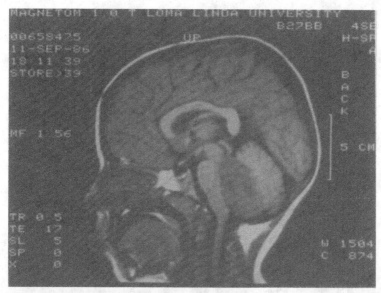

FIGURE 57 A nuclear magnetic resonance scan of the head of a child showing a large tumor, just behind the brain stem, that is already distorting the brain substantially. This is the kind of diagnostic information that provides a whole new dimension in medical care. This particular scan was carried out at a magnetic field of 1 T; the substantially higher fields that may eventually be available using high-temperature superconductors would be reflected in even more detail in these NMR scans. (Courtesy of Loma Linda University.)

example in the sunflower seed pattern. It is also a form of three-dimensional Penrose tiling—but most of all it is highly gratifying that in as mature a branch of physics as crystallography such complete surprises can still occur.

E. Tailored Materials

In a very major way, we have entered the era of tailored materials. We can now fabricate solids and surfaces having desirable properties that in the past were simply unattainable. A good example is the area of ion beam mixing. Many alloys that would be predicted theoretically to have desirable properties—the A-15 alloys in conventional superconductivity, for example—cannot be fabricated with any of the standard techniques. By laying down alternate or sequential layers of the desired materials in the relative amounts desired in the final alloy, and then subjecting the resulting layered structure to ionic bombardment, the heating along the ionic track mixes the components intimately and there is no opportunity for segregation or separation before the alloy freezes. This is of growing importance in the fabrication of corrosion- and wear-resistant as well as catalytically active surfaces.

Figure 59 is a micrograph of a superlattice fabricated by means of molecular beam epitaxy by Otsuka of Purdue. It is typical of the microscopic structures that can now be produced routinely. For the first time it has been possible to

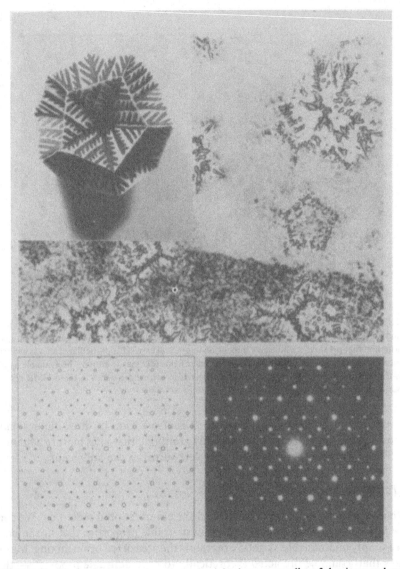

FIGURE 58   The microphotograph on the upper right shows a new alloy of aluminum and manganese which solidifies with an icosahedral symmetry—a symmetry strictly forbidden for crystals. The figure at the upper left shows a model of the crystal structure of this new alloy including the snowflake grains which are characteristic of it. At the lower right is the Laue diffraction pattern for this particular alloy and at the left is the pattern that would be calculated for an ideal icosahedral quasicrystal. (The microphotograph model and diffraction pattern are courtesy of R. Shaefer; the calculated scattering pattern is courtesy of LeVine and Steinhardt.)

FIGURE 59 A three-layer superlattice constructed using molecular beam epitaxy is here enlarged by 800 000 diameters in a dark field electron micrograph. The light bands are zinc-selenium and the major dark ones are zinc-manganese-selenium; each thin dark stripe is a single unit cell of manganese-selenium. (Courtesy of Nobuo Otsuka, Purdue University.)

actually fabricate the quantum-wells that occupy the first chapters of all quantum mechanics textbooks and experimentally study their characteristics. In such structures quantum-size effects are observed if the thickness of the layers in which the charge carriers are confined is smaller than the Bohr radius. As a consequence, the electronic properties of the layered quantum-well materials are entirely different from those of bulk material having the same composition. Quantum-well structures are already in use in new laser structures, nonlinear optical elements having very large susceptibility at room temperatures, optical modulators and so-called self-electro-optic devices (SEED) based on the quantum-confined Stark effect (QCSE).

In addition to fabricating microstructures, we have learned how to use ion and electron beams to carve the desired structures from bulk material. The level of precision attained here is illustrated in Fig. 60 from work of Wolf and his collaborators at the Cornell National Research and Resource Facility for Submicron Structures.

F. Temperature Behavior of Materials

One of the major challenges currently facing condensed-matter physicists and materials scientists is that of improving high-temperature performance. It is a matter of record that in essentially every instance the efficiency of industrial processes is ultimately limited by the high-temperature performance of some

FIGURE 60 Modern electron beam techniques have advanced to the point where submicron molecular-level fabrication has become one of the frontiers of technology. In this figure, a composite photograph, the upper panel shows an intact tobacco mosaic virus taken at the same magnification as the pattern below it which was etched directly into a sodium chloride film using a 100 keV focused electron beam. The width of the lines in the pattern is less than 2 nm and the degree of control of the electron beam at this scale is well illustrated by the precision with which the letters have been etched. The nanometer scale for the entire micrograph is shown at the lower right; the magnification is approximately 500 000. (Courtesy of E. D. Wolfe of the National Research and Resource Facility for Submicron Structures, Cornell University.)

component. Indeed it has been estimated in a recent Oak Ridge National Laboratory study that each degree Fahrenheit by which the operating temperature of the average US industrial process could be raised would be reflected in a one-billion-dollar annual economic return. Improvement in the allowable operating temperatures of engine components also will provide handsome return in improved efficiencies. Figure 61 shows recent and projected performance data for jet engines reflecting the evolution of new turbine blade material; on the basis of current research it would appear reasonable to project at least a 50% increase in allowable engine temperatures—hence efficiencies—over the next two decades.

At the other extreme in the temperature behavior of material, Fig. 62 shows the improvement in the delineation of the liquid-gas phase transition in ^{4}He at the lambda point (4.2 K). While the resolution has improved from left to right by a factor of 10^6 the shape of the transition is invariant. This figure is in many ways typical of the revolution that has occurred in the precision attainable in physical measurements during the past decade or so.

9. OPTICS

With pressure from military objectives, substantial progress has been made in recent years in the fabrication and use of large-scale optical elements. Pressure from the communication and information-handling community has led to comparable progress in the design and fabrication of electro–optical and integrated optical devices. Optical fibers have reached a remarkable degree of development where they now have channel capacities over 10^3 greater than coaxial cables and require repeater amplifiers at ~ 100 km intervals instead of the 3–4 km intervals characteristic of coaxial systems.

Two of the most active and interesting new areas in optics, however, have been those of microscopy and of adaptive optics—optical elements that change their characteristics in real time to achieve specific purposes.

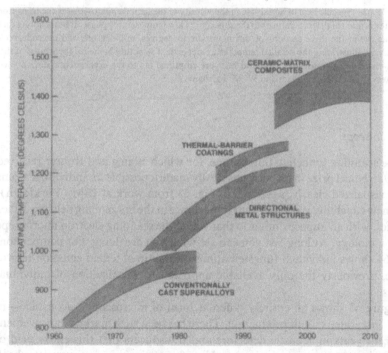

FIGURE 61 Current and projected operating temperatures for jet engines, extrapolating on the basis of new turbine blade materials. The lowest band on the graph indicates the temperature increase that has been achieved through improvement in nickel-based superalloys—the current standard turbine material. In the years to come, alloyed turbine blades made of metal strengthened through the use of directional crystal structures, blades protected by coatings of ceramics or special alloys, and ultimately ceramic-matrix composites, will allow an increase in turbine inlet temperatures and correspondingly higher engine efficiencies. (Courtesy of J. P. Clark and M. C. Flemings.)

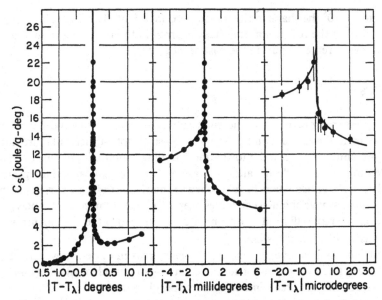

FIGURE 62   Specific heat of liquid ^{4}He in the vicinity of the lambda point—that is the transition from the normal (high temperature) to the superfluid (low temperature) phase. It bears noting that the abscissae in the three sections of this figure are in degrees, millidegrees, and microdegrees, respectively, illustrating the detailed temperature-dependent structure in the neighborhood of this second-order phase transition. The solid lines are empirical fits to the experimental data points. (Courtesy of M. J. Buckingham and W. M. Fairbank.)

A. Microscopy

The scanning tunneling microscope, for which Binnig and Rohrer received the 1986 Nobel prize in physics, has finally made it possible for individual atoms to be visualized clearly, as shown in Fig. 63 from work at IBM (Yorktown). This new capability is of enormous importance in the burgeoning field of surface physics, with an impact similar to that which the scanning electron microscope had on biology. At long last you can see what you are doing! For comparison, Fig. 64 shows individual tungsten atoms on the tip of a field emission microscope, previously the only available approach to visualization of individual atoms.

Figure 65 shows an entirely different form of microscopy—acoustical—in use in integrated circuit production. There has been such marked improvement in recent years in acoustic detectors and transducers that this entire field of physics is experiencing a renaissance. Use of acoustics in microscopy dates back to a suggestion of Sokolov in the Soviet Union in 1949, but the technology of the time was simply inadequate. The top panel of Fig. 65 is a high-magnification acoustic image of the base contact of a bipolar silicon transistor taken at 4.5 GHz; the center panel is the corresponding optical micrograph; and the lower panel is the corresponding scanning electron micrograph.

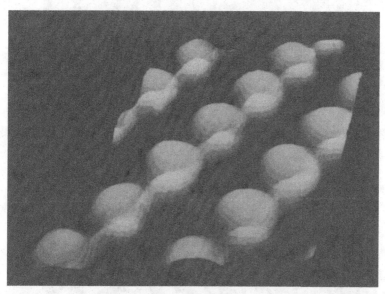

FIGURE 63 A scanning, tunneling micrograph of gallium arsenide along its 110 crystal plane. The large spherical objects are the gallium atoms and the smaller lighter ones the arsenic atoms. (Courtesy of P. Chaudhari, IBM, Yorktown.)

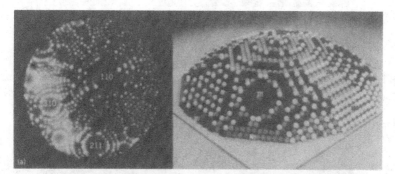

FIGURE 64 The left panel shows a helium field-ion micrograph of a tungsten crystal of nearly hemispherical shape with a radius of about 360 Å. The central dark area is the 110 plane. On the right-hand side is shown a model corresponding to the micrograph on the left, again labeling the prominent crystallographic features. The field-ion microscope was invented by E. W. Mueller in 1951 and for several decades was the only device capable of routinely revealing the atomic structure of a crystal surface. (Courtesy of T. T. Tsong, Pennsylvania State University.)

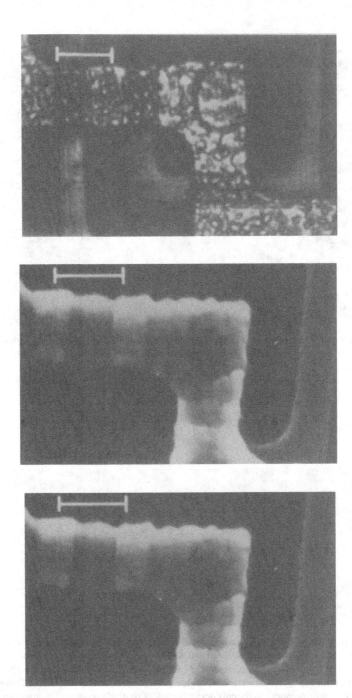

FIGURE 65   A comparison of images of the base contact of a transistor in a large-scale integrated circuit obtained with various microscope systems. The upper panel is a high magnification acoustic image. The central panel is an optical micrograph of the same area and the lower panel is a scanning electron micrograph. The scale bars in each of the three panels represent a distance of three microns. (Courtesy of B. Hadimioglu and C. F. Quate, Stanford University.)

One of the great advantages of acoustic micrography is that it permits focusing, as here, on a surface feature or, equally well, on features well below the surface such as individual fibers in composites that are inaccessible to other forms of microscopy.

B. Optical Phase Conjugation

The phenomenon of optical phase conjugation was first discovered by Zel'dovich and his colleagues at the Lebedev Institute in Moscow. An intense ruby laser beam had been smeared by passage through a frosted plate and then transmitted down a long tube filled with high-pressure methane. What was observed was that stimulated Brillouin scattering in the gas scattered photons back through the frosted glass with the methane acting as a most unusual mirror; to their amazement Zel'dovich *et al.* saw that the reflected wave, after its second passage through the frosted glass, formed an essentially perfect distortion-free image of the source. The distortions introduced by the first passage through the frosted glass had been cancelled out in the second. This was in 1972.

Since then, active research on this phenomenon has been driven by the desire to use it in eliminating distortion in optical transmission through turbulent media such as the Earth's atmosphere, and in optimizing performance of optical devices generally.

Figure 66 shows the use of optical phase configuration to compensate for atmospheric turbulence over a 100 meter optical path. The three left panels are 1/2000 second exposures of an effective point source dye laser showing marked time-dependent distortion. When the beam was phase-conjugated and sent back through the atmosphere the near perfect beam spot of the right panel was restored. Both stimulated Brillouin scattering as in the original Zel'dovich observation, and the now familiar four-wave mixing using nonlinear media, can be used to achieve the required phase conjugation. Figure 67 is a second illustration of the use of phase conjugation to restore a more detailed image. In this case, the image of the cat was transmitted through a frosted glass; in the top

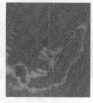

FIGURE 66 The images produced by a laser beam after traversing a 100-m outdoor range comprise very highly distorted beam spots as shown in the three 1/2000-sec exposures on the left-hand side of this figure. When the beam was reflected from a phase-conjugate mirror and sent back through the 100-m atmospheric path the near perfect quality of the beam spot was restored as shown in the right panel in a 1/2000-sec exposure. (Courtesy of G. J. Dunning and R. C. Lynn, Hughes Research Laboratories.)

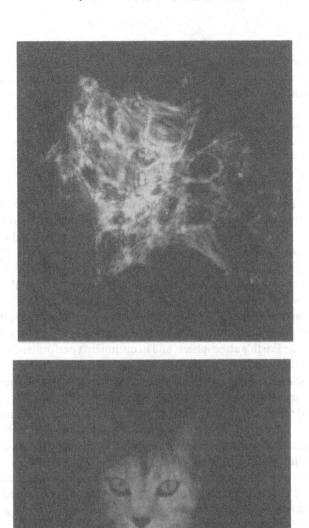

FIGURE 67   A further illustration of the use of a phase-conjugate mirror to compensate for distortions in optical systems. In both the upper and lower photographs the image of the cat was distorted by transmitting it through frosted glass. Reflection of this image back through the same piece of glass by an ordinary mirror yields the unrecognizable image shown in the top photograph whereas reflection by a phase conjugate mirror removes much of the distortion. (Courtesy of J. Feinberg, University of Southern California.)

panel it was reflected back through the glass by a plane mirror while in the bottom panel it was reflected back from a phase-conjugate mirror. The removal of distortion from the time-reversed beam is striking!

10. CHAOTIC SYSTEMS

One of the very surprising developments in recent years in physics has been the recognition that there is order in chaos—that seeming randomness may have an underlying geometric form. It has been discovered that even very simple deterministic systems with very few elements have the capacity to generate random behavior of a fundamental nature—to generate chaos. In principle, the future of such systems is completely determined by the past, but in practice very small uncertainties in initial conditions become amplified to such an extent that although the behavior may be predictable for short periods, over longer periods it becomes completely unpredictable.

The fact that there is order in chaos allows us to return to many problems such as turbulence, atmospheric phenomena, noise generation, and the like—problems previously considered intractable to physics although nonetheless of vital importance—with new hope of progress.

Fundamental to this work has been the concept of attractors—geometric forms that characterize the long-term behavior of a system in its phase space; an attractor is the point, line or surface in that state phase space toward which the system moves over the long term. The use of high-powered computers has made it possible to explore such behavior. Figure 68 shows the evolution of a particular attractor, the Lorenz attractor, in a fluid system with only three degrees of freedom; it was the first example of a chaotic or strange attractor and was discovered by Lorenz in 1963. The successive views in the figure trace the evolution in phase space of a system, beginning with 10 000 measurements of its initial conditions—all of which were sufficiently similar so that the 10 000 phase points initially were superimposed at a common point in this phase space. Because of the microscopic differences and fluctuations in these 10 000 measurements, however, as each point moves under the action of the equations governing the motion of the system, the tightly spaced cloud of points stretches into a long, thin thread which then folds over onto itself many times until the points are spread over the entire attractor. At that point prediction becomes impossible; the final state can be anywhere on the attractor. For a predictable attractor, in contrast, all the phase points would necessarily move together and would remain closely clumped in the final state.

The study of chaotic systems is still a young field of physics and one with enormous potential. Figure 69, a very simple physical situation involving both laminar and turbulent flow, as well as cavitation, is the sort of situation that even a few years ago would have been considered beyond the scope of realistic physical calculation but which now can be attacked effectively. Obviously here, as in

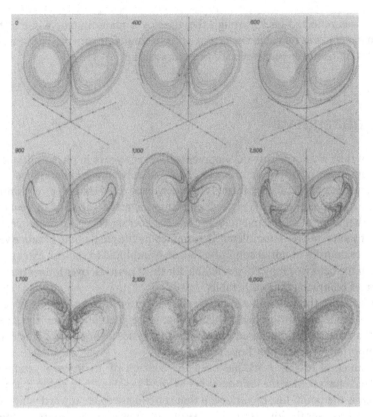

FIGURE 68   A supercomputer simulation of the development of a chaotic situation in a physical system to which the Lorenz attractor is appropriate. The nine views here are at units of 1/200th of a second. (Courtesy of J. P. Clutchfield, J. D. Farmer, N. H. Packard, and R. S. Shaw.)

Fig. 68, the availability of large supercomputers has made not only a quantitative but also a qualitative change in the scope of problem that has become tractable.

11. GEOPHYSICS

Turning now to flows so slow as to be completely imperceptible to humans, Fig. 70 shows the distributions of the world's land masses at four different epochs. It is now difficult to remember that as late as 1960 only a few mavericks in the geophysics community were prepared to give serious consideration to the concept of plate tectonics; now it is revealed truth! The change has reflected more than anything else the rapidly increasing sophistication of physical and geophysical measurements ranging from rock magnetism, to laser ranging, to acoustic probing of the Earth's interior.

Typical of these latter studies is that shown in Fig. 71, where seismic velocity profiles beneath the Sea of Okhotsk north of Japan have provided a striking

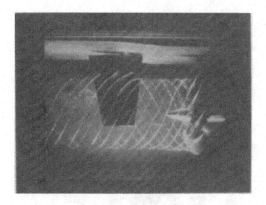

FIGURE 69   An experimental study of the flow in the vicinity of a propeller and rudder system showing cavitation at the propeller tips. (Courtesy of Larkin Kerwin, National Research Council of Canada.)

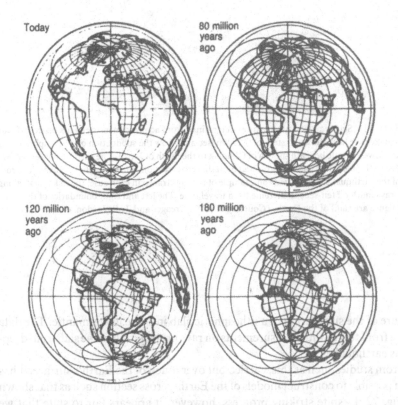

FIGURE 70   Four views of the Earth's surface showing tectonic migration over the past 180 million years beginning with the break-up of the supercontinent Pangea. The positions have been determined from paleomagnetic data and from examination of the layers of oceanic crust formed between the times indicated. (Courtesy of J. C. Maxwell, University of Texas, Austin.)

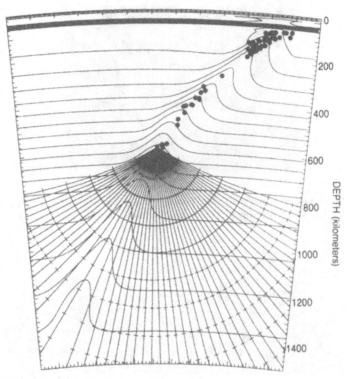

FIGURE 71   Seismic velocity structures determined from acoustic measurements on deep-focus earthquakes are here shown in steps of 0.25 km/sec and show the subduction of a lithospheric slab into the lower mantle beneath the Sea of Okhotsk to the north of Japan. The earthquake centers are marked with black dots and are projected onto this cross section. The radial lines emanating from one of the earthquake sites provide an example of the rays traced in each case; the tick marks along these rays mark off ten-second increments in travel time. The left and right boundaries of this true-scale figure are radii of the Earth. (Courtesy of K. C. Creager and T. E. Jordon.)

picture of the subduction of a lithospheric slab into the lower mantle. The data come from travel-time measurements on rays propagating in the slab from deep-focus earthquakes.

From studies such as this, carried out over much of the Earth's surface, it has been possible to construct models of the Earth's cross section such as that shown in Fig. 72. Despite striking progress, however, it appears fair to state that we currently have no real understanding of the mechanism that drives the tectonic plates in their stately odysseys around the globe. This remains a major challenge both to geophysics and to physicists of all persuasions.

12. ASTROPHYSICS

As we move away from our own planet, to take a longer look, the most striking change in recent years has been that in the number and width of the windows through which we view the universe. This is illustrated in Fig. 73 where the atmospheric absorption is plotted as a function of electromagnetic wavelength, together with an indication of the heights to which balloons, rockets, and satellites take detectors. The surface windows open in the visible region, in a few narrow windows in the infrared, and in the radio regime are shown clearly in this picture—as is the fact that effectively the entire electromagnetic spectrum becomes accessible to us as we reach satellite altitudes. We find that

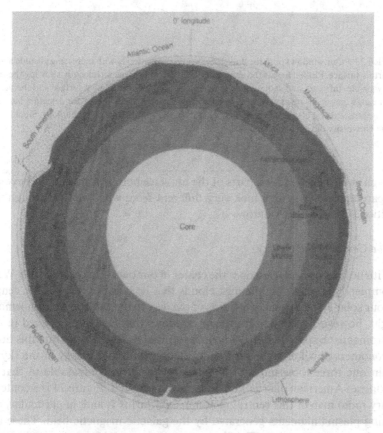

FIGURE 72   A cross section of the Earth at approximately 20° south latitude. The thickness of the lithosphere has been greatly exaggerated and the internal structure is not to scale. The major tectonic plates are bounded by oceanic ridges at which new crust is generated and by subduction zones where the older crust is consumed. Lines of volcanos commonly occur above down-going plates at subduction zones. Examples shown in this figure are the Andes of Western South America and the Tonga volcanic island chain in the Pacific. (Courtesy of J. C. Maxwell, University of Texas, Austin.)

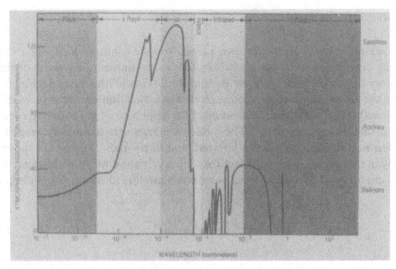

FIGURE 73   Our windows onto the universe open up dramatically with increasing altitude above the Earth's surface. Plotted here is the atmospheric absorption versus wavelength, showing the very narrow visible, infrared, and radio windows accessible to us on the Earth's surface and the vastly greater range that opens up as we use balloons, rockets, and satellites to take our detectors into and above the atmospheric blanket. (Courtesy of H. Alfvén, Royal Institute of Technology, Stockholm, and the University of California, San Diego.)

through these new windows parts of the universe both within, and far beyond, our parochial solar system look very different from what we see through the traditional visible-light window.

A. The Center of the Galaxy

With normal optical telescopes the center of our own galaxy, the Milky Way, is completely obscured by the dust clouds that lie between the galactic center and our solar system some two-thirds of the way to the galactic rim. Enormous insight, however, is coming from new windows that have been opened in the electromagnetic spectra. Both radio and infrared radiation penetrate this dust. Astronomers now know the position of the center of our galaxy in the sky to within one three-thousandth of a degree, a precision comparable to that of locating an American quarter at a distance of two miles. Figure 74 is a contemporary radio map of this central region in Sagittarius A and, in particular, the striking plasma filaments generated by the galactic magnetic field. These are truly enormous structures. The filaments are of particular interest and importance inasmuch as they delineate the galactic magnetic field in its central region. The combination, in recent years, of radio astronomy, optical astronomy, infrared astronomy, and even x-ray and gamma-ray astronomy has had revolutionary impact on our understanding of the structure of our own galaxy as well as of those in our galactic neighborhood.

Figure 75, in the left panel, shows a more detailed view of the galactic center as taken with the Very Large Array radio telescope. From measurements made at six-centimeter wavelength it has been possible to develop the detailed map shown on the right-hand side of this figure, showing the distribution of ionized gas in the region some five to six light years in radius around the galactic center. The arcs and filaments that appear in Fig. 74 are again evident, as also is a ring of dust and neutral gas some 12 light years in diameter. A compact radio source lies at the center of the ionized gas arcs and filaments.

Figure 76 shows this central region again to illustrate the improving resolution now available—with the upper and lower infrared maps now added to the radio maps. The upper panel is one obtained with the Infrared Astronomical

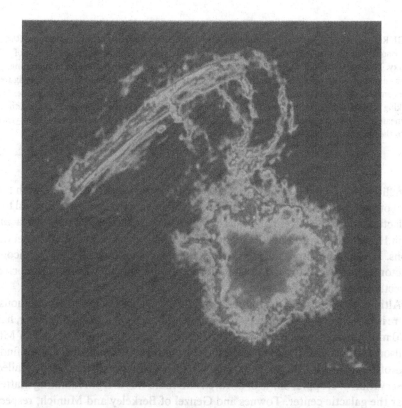

FIGURE 74 Shown here, at 20-cm wavelength, is a radial map of Sagittarius A at the lower right and an extensive network of filaments looping toward the upper part of the figure. The galactic center is the bright point—in the center of the Sagittarius A region. None of this structure, and indeed nothing in the region of the galactic center, is open to visible light observation, because dense intervening dust clouds in the plane of the galaxy block our line of sight. (Courtesy of the National Radio Astronomy Observatory.)

 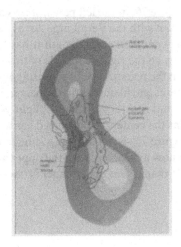

FIGURE 75   The left panel is a radio map made with the Very Large Array of the National Radio Astronomy Observatory, showing the ionized gas within five or six light-years of the nucleus of our galaxy. This map was made from signals collected at a wavelength of six centimeters; combining all the available measurements leads to the detailed map shown on the right-hand side of this figure. The compact radio source lies at the center of the ionized gas arcs and filaments, which appear in striking fashion in Fig. 74, out to a radius of some six light years, while the infrared observations delineate a large dust and neutral gas ring which extends out to a radius of some twelve light years from the galactic center. (Courtesy of the National Radio Astronomy Observatory.)

Satellite (IRAS) at 60 μm. At this wavelength interstellar dust shines with the heat of absorbed starlight and it is possible to delineate the dust rings around the galactic center. The lower panel was obtained with the Infrared Telescope facility in Hawaii and represents the highest resolution yet attained in such observations. It would be impossible to obtain such images without the new semiconductor infrared detector arrays and information-handling technology imported directly from physics research.

Although comparable details concerning our own galaxy are still ambiguous, there is rapidly growing evidence that our neighbor, the Andromeda galaxy, has a 70 million solar mass black hole at its center. Rickstone and Dressler of Mt. Wilson and Las Campanas Observatories have recently confirmed earlier findings of Tonry at MIT on this question. All this information comes from detailed observation of Doppler shifts of infrared and radio waves from radiating matter near the galactic center. Townes and Genzel of Berkeley and Munich, respectively, believe that there is compelling evidence for a similar black hole at the center of our own galaxy.

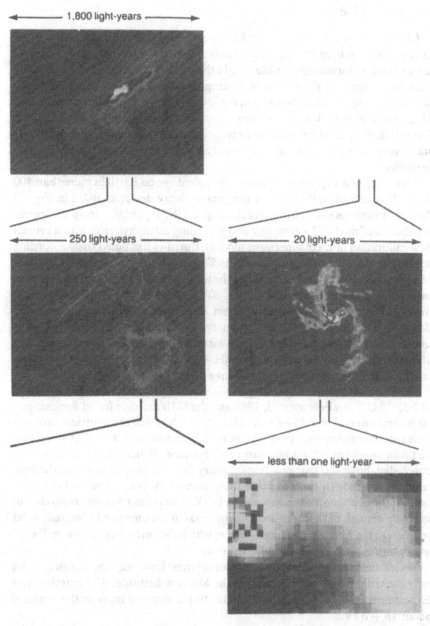

FIGURE 76 An illustration of the increasing resolution that is attainable through the new electromagnetic windows indicated in Fig. 73. The center two panels in this figure reproduce Figs. 74 and 75 while the upper panel is a photograph obtained with the Infrared Astronomical Satellite at a wavelength of 60 μm. The elongated oval region is some 700 light years along its major axis. The very high resolution infrared image shown in the bottom panel is one made with the infrared telescope facility on Mauna Kea in Hawaii. The bright clusters to the right of this view are the infrared source IRS 16—a very dense star cluster very close to the galactic center. (This composite figure courtesy of G. Wynn–Williams, University of Hawaii.)

B. Supernova 1987A

Clearly, the spectacular event of the year—and indeed of the century—in astrophysics has been the appearance of Supernova 1987A in the Large Magellanic Cloud, unfortunately visible only in the southern hemisphere except via its neutrinos. Figure 77 shows two photographs of the region of the southern sky involved, the upper taken before and the lower after the supernova event. These plates are unique in that, in the upper, it is possible to identify the specific star Sanduleak-69° 202 which underwent the supernova explosion, a super-giant B3 star—very young and very massive—and an outstanding candidate for such an explosion.

This is the first supernova visible to the naked eye on Earth in more than 400 years. It is unique in that two of the proton decay detectors listed in Fig. 23 detected neutrinos emitted in the early stages of the explosion—truly inaugurating the era of neutrino astronomy and providing astrophysicists with a crucial new calibration of their supernova models. Burrows has noted that, according to these models, the collapsing core of 1987A emitted 10^{58} neutrinos in a matter of a few seconds—a number ten times the total number of neutrons and protons in our sun! Of these neutrinos, about 3×10^{16} found their way into the 7000 cubic meter IMB proton decay detector; about 22 of them actually interacted in the detector. The neutrinos carried off energy corresponding to about 0.1 solar masses; the total kinetic energy of the material flung out in the explosion was 100 times less than this, and the visible light emission carried 100 times less energy still. Neutrinos are the dominant players.

The Kamiokande detector in Japan recorded 12 neutrino events in 13 seconds at 7:35:35 U.T. on February 23, 1987 and the IMB detector found 8 events in a 10 second interval at 7:35:44 U.T. Unfortunately, since the neutrino emission time spread is unknown, the spread in detection times in both detectors cannot be taken as evidence for a nonzero neutrino mass. Bahcall and Glashow have concluded, however, that observation times do set a limit of 11 eV on the neutrino mass although there is some disagreement as to the exact value of this limit. If the neutrino mass is as small as 11 eV, then primordial neutrinos cannot make up enough of the so-called missing mass in the universe to produce a flat solution to the Einstein equation—one in which the current expansion will slow asymptotically to zero in the infinite future.

Terrestrial measurements on the neutrino mass have been inconclusive, with only the tritium end point data from the Moscow Institute of Theoretical and Experimental Physics consistently indicating a nonzero mass in the range of about 16 ± 6 eV.

Bethe and Brown (B^2) have carried out detailed theoretical studies on the mechanisms of supernova explosions with a number of surprising results. These calculations, published prior to Supernova 1987A, reproduce the available observations thus far with remarkable success.

Figure 78 shows the results of the B^2 calculations during the period of maximum compression of the core. Time is plotted vertically downward in this fig-

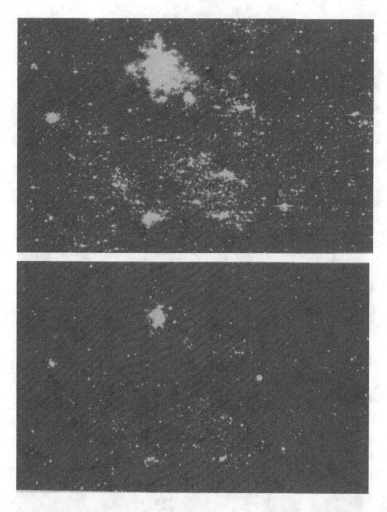

FIGURE 77 Views of the Large Magellanic Cloud before (above) and after (below) the appearance of Supernova 1987A. These are photographs taken with the 20-in. astrograph of the Yale Southern Observatory in Argentina with the upper photograph taken more than a decade ago and the lower one taken immediately after the appearance of the supernova. The upper photograph is important because, in it, the precursor of the supernova can be unambiguously identified. The most striking object in these photographs is not the supernova but the giant cloud of glowing gas known as 30 Doradus—or more popularly—the Tarantula Nebula. The supernova shows up clearly to the right of center in the lower photograph. (Courtesy of W. van Altena, Yale.)

ure, and each contour represents a shell of matter whose radial position is followed through a period of 12 msec in the computer simulation. The included mass—or total mass inside the contour—does not change as the shells contract and expand. Initially the core is of iron but the extreme compression of the collapse converts the innermost few kilometers into nuclear matter and ultimately into a neutron star or a black hole depending upon the total stellar mass

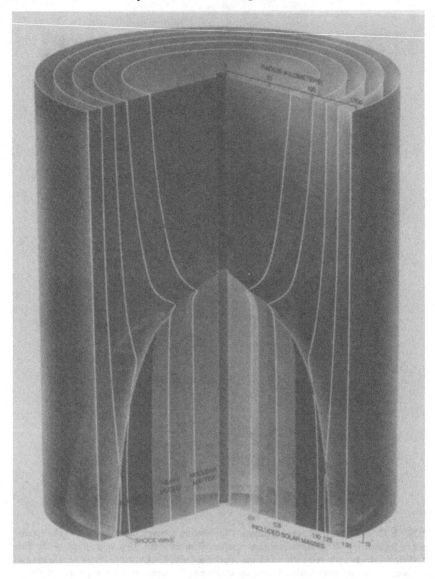

FIGURE 78   A schematic representation of the evolution of a supernova explosion in a massive star. Time is plotted vertically downward in this figure and the time interval depicted here includes the moment of maximum compression when the center of the exploding star reaches its highest density. Each contour represents a shell of matter whose radial position is followed through a period of 12 msec in the computer simulation. (Courtesy of G. E. Brown, Stony Brook.)

involved. Surrounding the innermost few kilometers is a shell made up of various heavy nuclei including iron. At maximum compression the rapid contraction stops abruptly, creating a shock wave which then travels outward at some 30 000 km/sec initially. In the wake of the shock, nuclei are broken up into individual neutrons and protons and, depending upon the mass involved, at some radius the in-falling material following the explosion meets the outgoing shock wave and for a significant period the shock remains stationary while the matter rains through it.

That a supernova explosion is the inevitable fate of any sufficiently massive star follows from the B^2 model and calculations by Weaver as shown in Fig. 79. It bears emphasis that *all* elements heavier than iron are assembled in the hot shock wave following the explosion and are blown out into the cosmos, where they mix with the interstellar gas before condensing into new stars and, possibly, planetary systems. There is a little stardust in each of us!

The coming together of astrophysics, elementary particle physics, and cosmology has led to what can only be considered remarkable insight into the early development and subsequent evolution of our universe. Figure 80 lists some of the major milestones in this 20 billion year saga.

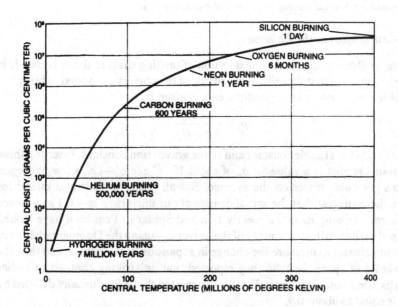

FIGURE 79 The evolution of a massive star—one many times the mass of our sun—is a steadily accelerating progress toward higher core temperature and density. In this figure the Bethe–Brown model of stellar development and explosion has been calculated for a star of 25 solar masses by T. A. Weaver of the Livermore National Laboratory. What is plotted is the central density against the central temperature and the various stages and durations of the dominant thermonuclear burning reactions are indicated along the curve. (Courtesy of G. E. Brown, Stony Brook.)

Major events in the universe's history

Cosmic time	Epoch	Redshift	Event	Years ago
0	Singularity	Infinite	Big bang	20×10^9
10^{-43} second	Planck time	10^{32}	Particle creation	20×10^9
10^{-6} second	Hadronic era	10^{13}	Annihilation of proton-antiproton pairs	20×10^9
10 seconds	Leptonic era	10^{10}	Annihilation of electron-positron pairs	20×10^9
2 minutes	Radiation era	10^9	Nucleosynthesis of helium and deuterium	20×10^9
1 week		10^7	Radiation thermalizes prior to this epoch	20×10^9
70,000 years	Matter era	10^4	Universe becomes matter-dominated	20×10^9
300,000 years	Decoupling era	10^3	Universe becomes transparent	19.9997×10^9
$1-2 \times 10^9$ years		10–30	Galaxies begin to form	$18-19 \times 10^9$
3×10^9 years		5	Galaxies begin to cluster	17×10^9
4×10^9 years			Protogalaxy collapses to the Milky Way Galaxy	16×10^9
4.1×10^9 years			First stars form	15.9×10^9
5×10^9 years		3	Quasars are born; population II stars form	15×10^9
10×10^9		1	Population I stars form	10×10^9
15.2×10^9 years			Parent interstellar cloud to the solar system forms	4.8×10^9
15.3×10^9 years			Collapse of protosolar nebula	4.7×10^9
15.4×10^9 years			Planets form; rock solidifies	4.6×10^9
15.7×10^9 years			Intense cratering of planets	4.3×10^9
16.1×10^9 years	Archeozoic era		Oldest terrestrial rocks form	3.9×10^9
17×10^9 years			Microscopic life forms	3×10^9
18×10^9 years	Proterozoic era		Oxygen-rich atmosphere develops	2×10^9
19×10^9 years			Macroscopic life forms	1×10^9
19.4×10^9 years	Paleozoic era		Earliest fossil record	600×10^6
19.55×10^9 years			First fishes	450×10^6
19.6×10^9 years			Early land plants	400×10^6
19.7×10^9 years			Ferns, conifers	300×10^6
19.8×10^9 years	Mesozoic era		First mammals	200×10^6
19.85×10^9 years			First birds	150×10^6
19.94×10^9 years	Cenozoic era		First primates	60×10^6
19.95×10^9 years			Mammals increase	50×10^6
20×10^9 years			*Homo sapiens*	1×10^5

FIGURE 80   Remarkable progress has been made in delineating the milestones in the past history of our universe. Some of the major ones are collected in this figure.

C. Missing Mass in the Universe

I have already mentioned the question of missing mass in the universe. It is easy to calculate that the critical density ρ_c of the universe, required to bring the Hubble expansion to an asymptotic end, is given by

$$\rho_c = \frac{3}{8\pi} \frac{H_0^2}{G},$$

where H_0 is the Hubble constant and G the gravitation constant. Inserting these constants results in a value for ρ_c of about 10^{-29} g/cm³—i.e., a few hydrogen atoms per cubic meter on the average. But all measurements and inferences from them suggest that the actual density of our universe ρ is $\sim 0.1 \, \rho_c$—hence the term "missing mass". Recently Loh and Spillar at Princeton have looked instead at the intrinsic geometry of the universe—using the Doppler redshifts of distant galaxies to measure the change in expansion with distance and thus the curvature of space. In 1986 they reported that, in striking contrast to earlier results, their data—shown in Fig. 81—suggested that ρ/ρ_c (usually denoted by Ω) is equal to about 0.9.

This is an appealing result, both because it renders somewhat moot the question of missing mass but also because the very attractive inflationary Big Bang cosmologies require $\Omega = 1$, all curvature having been inflated away during the first 10^{-35} sec after the Bang. Obviously a result such as that of Loh and Spillar is highly controversial and is one more readily accepted by physicists than by astronomers!

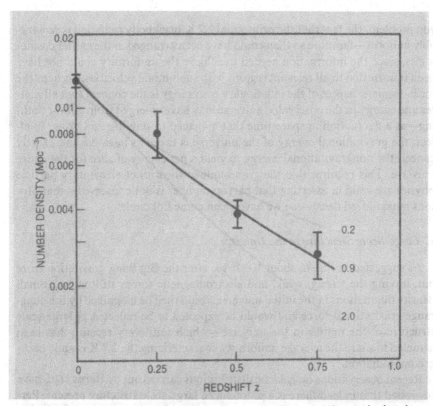

FIGURE 81 The number density of galaxies per unit volume of red-shift space is plotted as a function of the red shift and normalized so that the density at zero red shift is simply the ordinary spatial density as observed in the vicinity of our own solar system. The data points here are from the survey of Loh and Spillar of Princeton and the curves are for the indicated calculated values of Ω, the ratio of the actual to the critical density of the universe. The best fit is obtained for $\Omega = 0.9$, very close to the exact criticality favored by modern astrophysics theory but much larger than the density observable in the universe. (Courtesy of E. Loh and E. Spillar, Princeton.)

D. Inflationary Cosmology

The inflationary cosmology, developed originally by Guth, suggests that the observable universe is embedded in a much larger region of space that underwent extraordinary growth immediately after the Big Bang; it agrees with the previously accepted cosmologies for times greater than 10^{-30} sec but differs dramatically for earlier periods. The inflationary cosmology suggests that during this earlier period the universe expanded by a factor of about 10^{50} beyond what earlier cosmologies would have suggested; it was developed in order to avoid the facts that in the earlier cosmologies a number of stringent, unexplained, and arbitrary assumptions were necessary concerning the initial conditions and that all of them led inexorably to abundant production of unobserved magnetic monopoles. The inflationary cosmology also solves the so-called horizon problem, the fact that the primordial 2.7 K blackbody radiation is remarkably uniform—far more so than could have been arranged in the earlier cosmologies, since the information needed to achieve the uniformity could not have been transmitted to all relevant regions with subluminal velocities. Perhaps the most dramatic aspect of the inflationary cosmology is the concept that all matter and energy in the observable universe may have emerged from almost nothing—as a fluctuation in space-time that managed to stabilize and grow. In effect, the gravitational energy of the universe is taken as negative and exactly cancels the nongravitational energy to yield a net energy of zero for the entire universe. This requires that the grand-unified theories of elementary particle physics are valid in asserting that baryon number is not conserved—that protons must indeed decay—so we have again come full circle.

E. Large-Scale Structure in the Universe

As suggested in Fig. 6, about 10^{-43} sec after the Big Bang gravitation froze out, leaving the strong, weak, and electromagnetic forces still unified. Small density fluctuations in the initial universe would then be magnified by this long-range gravitational force and would be expected to be reflected in large-scale structure of the matter in the universe—which until very recently has been assumed to share the average uniformity characterizing the 2.7 K cosmic background radiation.

Recent observations on galactic distributions carried out by Burns *et al.* have suggested that indeed there is a pronounced large-scale structure present. Perhaps their most striking result was the discovery of the linear string of galaxies, more than a billion light years long, originating in the region of the Perseus and Pegasus constellations. It is—by a large margin—the largest structure known and is surrounded by three voids each roughly 300 million light years in diameter.

Such voids and structures are precisely what emerge from recent Cray supercomputer simulations that have followed 10^6 galaxies in an evolution model developed by Zel'dovich. Figure 82 shows the results of such a simulation.

13. GRAVITATIONAL RADIATION AND GRAVITATIONAL LENSING

Gravitational radiation was one of the early predictions of Einstein's general theory of relativity in about 1916, but this concept lay fallow until the 1960s when Weber mounted an ambitious program for its detection using large resonant-bar detectors. In 1969, he announced detection of gravitational radiation from the galactic center, but these results were not confirmed.

Although still not unambiguous, the first generally accepted evidence for the reality of gravitational waves comes from observations made on a binary pulsar, initially discovered in 1974 by J. H. Taylor and R. A. Hulse, then at the University of Massachusetts at Amherst. The upper panel of Fig. 83 is a schematic view of what this binary pulsar might look like, with one member of the binary pair

FIGURE 82 Using the theory of Yakov B. Zel'dovich, a Cray 1 supercomputer has been used to simulate the clustering of one million galaxies. This work was carried out by J. Centrella and A. Melott at the Lawrence Livermore National Laboratory. The top figure shows that the galaxies have clustered into flat pancakes and filaments while the voids in the lower figure are more spherical in shape. These simulations appear to provide a reasonable reproduction of the most recent observations on large-scale structure in the universe.

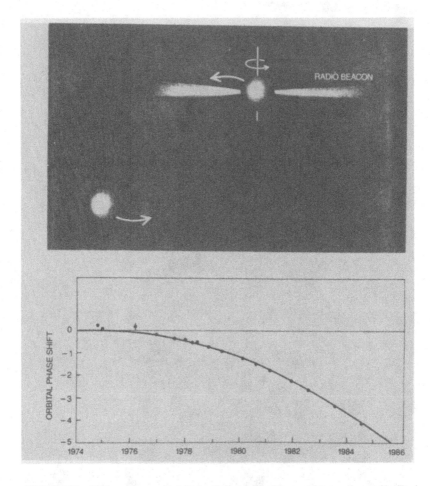

FIGURE 83 Evidence for gravitational radiation from a binary pulsar, as observed by Taylor and Hulse. The upper panel is a schematic illustration of the binary pulsar that they have studied since 1974 and the lower panel shows the systematic shift in the pulsar's orbital period over the intervening years—a shift that is almost precisely reproduced by calculations of the amount of energy and angular momentum that gravitational radiation would be expected to carry away from such a system. (Courtesy of A. D. Jeffries, P. R. Saulson, R. E. Spero, and M. E. Zucker.)

being a pulsar whose period has now been observed for a period of 12 years. The decrease in this period, as plotted in the lower panel, is measured by carefully timing the arrival of the pulsar radio signals and implies that the paired stars are gradually losing energy and spiraling together. The observed energy loss given by the data points is remarkably well reproduced by the curve of predicted energy loss through gravitational radiation; gravitational waves would appear to carry away just the observed amount of energy and angular momentum from the system.

The possibility of directly detecting gravitational waves is being explored in a number of laboratories. Figure 84 is a schematic drawing of the 4800 kg cryo-

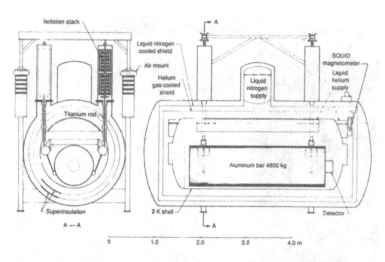

FIGURE 84   A schematic illustration of the 4800-kg cryogenic resonant gravitational wave bar detector at Stanford University. The SQUID magnetometer detects the strain on the bar that would correspond to the reception of gravitational radiation. The aluminum bar is maintained at a temperature of 4 K. Similar resonant detectors are installed at Louisiana State University and at the University of Rome for use in a three-way coincident system. (Courtesy of Joseph Reynolds, Louisiana State University.)

FIGURE 85   A photograph of the Stanford resonant-bar gravitational wave detector shown schematically in Fig. 84. (Photographed by Frans Alkemade, Courtesy of Stanford University.)

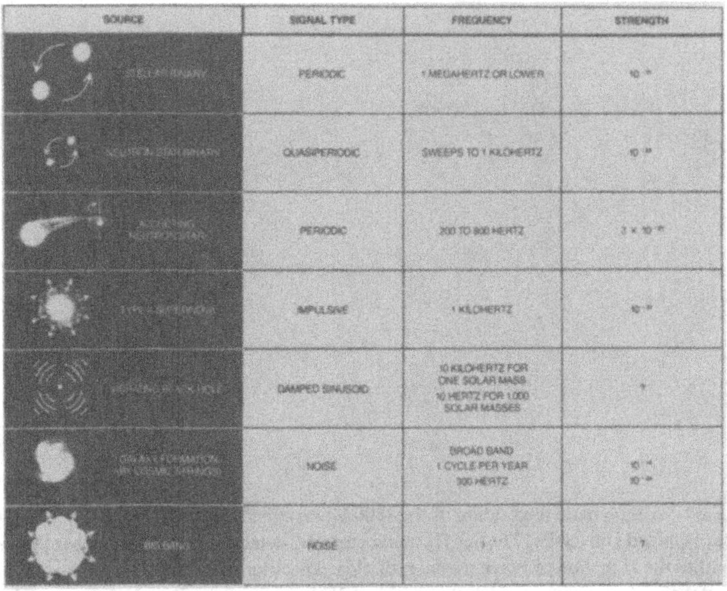

SOURCE	SIGNAL TYPE	FREQUENCY	STRENGTH
STELLAR BINARY	PERIODIC	1 MEGAHERTZ OR LOWER	$10^{-?}$
NEUTRON STAR WHARF	QUASIPERIODIC	SWEEPS TO 1 KILOHERTZ	$10^{-?}$
ROTATING NEUTRON STAR	PERIODIC	200 TO 800 HERTZ	$3 \times 10^{-?}$
TYPE 1 SUPERNOVA	IMPULSIVE	1 KILOHERTZ	$10^{-?}$
COLLAPSE TO BLACK HOLE	DAMPED SINUSOID	10 KILOHERTZ FOR ONE SOLAR MASS; 10 HERTZ FOR 1,000 SOLAR MASSES	?
GALAXY FORMATION OR COSMIC STRINGS	NOISE	BROAD BAND 1 CYCLE PER YEAR; 300 HERTZ	$10^{-?}$ $10^{-?}$
BIG BANG	NOISE	?	?

FIGURE 86   Possible sources of gravitational radiation. (Courtesy of K. Thorne, Caltech.)

genic resonant-bar detector at Stanford University; Fig. 85 is a corresponding photograph. This detector is one of three similar units, the other two installed, respectively, at the Louisiana State University and the University of Rome for ultimate use as a three-way coincidence detection system. These detectors are sensitive to mechanical strains of order 10^{-18} and already represent a triumph of instrumentation; a further factor of 10^5 is claimed possible before fundamental quantum limitations are encountered. For comparison, a supernova at the galactic center, converting 1% of a solar mass into a gravitational radiation pulse 1 msec in duration, would produce a strain of 3×10^{-18} on the Earth. Although no unambiguous gravitational signals have been detected as yet, the development of these resonant detectors has led to new high-phase-stability superconducting-cavity microwave sources, to gravity gradiometers of unprecedented precision, and to vibration isolation techniques yielding isolation values larger than 10^{10}.

The sensitivity of modern gravitational wave detectors is such that it would be expected that gravitational waves from a number of different sources could be detected. The detection of gravitational radiation remains, however, a technological *tour de force* of the first magnitude. Figure 86 shows the sizes and types of signals to be expected from a variety of possible sources. Waves traveling from a binary star system, for example, would shift all masses by only 10^{-21} meters per meter of separation.

A number of groups, worldwide, are attempting to improve gravitational wave detection sensitivity by long-arm (several kilometer) laser interferometric

GRAVITATIONAL WAVE DETECTORS

Michelson

Fabry — Perot

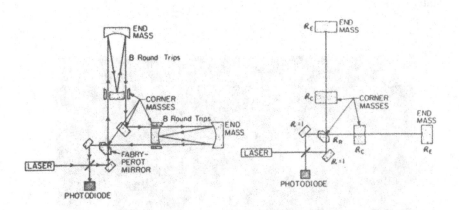

FIGURE 87   Schematic layouts of two types of interferometric gravitational wave detectors. (Courtesy of K. Thorne, Caltech.)

systems—two of which are shown schematically in Fig. 87. Calculations suggest that these should be between 100 and 1000 times more sensitive than current bar detectors. In both cases it may be possible to utilize the squeezed state technique of Fig. 44, as recently suggested by Caves, to further improve the detection efficiency, but this has not yet been attempted. Beyond question, however, gravitational physics has finally become experimental physics!

Perhaps one of the most dramatic consequences of general relativity, originally predicted by Einstein in 1936, was the possibility of gravitational lensing—for example, by the large gravitational field of a galaxy which might be in the line of sight to a more distant astronomical object, so distorting the space-time in its vicinity that light from the distant object in passing through the distorted region might be deflected and lead to apparent multiple images of the distant object. It is remarkable that several examples of such lensing have now been discovered. The first example found is shown in Fig. 88—the double image of a distant quasar. Were the intervening galaxy sufficiently massive it could conceivably bend the passing radiation sharply enough to entrap it, but in this case the mass of the intervening galaxy is not large enough and the quasar light is simply deflected in its passage to us.

14. CONCLUSION

As is obvious from all of the above and from the specialized papers that follow in this volume, the past three years have been among the most stimulating and exciting in the history of physics. It is encouraging that new discoveries such as high-temperature superconductivity continue to surprise us; the rate of such discoveries is, if anything accelerating. Physics, worldwide, is in an excellent state.

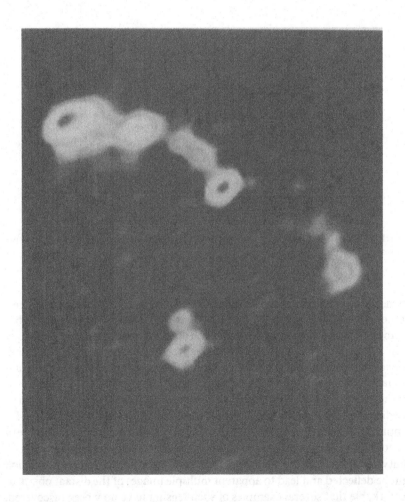

FIGURE 88   A striking example of what appears to be the gravitational lensing action originally predicted by Einstein in 1936. The two ring images in the upper and lower central region are those of a single object—Quasar 0957 + 561—whose radiation has been deflected by the gravitational field of a galaxy (that appears here as the faint patch just above the lower image) that lies along the line of sight to the quasar. The observations on this quasar were made by P. E. Greenfield, D. H. Roberts, and B. F. Burke. (Courtesy of the National Radio Astronomy Observatory.)

In the time and space available, I have necessarily focused on the more fundamental aspects of physics in this paper; this in no way reflects any diminished interest on my part in the applications of this physics or in the importance of these applications. They merit a separate review of their own and I have attempted such a review in the November–December 1986 issue of *American Scientist* on the 100th anniversary of the founding of Sigma Xi.

Let me only conclude here that physics has succeeded in changing, dramatically—and for the better—the conditions of our life, how we view our universe, and our place in it. It remains one of the greatest adventures in which mankind has engaged. It is a triumph of the human intellect. Those of us who are privileged to be physicists at a time of such excitement and discovery are, indeed, fortunate.

15. ACKNOWLEDGMENTS

In a review of this scope, even to attempt to reference all the persons and publications from which I have gleaned input would result in a listing almost as long as the paper itself. I am indeed grateful, however, to the Chairmen of the IUPAP International Commissions and to many of my old friends in physics who once again have responded generously to my request for input and suggestions. To all whose work and results I have included herein—my thanks; and to all those whose work I have, for whatever reasons, omitted—my apologies. For almost every example that I have included, I was forced to pass over several that would have been equally effective and interesting.

I am particularly indebted to the editors of *Scientific American* and *Physics Today* for permission to reproduce illustrations that appeared originally in their publications and to the many persons who have provided me with figures and photographs. I have attempted to identify all of them in the figure captions.

Finally, I would be remiss indeed were I not to acknowledge all the assistance that I have received from Mrs. Dorothy Berenda, Mrs. Rita Bonito, Ms. Lisa Close, and Mrs. Mary Anne Schulz in preparing this review.

The Roles of Government / PIERRE AIGRAIN

It is very difficult to take the floor while people are still remembering the brilliant performance of Professor Bromley this morning. Don't expect anything quite as enjoyable from me. There are several reasons for this. In the first place, I cannot compete with the brillance of Professor Bromley. Secondly, he demonstrated very clearly that physics is fun, and, unfortunately, governments are not! But there is a further reason, stemming from the fact that I accepted the talk's proposed title—"the roles," with an "s," of government. I think that the "s" is indeed necessary, but it complicates my task.

The concept of government is not uniform around the world, not even between the countries that are represented in IUPAP. To give some examples, it is clear that, in controlled-economy countries, the governments, at least in principle, are leading everything, so the role of government in such systems is theoretically the role of everything in science and technology—clearly much too wide a task to describe. Even in market-economy countries you have very large differences. At one extreme are highly centralized governments, like the situation in my own country, France, about ten years ago. (There is now a slow move toward more decentralization.) At the other end of the scale you have Switzerland, which is a confederation of independent cantons. In between, you have the USA or West Germany, which are federal countries; and there you have a problem. When you speak about government, do you speak about the central government, which is the federal government in a federal country, or do you speak about the local governments? I think you have to include the local governments in the picture, because of their role, in many cases, in matters relating to science policy, especially in physics and technology—physics-based technology. (I take advantage of this moment to mention that I think that one of the great evolutions during these last decades has been the growing influence of technology on basic physics. Basic physics is an enormous user of advanced high technology, and a lot of things would not be possible in basic physics if it did not rely on computers, integrated circuits, and a lot of other things. So we are not

PIERRE AIGRAIN *has had a distinguished career as scientist and administrator. He has served as Scientific Director of the French Army Ministry, as Secretary of State in charge of research in the Prime Minister's Office, and as chairman of a governmental advisory committee on Franco/Chinese economic and industrial cooperation. Since 1981 he has been with the Laboratoire Thomson in Paris, and in 1988 he was elected to the French Academy.*

speaking only about the effect of basic science on technology; the opposite flow is just as important.)

It is clear that local governments are playing a growing role. In fact, they are often putting a lot of money into basic science and, more often still, into what they think is going to help high-technology companies in their region. However, they also put money into high technology itself. I am not sure that I would like to be a taxpayer in the American state that is going to win the site for the SSC! So I decided to take the following definition of government for the purpose of this talk. It is the kind of things that are done by both central and local governments in a country like the United States, which I will often use as an example because it is our host country today. The other country I will sometimes use as an example is, of course, my own country, France, because it is the one I know best.

Now I had a choice between three approaches. The first one would have been to try to describe what the science policy had been (and, by the way, it has been changing all the time) and what the high-tech policy had been in all the countries of the world—and the roles that governments all around the world have played in these matters. Anybody who is interested can find literature—piles of literature meters high—on this subject. For the Western countries, the OECD Reports on Science Policy are available, and they are already quite a thick pile. A lot of work has been done by various university groups in various places on science policy and technology policy in many other countries—impossible to summarize in 40 minutes, or 40 hours! I shall not try to do it.

I was tempted to give a second type of report, which could be very useful. It was to try to draw up a list of all the things governments have done and which have proved bad for science and technology. That would be a tremendously useful piece of work. It could be done because I think so many bad things have been done everywhere that probably the list would be fairly complete. It is probably hard to invent a stupid thing to do which has not already been done somewhere, sometime! So you could have a rather complete description of the pitfalls to avoid. Well, I am not going to do that either, or at least I am only going to develop it partially, for at least two reasons. First, doing such a complete piece of work would require an enormous amount of pathological work. I would suggest that this could be a program of research for one of the universities working in science policies—the University of Sussex, for example, in the U.K., might undertake it. Second, if I had been even half successful in this, I might have found that I was not allowed to enter about 45 countries any more, including my own, and that would, of course, have been very annoying for me! So the third solution was the only one available. It is, more or less, a normative approach—to say what governments *should* do, and, with a few examples, what they should avoid doing, in order that physics—and in many cases this bears on the other sciences too—may develop in the most efficient way and, more important (because I think this is what is expected of me) so that physics-based high tech can develop with maximum efficiency.

Strangely, it is much easier to convince governments to do things in chemistry than in physics, for the simple reason that the industries which are the biggest users of the results of chemistry research are called the chemical industry, and any politician can recognize that there is a link between chemistry and the industrial branch. It is much more difficult to understand the link, though it is just as important, between physics and, say, the electronics industry—or, in fact, between physics and the chemical industry, which is an enormous user of physics results.

The most important thing that governments can and should do, and in general do not do properly, is to foster the education of physicists (amongst others). I do not think that high tech, physics-based high tech, can properly develop and flourish if one does not have a large supply of people, scientists and engineers, who are well trained in physics. I think it would be nice if you also had a lot of politicians who had at least some education in physics, and I am not sure that enough is being done in this direction by governments. It is clear that education is to a large extent paid for by governments—central and local governments—sometimes in a direct way by tax deductions and so on, which anyway are government money. It is not clear that the problem of how you improve teaching in physics has been taken up properly. And I think that the most important part of that is teaching physics in secondary schools.

I have said that physics is fun; and the fact that it is useful may come from the fact that it is also fun. But who amongst us found it fun when we were receiving our secondary education? Which of our children really found it fun in high school? Are you sure that, even though they learned a lot of things, what they learned was really physics or some second-rate mathematics applied to useless problems? I think that we have a real problem if we want to have enough people who understand physics, either to practice it, to use it in engineering practice, or just simply to tolerate it in their everyday life. We must make secondary education in physics attractive. Even in a highly centralized country like France where secondary education is completely controlled to the last details of programs by the central government—even there, when the suggestion arises, "Oh, it would be nice if the students knew a little more science," what is done is just to increase the number of hours of science teaching. Unfortunately, teachers in France, as in most other countries, are underpaid. (They used not to be in France, but the salary scale has slipped progressively compared to other jobs they can do.) If you increase the number of hours' teaching, the problem you meet with is that you need more teachers, and you have to select these teachers from those available; they are usually not the best. I think the big problem is to develop programs that can make the whole thing attractive. If students do not know much physics when they come out of secondary schools, but if they have found it interesting, then they will be able to learn more physics later in higher education where things are on the whole rather better. Of course, we could look at more drastic solutions. One of them was suggested in the round-table I was in yesterday. It would be, for example, in the United States, to introduce several hours a week of teaching of law. That would make law very unattractive, which would, I think, improve the situation in the USA in two ways. First, there would be more scientists; and, second, there would be fewer lawyers. I think that would be a boon in both ways! In any case, education is probably the first and most important thing governments should do. The methods may vary in different countries because of cultural and historical reasons, but it is extremely important.

The second thing which is paid for by governments almost a hundred percent, and on which governments should thus be extremely careful, is basic research. I am speaking about high tech, and I say that high tech can only flourish if surrounded by the right basic research environment. And by "right" I mean good quality. Volume is important; the research is becoming harder and harder. Bromley was showing it clearly this morning; of course, in his field it is especially true. It is harder and harder to make good research without significant investments. And that is true not only for the traditional "big science" fields such as elementary particles, nuclear physics, and astronomy; it is becoming true also for the so-called light science field. I have just spent nearly one-third of my time for a year trying to untangle the rather messy situation we were in with regard to the European synchrotron radiation facility, where the problems of administrative arrangements, finance, and so on had not been completely solved. It takes time to solve them. Yet the European Synchrotron Radiation Facility, like the one planned in this country at the Argonne Laboratory, like the one planned in Osaka, Japan (and possibly there are also some in the USSR—I am less informed there) are not big-science machines; they are equipment designed to help physicists, chemists and biologists to do work which years ago could be done with a simple x-ray machine, except that you couldn't do the work with the same accuracy, some problems were also just impossible to attack, and so on. So it is true that much of science requires heavy equipment.

It is also true—and this creates a serious problem for people who have to choose—that a lot of very good science can still be done with relatively small resources. I was speaking with my old friend Praveen Chaudhari, and noticing that the Zurich laboratory of IBM has gotten a Nobel prize recently for the tunneling microscope, which certainly did not require very large investments to be made, and that the more recent work in the same laboratory by Müller and Bednorz, leading to the new superconductor family, did not require very heavy equipment. One problem with these new superconductors is that they are too easy to make, and the result is that an enormous amount of work is being done all around the world, in all universities; it fits perfectly the description that another old friend of mine used: he called it xerographic research—namely, reproducing what has already been done, getting clearer and clearer copies, but really with no new information in it! And I think that the problem of a good science policy is to manage to give what is needed to good science and at the same time, if possible, to eliminate or reduce xerographic research, which, strangely, is often done by people who could do very much more, who could be very active in another field but who, because they have achieved a reputation in a narrow field, interesting but narrow, may have a tendency to stay in it forever. Changing fields or subfields in physics is as difficult as changing countries or changing wives—a thing which I found so difficult that I never tried it! So sometimes a mechanism has to be put in place in order that the right kind of basic research is done. And the right kind of basic research is *good* research. The subfield is relatively unimportant, even though of course you cannot expect to have a high-tech electronics industry if you do not have some high-caliber work being done in solid-state physics. Of course, having an industry in electronics and chips may turn out to be extremely useful to the people who are doing particle physics. Some of the detectors mentioned by Bromley this morning contained, if I remember correctly, a half million Application-Specific Integrated Circuits (ASICs) specially designed for that application in CERN.

So here you have the first two things—education and basic science. And I think that if the government would fully exercise its role in these two areas, leaving other things alone, it would probably already do a pretty good job compared to what is occurring in most places. Of course, this may be a little exaggerated, because we have to think about one other area where it is clear that governments are necessarily strongly involved. This has to do with high-tech products for government use—the most important ones, of course, being in the field of defense. In this case governments are the customers; they are the market—not only, by the way, the government of the country in which the work is done; there is quite an export trade of government electronics, for example, or defense products. One may like it or not like it; but it is a fact, and it is clear that the customer who is in the market as a result is very important in deciding what should be done, deciding the specifications of the product; and there what one can ask of governments is that they be competent customers. In most cases the Defense Department has been rather good—but not always. There is a natural tendency for a government agency that has lots of money, and which is technologically minded, to try to ask for development of devices that will do three things at once. In France we have a word for that; we speak of "the cane coffee pot combination." There is a story—of course imaginary—that a Defense Department organization had asked for the development of a cane for wounded soldiers which could also be used as a coffee pot. Of course, it would have been a very bad cane and a bad coffee pot. But these are the kinds of exaggerations and pitfalls that governments, as users of high-tech products, may fall into. It does occur in many other areas: for example, in the field of education-related products where, unfortunately, there is not usually in the Administration a group of technology-minded people; and so what they order may be completely inapplicable to the goal perceived.

We enter now the much more delicate problem of the many hidden but very important roles of governments. They often take the form of fiscal policies. Sometimes this fiscal policy has never been designed to help technology and yet proves useful for technology. This, I think, was the case in the USA for the great fiscal advantages given to the R&D limited partnership, maybe seven or eight years ago. This created an influx of money into high tech from private sources, amounting in the best year to $4 billion per year, then tapering off a little. I understand that the recent fiscal reform in this country has eliminated almost all of it. There are many problems of this kind. The initial decision with respect to

R&D partnerships was planned to be a help for high tech—and nothing else. But it disappeared because of possibly very good reasons having nothing to do with R&D. A sweeping reform, made in another area, simply eliminated what had turned out to be an extremely efficient thing. Now, this example is well known in the United States, but I could give hundreds of similar examples of actions that have proved extremely useful because they had enormous leverage.

Through fiscal policies it has been possible to channel a lot of money into high tech, while at the same time keeping the necessary connection with the market, without which high tech can lead to very interesting products that nobody wants—or which are only gadgets. Many worthwhile developments have occurred, but have later disappeared because nobody realized, at the time when a new law was passed, that it would have this effect. I don't want to criticize exclusively the American government—that is not my purpose—so I would like to give an example in the case of France, dating from about ten years ago. The local authorities—I mean municipalities, cities, and so on—get their money from local taxes which are based on a few things—in particular real property of various kinds—a tax called *patente*. It was felt that this was rather antiquated—which was true. The definition of the excise was a hundred years old, so it was replaced by a new one called "*taxe professionnelle.*" The purpose was neither to help nor to kill anything to do with high tech; but the base on which this new tax was paid included salaries and the equipment present on the site—none of which amortized for what it had been bought for. The result was that the tax proved very heavy on industrial research labs (it is not paid by university labs, fortunately). Let me give just one example. Very often an industrial research lab may share facilities on the same site with things that are not research; one such laboratory with which I am familiar is a Philips-owned lab in Paris—LEP. The LEP suddenly found that their tax was increased by a factor of 21 from one year to the next. Well then, of course, measures were taken to provide that, when the increase exceeds a factor of three or four, it will be limited to a factor of three or four. But this is still a very large factor, especially since the old law excluded a number of R&D operations from taxation. So I think that governments should be much more careful.

Very often simple fiscal systems, with little cost to the taxpayer, can have a tremendous impact on science and technology, but only if the arrangements last for years. Too often, measures that had proved very good have been cancelled, even before they had come to fruition. It is like planting an expensive fruit tree and then chopping it down the year it is going to bear fruit for the first time. These are things that governments have been doing out of sheer incompetence combined with an obvious desire to do well—because one thing I would like to mention, having been in touch with these people, is that they are all very well-meaning; they are trying to do well; they work very hard in order to try to help, and then sometimes it doesn't turn out the way that they expected. So I believe that it is in this fuzzy area of indirect action that governments could and should do more. I believe that any measure of a fiscal or similar nature should be studied somewhere in government for its impact on high tech—and this is usually not done. You should not spend a lot of money to develop high tech and do things that will stop the development of high tech—but this is being done all the time. It is like driving a car with the accelerator and the brake applied at the same time—you don't go far, you use a lot of gas, and you destroy your brakes and possibly your engine at the same time.

The other thing that I would like to indicate is the obvious importance of contacts between basic research and high tech—all types of contacts, among them the maximum facility for personal motion from one to the other and *vice versa*. If you believe that high tech is founded on basic research, you have to be involved in some basic research. This is necessary even if the research being exploited was done twenty years ago. And this is why a small country, or a medium-sized country like my own, cannot do without basic research. I know very well that France cannot hope to produce more than, say, seven percent of the basic research results in the whole world, because our population is only about seven percent of the total population of the developed world. But this seven percent is important, not because if it were not there we would be left with only 93 percent; it is important because we would be left with nothing. The seven percent is essential because it is the way to access the 93 percent that is not being done in France. This implies also that people who are trained in basic research should be able, if they feel the vocation to do so, to transfer to industry. It implies that people in industry whose background is in technology should be able to move into research. But, even if they are just specialists in technology, such people should be able with maximum facility to move into universities and similar places. This is very unequal in various countries. Very often there are serious limitations to this possibility of motion—limitations always established for the best of reasons, sometimes 100 years ago or 50 years ago by people who did not realize that this might have any bad effect on science and technology.

I will take just one example. Professors in France are civil servants; the corps of civil servants is under a law of 1936 which has been modified marginally but is still in effect. This law of 1936, even though it says that a professor of physics, for example, can consult freely, which is good, also says that he cannot have any commercial activity, that is, he cannot be on the board of a company, even a company he has founded himself—and this has been a serious obstacle to the creation of start-up companies by university professors in our country. We still have such a law on the books, and obviously it is completely stupid from the point of view of the development of high tech. And even though successive governments have come to the conclusion that it should be changed, it takes so much time to pass a law through parliament that the government is usually replaced before the law is passed—and the new one starts again! So we should be a lot more careful about this problem of obstacles to personnel motion one way or the other. In the case of this country, the United States, the situation is not too bad. In the case of many European countries, with the possible exception of The Netherlands (they are on the whole doing pretty well in high tech), I think that the situation is much poorer. I am really sorry about that; and I am not sure that the situation is very good from that point of view in the eastern countries either. I think that probably a lot could be improved in this matter.

These are a few basic thoughts I wanted to present about the roles of government: not only the direct and very important roles in education and basic research, of course, and in the procurement of research support for projects that are for government use, but also the very important role of government in taxes and other rules, which have an influence on the development of high technology in government physics, though for indirect reasons. They are so important precisely because, being indirect, they need to avoid other pitfalls, such as one might have if a decision by civil servants in a Ministry had the effect of determining what should be on our plate tomorrow, which would, I think, be the end of French gastronomy! Or spending a lot of money for what the market itself would be willing to pay for, so that it is completely useless to pay for it out of taxpayers' money. Government can be very efficient if it is properly used, but that requires more and more technological competence at government level. And here I would like to turn to the physicists. Are you sure that you physicists here—people who at least are still trying to be physicists by reading the literature—are you sure that you are always ready to give that kind of advice, and participate in the backbreaking decision process of government? And if you are not, should you have a right to complain if governments then make mistakes? That is the last thought I would like to leave with you.

The Role of Industry:
Knowledge and Skills/ H. B. G. CASIMIR

It is evident that the spectacular development of our industrial technology would have been impossible without the basic research that preceded it and without the so-called applied research that accompanied it. It is equally evident that the spectacular progress in our understanding of nature would have been impossible without the products that have become available, thanks to industrial technology. The fact that there often exists a love–hate relation between science and industry does not contradict this simple truth. Industry may occasionally scoff at long-haired scientists and find their preoccupations useless and futile; scientists may look down on the mercenary ambitions of industrialists, but both parties know they need each other. In order to clarify the issue, I would like to introduce the notion of a science technology spiral. Let us begin by looking at science and technology as two independent streams. Both streams present a complicated structure. Some developments may come to a stop, often a temporary one. There are parallel lines, also apparently divergent ones, that later may come together.

But the progress of technology uses the results of scientific research, often well-established, almost antiquated results. Euclidean geometry is perfectly adequate for the mechanical engineer and the architect. Maxwell's theory of the electromagnetic field satisfies the needs of electrical engineering. However, technology does not use the most recent results of scientific research. There is always a time lag of some five to twenty years and during our century that time lag has not become any shorter. That statement is one of my hobbyhorses and goes definitely against popular opinion, so let me give some examples. Between the discovery of the electron and the appearance of the first electron tubes, less than ten years had passed. Between the introduction of the notion of positive holes by Peierls, and the discovery of hole survival in p-type material (which is the essence of the invention of the transistor), twenty years had passed. That was also the first really important application of the notions of the new quantum mechanics. Five years after Röntgen's discovery of x rays there was already a flourishing industry of medical x-ray equipment. Nuclear magnetic resonance, found experimentally just after World War II, was only recently harnessed for

H. B. G. CASIMIR *spent most of his career with Philips, Eindhoven (Netherlands) where he was Director of Research for many years until his retirement. He was President of the European Physical Society from 1972 to 1975, and has also served as chairman of the IUPAP Commission on Physics for Development.*

medical purposes, and the whole domain of high-energy physics, particle physics, and so on has not yet found any direct applications. Nor are any in sight—
for which I am grateful. Neither are the most refined results of fundamental
research applied. The general theory of relativity is a wonderful creation of the
human mind; I do not know of any application. The minute changes that result
from the more refined forms of quantum electrodynamics are so far of no technical importance either.

Scientific research, on the other hand, uses the results of technology and here
there is hardly any time lag at all. In this respect astronomers are very expert.
The first Sputnik had hardly been launched before the astronomers began to
make plans for extraterrestrial observations. In radio astronomy they use the
most advanced low-noise amplifiers, etc. But physicists are no less versatile. The
days of string and sealing wax are definitely over, regret it or not. Sometimes
science is even creating its own technology and thus furthering technological
progress directly. The thermos bottle, invented by Dewar as a tool for his low-
temperature work, is a humble but striking example.

What is the influence of wars (including cold ones)? In general, scientific
progress at the very frontier is retarded, but, on the other hand, the time lag I
mentioned is shortened. The development of centimetric and millimetric electronics during the war is a case in point. It did not use notions that were unknown in principle at the beginning of our century. A man such as my compa-
triot H. A. Lorentz could immediately have made major contributions. After
the war, the new technology was used to great advantage for research purposes
which led to several Nobel prizes!

So far, so good, but where do the industrial research labs come in? I think
their primary task is to be a kind of matching device between science and industry. After all, the results of academic research are usually not fit for immediate
application. Academic research may have reached an understanding in principle, but technology is interested in precise quantitative data. And it asks for
materials answering certain well-defined requirements. In order to make results
of "pure" science applicable, a lot of work still has to be done. I do not like the
phrase "applied research". I think it is misleading. You can distill whiskey,
drink whiskey, and distill drinkable whiskey, but you cannot distill drunk whiskey. You can do research, apply the results of research, and perform applicable
research, but you cannot perform applied research. But my real objection is that
the phrase applied research suggests a greater difference between academic
research and industrial research than exists—or in any case greater than ought
to exist—in reality. The point is that most academic scientists do not work at the
extreme frontiers of scientific thought. They too are tying up loose ends, clarify-
ing, simplifying, looking for new examples, etc. True, the motivation is different, but even that is only partly true. For the manager of a company the situation
is clear: The research laboratory is there to help him to stay in business. (In our
economic system that is tantamount to saying that it should help him to make
money; my formulation would also hold in a different system.) But the research
man working in industry has a double motivation. On the one hand, he wants

recognition inside the company; on the other hand, he also wants to be recognized in the world of science, and in that respect he has the same ambitions as his academic colleague. Moreover, whether in industry or in an academic laboratory, research work of good quality can only be done if the research man is interested in the problem as such. Whether a man studying the electron optics of a TV picture tube is interested in television or not, does not matter. He will not accomplish much unless he is interested in the mathematics of the problem for its own sake.

Occasionally, work in industry leads to important contributions to fundamental science. I shall give a few examples—which does not mean that there are no other cases equally or even more important. I mention the work of Langmuir at General Electric on surface chemistry, the discovery of electron diffraction by Davisson and Germer at Bell Labs, and I hope you will forgive me if I also mention the work of Verwey and Overbeek at Philips on the stability of colloids, work that also inspired my own contribution to the theory of long-range retarded forces. But, as I said, there are many more examples. Several recent Nobel prizes bear witness to this. Now such work is certainly important for the reputation and for the morale of the laboratory concerned. Yet in my opinion it is not the main justification for the existence of industrial laboratories. Similarly, the academic laboratory is not primarily concerned with making new technological inventions, although occasionally it does.

Industrial research labs start from existing basic knowledge. They add what is required to make it applicable, thus contributing to the basic knowledge. They may also formulate questions to be answered by academic science. But *industry* does more: it also develops specific skills and it finds partly by scientific reasoning, but largely by empirical and often somewhat alchemistic methods, accurate prescriptions that are to be followed in manufacture.

Time and time again, I have been astonished to see how a delicate piece of equipment, for instance a camera tube for television that was conceived in the research lab and of which a few samples were made by highly competent scientists assisted by extremely skillful technicians, were eventually produced in the factory in large numbers and with better, more reproducible characteristics, of course after unavoidable birth pangs and teething troubles.

Enough of generalities; let me discuss a few examples. In these days when superconductivity has become a hot subject it may be of interest to have a look at Kamerlingh Onnes. In a way, he was a man who did big science "*avant la lettre*". In his laboratories he gradually built up what, by the standards of those days, was an impressive system. He installed well-equipped workshops. He constructed liquefiers for air, hydrogen, and helium. He drew upon industrially available apparatus as much as was possible but, to begin with, he had to produce his own electricity! In his early days there was no electric power station at Leiden, so he bought two engines—a British one, running on coal gas, and a German one, running on liquid fuel. He bought compressors in Switzerland but the liquefiers had to be built in the laboratory workshops. The air liquefier was the first to be replaced by an industrially manufactured one, but when I worked in the K.O. Laboratories, in the middle and late 1930s, the hydrogen and helium liquefiers were still very much home-made, and liquid helium was still a sacrosanct liquid, available only once a week and then in limited quantities.

Today, thanks to industrial developments, that situation has drastically changed: liquid helium has become a generally available commodity.

When superconductivity was discovered in 1911, Kamerlingh Onnes thought at once about applications. But his early high hopes were crushed because of the existence of a critical magnetic field, and as long as liquid helium and even liquid hydrogen were rather esoteric liquids, industry was not interested anyway. It is a rather ironic situation that now that the low temperatures are far more accessible, soon they may no longer be required for applying superconductivity.

As another example, 1932 was the "annus mirabilis" of nuclear physics, in which the neutron, heavy hydrogen, and the positive electron were discovered. It was also the year in which Cockcroft and Walton for the first time observed a nuclear reaction brought about by protons accelerated by a high-voltage generator. It is curious that in two recent books, Richard Rhodes's *The Making of the Atomic Bomb* and Luis Alvarez's autobiographical *Adventures of a Physicist*, this is not mentioned at all. It is undoubtedly true that the USA soon produced more powerful accelerators: the Van de Graaff, the cyclotron, and so on, but Cockcroft and Walton were the first. The suggestion that the Cavendish Laboratory under Rutherford went beyond the string and sealing wax only *after* the cyclotron had been successful is incorrect.

The Philips research labs were among those that took an early interest in accelerators for nuclear physics. In 1938 a better engineered version of a cascade generator was built at Eindhoven and installed at Cambridge. It gave valuable service throughout the war. As a matter of fact two generators were delivered but, as I remember it, only one was kept in operation, the other one being used as a stock of spare parts. It was later donated to the University of the Witwatersrand at Johannesburg, where during many years it functioned well and in any case contributed to physics teaching. Finally, the condensers began to break down. Philips people claim it was because they had not been designed for use at 1750 meters above sea level.

After World War II Philips built many more cascade generators, but when the High Voltage Corporation put Van de Graaff generators in pressure tanks and went to voltages of several million volts, the Philips generators were no longer competitive. Philips also built a number of cyclotrons, both synchrocyclotrons and isochronous ones. It was an interesting feature of this activity that it brought us into contact with leading figures in basic science. I remember how Irène Curie came to Eindhoven to discuss the specifications of a cyclotron we were building for her. After her death in 1956, at the age of fifty-nine, her husband (Frédéric Joliot) took over, but he survived her by no more than two years. My impression was that neither of the two felt entirely at home with such a machine, and Joliot spoke somewhat wistfully about the thirties, when he and Irène did their outstanding work with rather simple apparatus and when the circle of physicists was so much smaller.

Also, the cyclotron work at Eindhoven came to a stop. The commercial department decided—probably wisely—that there was no continuity there. And in the meantime fundamental research began to demand ever bigger installations, the building of which no industrial outfit would be willing to undertake as a turnkey project, or on a no-cure no-pay basis. That is characteristic of big science. But these big projects had to draw heavily upon existing technologies, just as Kamerlingh Onnes had done in the old days. Philips, for instance, delivered large quantities of ferrites for the accelerating sections of the Brookhaven cosmotron and for the proton synchrotron at CERN. This brings me to another domain, a domain where industry has not only delivered scientifically useful products but has also contributed much to understanding—the field of materials sciences.

Of course, the most striking post-WWII example is the study of semiconductors. Here industrial laboratories, headed by Bell Labs, were leading. It is true that they could build on the work of Bloch, Peierls, Wilson, Bethe, and a few others, but much of our detailed knowledge of the band structure, the nature of donors and acceptors and so on, is due to work in industry. Industry also did an amazing job in mastering an ever more complicated technology. I recall the days when it was quite a task to produce with good yield a chip with 25 components!

Chips make today's computers possible. By the way, I think that the choice of this name for an integrated microcircuit is particularly unfortunate. Why add one more meaning to the sixteen that are listed in Webster's Dictionary? (I am speaking about the second edition that appeared in 1958, not about the next one that was burned page by page by Nero Wolfe.) The only thing I can say in favor of this term is that it is almost universally and internationally accepted, a feature it has in common with the SI system of units, which I dislike even more. But after all, what's in a name? And I could not think of anything that better demonstrates how indispensable industrial products are for the advancement of science than computers. I know that this is so, but I have not mastered their use myself—I am not up-to-date. Neither am I up-to-date in my style of presentation. Well, that is the risk the organizers of this meeting took when they invited a man of my age to speak about present aspects of science and industry!

The electronics of the present is solid state electronics. It was preceded by vacuum electronics, and we should not forget that radio and television, and the first digital computers, were based on the vacuum tube. The main principles of electronic circuitry were established in the age of vacuum electronics. And vacuum electronics is still with us, for high power transmitters, for picture tubes and camera tubes, for image transformers, and for photomultipliers.

Sometimes it seems to me that the tendency to replace tubes everywhere by solid state devices is partly a matter of fashion, and I suggested once that one might market good old-fashioned valves as a new kind of "solid state device with vacancies in a collective mode," but my suggestions in the commercial field were usually not taken seriously.

Vacuum electronics provided ample scope for work in the industrial research laboratories. They contributed much to the understanding of thermionic emission. Yet the final prescriptions to be followed in the manufacture of cathodes were partly the result of trial and error, and here we have to speak about skills rather than about knowledge. Much work of a mathematical nature was done on the motion of electrons, the influence of space charge, and so on, but the art of making and accurately positioning fine grids and other electrodes comes under the heading of skills.

The study of surface phenomena, photoemission, secondary emission, and cold emission, is closely related to similar work in solid state electronics, and I believe that there are some indications that knowledge gained in solid-state investigations may lead occasionally to important innovations in vacuum electronics.

Earlier, I mentioned magnetic material delivered by Philips to Brookhaven and CERN. Now magnetism is certainly a subject where there has been a constant and fertile interaction between academic research, industrial research, and industry. Let me tell you something about ferrites. Ferrites are compounds with the general formula $Me^{II}O \cdot Fe_2^{III}O_3$ where Me^{II} is a divalent metal—or mixture of metals. Iron ferrite $FeO \cdot Fe_2O_3$ (or, written more shortly, Fe_3O_4) is the first magnetic material known as such to man, for this is the formula of the famous lodestone. I believe the Japanese were the first to prepare ferrites that were magnetically soft, had large permeability but little remanence. But extensive research and development work was done at Philips, where the crystal structure and its relation to the magnetic properties was carefully studied and where a whole range of useful compounds was made. Now the French theoretician Louis Néel had developed the theory of what he called ferrimagnetism, a phenomenon that can also be described as uncompensated antiferromagnetism. Let me remind you of the main idea, which is, as good ideas should be, simple, elegant, and very ingenious. How does normal ferromagnetism come about? Let us assume that a substance consists at least partly of magnetic ions, of ions having a magnetic moment. Now if there is an interaction between neighboring ions that tends to orient them parallel, this can at sufficiently low temperatures lead to spontaneous magnetization; at higher temperatures thermal motion will prevent this. There exists a critical temperature, the Curie temperature, below which spontaneous magnetization sets in. But it can also happen that the interaction between neighbors tends to orient them antiparallel. Then at low temperatures the magnetic moments will be arranged regularly, up down, up down, and again this kind of long-range order will begin to appear below a critical temperature, the Néel temperature.

In ferrimagnetism, the situation is a bit more complicated. Suppose there are two kinds of magnetic ions A and B, and suppose all interactions between neighbors are negative—that is to say, tend to orient the moments antiparallel. Now the arrangement of the ions can be such that the interaction between the A-ions and the B-ions outweighs the interaction of the A-ions among themselves and also the interactions between the B-ions among themselves. Then—again below a certain critical temperature—the A-moments will be parallel to one another, the B-moments will be parallel to one another, but the total A-moment will be opposite to the total B-moment, and the total resulting moment will be given by

$$M = \sum \mu_A - \sum \mu_B .$$

Now this theory of Néel was an important guide in developing new ferrites. But, looking at it the other way round, the Philips work served as a beautiful confirmation of Néel's ideas.

I like to illustrate Néel's theory in the following frivolous way. Suppose that two quiet gentlemen are living in an apartment building, one on the first and the other on the third floor. Two saxophone players who like to practice after midnight are living on the second and fourth floors. The two quiet gentlemen do not like each other. Neither do the two saxophone players. But the quiet gentlemen present a united front against the united saxophone players. Agreement by shared antagonism: that is the crux of ferrimagnetism.

Let me now tell you two of my favorite stories; I think they are instructive. When Philips began delivering ferrites to Brookhaven, according to our own measurements the material just met the specifications we had agreed upon. But according to the Brookhaven people the material was considerably better and had lower losses. One of our best engineers checked up on our measurements and found the Brookhaven people were right—an error caused by nonlinearity and higher harmonics had crept into the Philips measurements. The Brookhaven people were pleased—it is always nice to be right—and they got a better material than they had bargained for. And we probably got an undeserved reputation for being dumb but honest. But suppose the error had had the opposite sign. Then one would certainly have blamed greedy industry for trying to get away with inferior material by manipulating measurements.

Second story: Besides the cubic ferrites there exist hexaferrites, compounds of the type $BaO \cdot 6Fe_2O_3$, which happen to have a hexagonal crystal structure. (Instead of barium, one can also take strontium, or a mixture.) These hexaferrites make useful permanent magnets. Now the history of the discovery of this permanent magnetism is amusing. A lab technician had been instructed to make a preparation that would almost certainly have been a soft magnetic material, but in converting atomic weights to grams he made an error. The research scientist who had ordered the preparation noticed its peculiar properties, had it analyzed and then began a thorough investigation. The moral of the story is not that one should encourage people to make mistakes, but that, once a mistake has been made, it is worthwhile to consider whether one can understand it and what one can learn from it.

Let me try to summarize this somewhat rambling talk in a few sentences. Progress in fundamental research depends on the knowledge and skills of modern industry. That knowledge and skill exist and will continue to grow, thanks to academic research. Academic research aims primarily at discovering or creating new phenomena and at understanding them. Industrial research aims at applicable results. But nonapplicability is neither a necessary nor a sufficient criterion for inherent beauty and value.

International Space Science/ HANS MARK

1. INTRODUCTION

Science always has been an international enterprise. From the earliest days of organized scientific effort that began some 300 years ago in western Europe, a strong international communications network has existed among people engaged in science. Fermat, the Bernoullis, Newton, Leibniz, Galileo, Kepler, and many others carried on an extensive international correspondence with their colleagues. None of them felt that national boundaries were important in pursuing the quest to understand nature.

Perhaps the most interesting historical example of the international nature of the scientific enterprise, even in periods of great conflict, is the story about Sir Humphrey Davy's tour of Europe in 1812 and 1813. You may remember that the Napoleonic Wars were just coming to a climax at that time and England and France were bitter enemies. Yet, Davy and his young and brilliant assistant, Michael Faraday, were able to travel freely throughout Europe and deliver their scientific lectures even in Paris.

It is also true that nations have not been reluctant to use scientific talent from foreign countries for purely "national" and often highly classified projects whenever it served their purposes. This is perhaps the other side of the coin that illustrates the truly "international" nature of science. The stories regarding the contributions made to the U.S. nuclear weapons program by refugee scientists during the Second World War are well known. After the war, the contributions of German engineers and scientists to both the American and the Russian space programs are equally familiar. Nor is this anything new. There is, for instance, the case of the Russian-born chemist and Zionist leader, Chaim Weizmann. Weizmann's scientific contributions to the British war effort in the area of explosives during the First World War led to a statement of British support for the Zionist cause in 1917 (The Balfour Declaration). Finally, there is the story of Benjamin Thompson. Thompson was born in Massachusetts in 1753. He made a fundamental error in judgment in supporting the British during the American

HANS MARK *has divided his career between academia and government service. He has held teaching and research appointments at MIT and Berkeley, but became deeply involved in the United States space program, becoming Director of the NASA Ames Research Center and later Deputy Administrator of NASA. He has also served as Secretary of the US Air Force. He is now Chancellor of the University of Texas system.*

Revolution, which resulted in exile for him after the war was over. Thompson spent the rest of his life in Europe. There, he had an extraordinary career in politics, military affairs, engineering, and science. He is better known today as Count Rumford, a title bestowed on Thompson by the King of Bavaria for service rendered to that state. Rumford made two important scientific contributions while serving as a military advisor to the King. One was that he laid the foundations for the first law of thermodynamics by demonstrating the continuous generation of heat from mechanical work while boring cannons. Second, he discovered the phenomenon of heat convection while trying to find out how to make uniforms that would keep Bavarian soldiers warm. Rumford was obviously a valuable officer of the state and it did not bother the King that Rumford was an American.

The fact that all of this should be the case is really not too surprising. Pure science is practiced really well by a very tiny fraction of the population. The people who are devoted to pure science are not moved by the same political considerations, generally speaking, as the population at large, and they generally are not close to the government. They are people who are extremely important citizens of the nation but they are almost never directly involved in the process of making policy.

I wanted to talk a little bit about history to make the following point: I believe that doing science on an international basis is extremely important but it is not a way of conducting foreign policy. For example, those who believe that doing international scientific programs has anything to do with the preservation of peace have simply not looked at the record. Heisenberg's visit to Bohr in 1941 did not prevent Nazi Germany from brutalizing Denmark. Victor Weisskopf's graduate studies in Göttingen did not discourage Nazi Germany from invading Austria in 1938. The fact is that scientific projects do not help to cement international relations in any significant way. At least, that is what the historical record says. Those who believe otherwise overestimate by a large factor the importance of scientific work in the conduct of international affairs. It is a self-delusion that affects us as scientists, an honorable one perhaps, but a delusion nevertheless.

If what I am saying is true, why should we do international science at all? Why not just adopt a very chauvinistic view and do only "national" science? The reason for doing science on an international basis is that it will lead to *better science, not better international relations.* As far as we can tell, genuine scientific talent is very rare and it seems to be equally distributed among people all over the world. This rarity of talent and its equal distribution are the real reasons for seeking international scientific collaboration. It is important to do things for the right reason because only then will the international programs that we seek survive the inevitable political fluctuations in international relations.

2. TWO INTERNATIONAL SPACE PROGRAMS

Some of the things I have just said may sound slightly unorthodox. Nevertheless, I think they must be taken seriously. Why is space science a particularly good area for international collaboration? Is there anything that makes space projects particularly attractive as candidates for international collaboration? My feeling is that space science projects are good but not necessarily better than other science programs. Because they tend to be expensive, resource sharing may be a consideration. However, this is also true for large particle accelerators and is becoming true for other scientific areas that once were considered to be "small science." The recent proposal to sequence the human genome is perhaps the best example of what used to be considered small science that might now become large enough to be considered as an international project. The resource-sharing argument is certainly a valid one and it has proved to be good over the years.

Another related argument is the existence of a unique capability in the hands of one party or another that would make international collaboration particularly attractive. This reason is often compelling and has led to several space science collaborations in which I have personally participated. If I may, I would like to describe briefly two international space programs that I helped to initiate and execute, in order to perhaps draw some lessons about how international space projects are best conducted.

A. The Soviet–American Space Biology Program

As some of you may remember, the United States flew a biological research satellite called "Bio-satellite" in the Summer of 1969. It turned out that this satellite was extremely expensive and the flight was deemed only partially successful. As a result, the United States decided to discontinue flying biological experiments in space until the shuttle became available. At the same time, scientists in the Soviet Union were initiating a long-term series of satellite flights in which biological experiments would be conducted. In 1969, when I was serving as Director of the NASA–Ames Research Center, a group of people at the Center suggested to me that we should develop a collaborative space biology program in which American payloads would be flown on Soviet satellites. The Russians had a unique capability for flying biological payloads, and our people felt we should somehow use it. We also had a unique capability, which was the strong group of biological scientists NASA had established at the Ames Research Center under the leadership of Dr. Harold P. Klein. We felt that putting these two groups together would be beneficial to both parties and that good science would result. I therefore strongly encouraged the establishment of the program.

The program has now been in operation for over 15 years. We have flown American payloads on six Russian satellites starting in 1975. (The flights were conducted in 1975, 1977, 1979, 1983, 1985, and 1987.) Papers based on the results of these flights have been published in international scientific journals and several really significant results came out of the collaborative work done by the two scientific teams. The most important of these are in bone formation and in the response of the cardiovascular system under zero gravity conditions.[1,2] On the Soviet side, the program was led by Dr. Oleg Gazenko, and on the

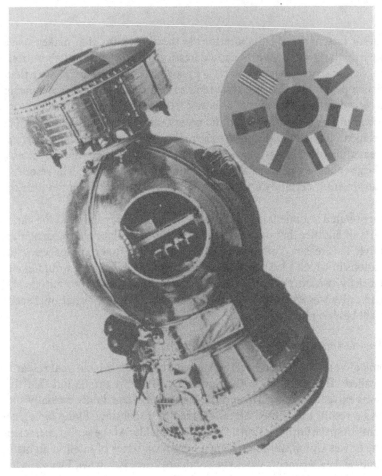

FIGURE 1   This is a picture of the modified "Vostok" spacecraft used by the Soviets for their space biology experiments. American payloads have flown on this type of spacecraft six times since 1975. Note the American flag on the logo for this particular flight.

American side by Drs. Harold P. Klein and Joseph C. Scharp. It is an interesting fact that the joint Soviet–American space biology program has survived all of the political fluctuations in the relations between the Soviet Union and the United States since the program was initiated in 1971. It is equally clear that the existence of the joint Soviet–American biology program has had no impact whatsoever on Soviet–American political relations.

B. The Infrared Astronomical Satellite (IRAS)

The Infrared Astronomical Satellite is perhaps the most successful international scientific program that the United States has ever mounted. It was initiated in 1977 by a group of infrared astronomers headquartered at the NASA–Ames Research Center and working in collaboration with other American scientists. This group had experience performing infrared astronomy above the atmosphere using an airplane-mounted telescope (the Kuyper Airborne Observatory). Since the Ames group was not in the scientific mainstream of space astronomy, the proposed international collaboration seemed to be a good way to gain the necessary support to push the project through. I strongly encouraged this approach to the development of the project.

Eventually, an international arrangement was agreed upon to develop an orbiting infrared telescope involving the United States, The Netherlands, and Great Britain. The work was divided in the following ways among the three partners: the United States would provide the infrared telescope (a responsibility of the Ames Research Center) and would also provide overall project management (a responsibility of the Jet Propulsion Laboratory); the Dutch would build the spacecraft through a contract with the Fokker–VFW Corporation; and finally, the British would provide the ground station and the initial data-

FIGURE 2 This is a picture of the Infra-Red Astronomical Satellite (IRAS). The telescope was built by the Americans, the spacecraft by the Dutch, and the ground-based data systems by the British. The IRAS was flown successfully in 1983.

processing system. It is really remarkable that this manifestly complicated arrangement worked as well as it did. I have already said that I participated in the initiation of this project when I was working at the NASA–Ames Research Center in 1977. Six years later, in 1983, it fell to me to participate in the decision to launch the IRAS Observatory while serving as the Deputy Administrator of NASA in Washington.

As you all may know, the Infrared Astronomical Satellite (IRAS) remained in Earth orbit for about a year and produced a wealth of new information in infrared astronomy. Among other things, 250 000 new infrared stars were discovered by IRAS. It is no exaggeration to say that the IRAS data opened up an entirely new field of astronomy.[3] While the scientific success of the program is beyond question, I could not see any discernible effect of the IRAS project on relations between Britain, The Netherlands, and the United States.

What did I learn during my own participation in international space science programs? Are there any lessons that I can share with you in order to make it easier for future collaborations to be established? I believe that at this point there are several statements about international projects which might be worth making.

First and foremost, it is important to keep the technical interfaces simple. This is generally good policy in any space program, but in one that crosses international boundaries it is more important than ever. A second important point is that good personal relations among the key individuals are even more important in international programs than they are in the ordinary space projects conducted by a single nation. Good personal relations always make it easier to work with people, but in this case the establishment of such relationships is also a good hedge against political fluctuations. One of the interesting things about the space biology program mentioned above is that once a good international team is established, an "us against them" philosophy develops where "us" is the international team and "them" is everyone else. The kind of *esprit de corps* that such an attitude fosters turned out to be extremely valuable in surviving the various downturns in Soviet–American relations over the years. Finally, it is important to remain very flexible. I learned this in both cases—the space biology program and the Infrared Astronomical Satellite program. There were the inevitable changes of plans in both programs and it is sometimes more difficult to negotiate such changes across national boundaries than it is if the program is purely national. A degree of flexibility above and beyond that required for the ordinary space program is therefore necessary.

I should close this section of my talk by saying that with all the difficulties one encounters in international space programs, the results are worth it. There is no question in my mind that we have done first-class science on an international basis that could not have been done otherwise. It is equally true that the science we have done has been of higher quality because we have done it with our friends and colleagues in other nations. These results, I believe, speak for themselves on the value of international collaboration on space science and in other scientific areas as well.

3. THE NASA SPACE EXPLORATION PROGRAM

Ever since NASA was established in 1958, there has been a very basic plan for the nation's space program. The fundamental points of this plan were and are to send people into space in order to learn how to live and work there, to do scientific experiments, and to conduct other operations as well. At the same time, a vigorous scientific research program using unmanned spacecraft and satellites would also be conducted. Since 1970, about 80% of NASA's space effort has gone to the program designed to put people into space and about 20% to scientific exploration. It is remarkable that this ratio has held constant within a few percent in spite of the large variations in NASA's budget. The peak scientific spending percentage was about 25% in 1973, which was the top spending year for the Viking program. Prior to 1970, scientific spending tended to be a smaller fraction because of the large amount of money spent on the Apollo program. While this is true, it is also true that the largest scientific spending in any single year was in 1967, which was also the peak of spending on the Apollo program. History has demonstrated that scientific work in NASA has fluctuated approximately with the NASA budget. When the NASA budget has increased, scientific spending has increased, and when the NASA budget has decreased, scientific spending has correspondingly been curtailed.[4]

The basic NASA plan for space exploration has been remarkably consistent over the years. I first heard about it in 1969 from Wernher von Braun at a meeting of a committee, of which we were both members, to determine what should be done after the successful execution of the Apollo program. Von Braun's envisioned sequence of events was very simple. First, we would have to develop a reusable space ship in order to transport people to Earth orbit and back in the most effective and efficient manner. Second, we would establish a permanently occupied space station in Earth orbit. Third, we would establish human presence on the moon and eventually put a permanent base on the Earth's large satellite. Finally, we would initiate human exploration of the solar system by making a trip to Mars. It is interesting that the essential elements of this plan were developed in the late 1940s and early 1950s by a group of visionaries (including Wernher von Braun) who developed the ideas in some detail. Two interesting documents from that period tell the whole story. One is a report issued by the Douglas Aircraft Corporation in 1946, authored by David Griggs, Louis Ridenour, and Francis Clauser.[5] While the main portion of this report deals with rocketry, there is considerable speculation on what can be done with such rockets and, of course, a trip to the moon is mentioned. The second document is an article that appeared in the 22 March issue of *Collier's Magazine* in 1952. It is called "Man Will Conquer Space Soon," and the cover picture shows the space shuttle and a space station, as well as the first destination—the moon. The article was coauthored by Wernher von Braun, Joseph Kaplan, Fred Whipple, Heinz Haber, and Willy Ley.[6]

We are now well on the way to the execution of this plan. We have built the space shuttle and, in spite of the current difficulty, we have learned how to use it. In his State of the Union Message of 1984, President Reagan identified a permanently occupied space station as the next important objective of our space program. In his address, he once again repeated the major reasons for building this space station, reasons that were first outlined in the 1952 Collier's article.[7] It is, perhaps, worthwhile to repeat them here.

First, the space station will be a maintenance facility in Earth orbit for the repair, retrieval, and refurbishing of unmanned satellites. Second, the space station will be a laboratory in Earth orbit. Particular attention will be paid to developing experiments that will demonstrate and enhance the capability of people to live and work in space. Third, the space station will be a staging base for more ambitious missions, the first of which has always been envisioned as the establishment of a base on the moon. To quote the 1952 Collier's article, "From this platform, a trip to the moon itself will be just a step as scientists reckon distance in space."

The objectives I have outlined have been generally accepted by successive American administrations since 1970. There is every reason to believe that future presidents will also commit themselves to continuing this overall conceptual plan.

4. THE LUNAR BASE—AN INTERNATIONAL ASTRONOMICAL OBSERVATORY

The space station that I have mentioned is due to be deployed before the middle of the next decade. If we are able to adhere to this schedule, then it will be possible to take the next step, which is the establishment of a permanent base on the moon by the year 2000. In the summer of 1984, a meeting was held at the Los Alamos National Laboratory to discuss the prospects of a lunar base and to develop the technical concepts that would have to be employed in order to achieve the objective. One of the most important aspects of this discussion had to do with the development of strong reasons for putting a base on the moon.

There is no doubt that a political imperative exists for occupying the moon. It is this factor that will influence future administrations to reach the decision to place a permanent base on the moon. It is really this same political imperative that motivates putting people in space in the first place. The participants in the Los Alamos study, myself included, recognized this political imperative, but we also felt that establishing a base on the moon, just to be there, was simply not enough. This point was also recognized in the recent report authored by Dr. Sally K. Ride on the future of the NASA program.[8] Dr. Ride and her colleagues advocated the construction of a lunar base as the next step and, in doing so, they simply adhered to the plan I have already outlined.

What has been missing is a strong reason for taking the step to put a permanent base on the moon. I submit that a strong reason actually exists, and it is the establishment of an international astronomical observatory on the moon. It was this rationale, presented to the Los Alamos group by Professor Harlan J. Smith of The University of Texas at Austin McDonald Observatory, that provided the basic motivation for many of us at the meeting.

What are the advantages of putting an astronomical observatory on the moon? Are they sufficiently great to justify the enormous expense that would be incurred if this project were to be executed? Let me provide a quick summary of the arguments that were made by Professor Smith at the Los Alamos meeting three and one-half years ago.[9]

The moon has no atmosphere and, therefore, the observatory would exist in a vacuum. This is good for low-frequency radio astronomy because the ionosphere that is present on the Earth would be absent. It is obviously good for diffraction-limited imagery because there would be no atmospheric interference. Phased coherent interferometry would be easier, again because there would be no distortion by any atmosphere. In the realm of electronic detectors, the absence of an atmosphere may be particularly important. One can imagine large arrays of naked cathode detectors used with large light-gathering mirrors. The mirrors themselves would last much longer in the vacuum than they do in the Earth's atmosphere because the optical coatings that must be applied would last practically forever.

The dark side of the moon is very cold and there would be very little interference from other sources of light. This would make it possible to look at ultra-faint objects which are many magnitudes smaller in intensity than those that can presently be observed from the Earth's surface. The moon would be good for spectroscopy because it would be free of the kind of sky emission and absorption due to the atmosphere that exists on Earth. Even during the lunar day there would be good observation conditions for various objects. Finally, because of the low temperatures that exist on the lunar surface, during nighttime there would be a natural cryogenic environment for telescope detectors designed for infrared and other long-wavelength observations.

The moon is a stable inertial platform as opposed to Earth orbit which is currently used for the placement of space telescopes. This is particularly important for things such as long-baseline interferometry, where a stable place to put the elements of the array is necessary. It is also very important for pointing and tracking. Accurate pointing of space telescopes in Earth orbit is a problem requiring very sophisticated technologies. These would not be necessary in the case of a lunar-based telescope.

Another advantage of the moon is its slow rotation rate. An observatory properly placed could see the whole sky over a reasonable period of time. Furthermore, the fact that nighttime on the moon lasts two weeks means that it is possible to think about very long exposures for faint object detection.

An observatory placed on the far side of the moon would avoid absolutely all interference from radiations of all frequencies coming from the Earth. At the present time, light pollution is a major problem at many of the world's astronomical sites. There is also stray light from atmospheric scattering of sunlight and starlight which would be absent on the moon. Finally, in the case of Earth-based radio astronomy, interference from television and other microwave signals has become an important limiting factor for that particular scientific endeavor.

The gravitational field of the moon would also be an advantage. One of the difficulties of conducting astronomy from Earth orbit is that all spacecraft on which the telescopes are mounted are sources of gas. In the vacuum of space the spacecraft outgas. These gases emitted by the spacecraft tend to condense on optical surfaces, and especially on the surfaces of cooled detectors and mirrors. The small gravitational field on the moon would eliminate this problem since the materials that outgas from telescopes and other instruments would simply fall to the ground in due course. This is the advantage when compared to space astronomy but there is also one when compared to astronomy done from the ground. The smaller lunar gravitational field would make it easier to erect large structures on the moon and to accurately control their pointing.

One of the particularly interesting advantages of performing radio astronomy from the moon is the existence of a large number of craters that could be employed in the same manner as the Arecibo Radio Observatory is now used. As you all know, Arecibo is a natural crater in which a radio telescope reflector has been placed. Arecibo is 300 meters in diameter. Based on our observations, there are craters on the moon with approximately the right shape for a parabolic reflecting telescope which have diameters up to 10 and 15 kilometers. Putting radio telescopes in such craters would increase the radiation gathering power of the instrument by many orders of magnitude. [10]

These are some of the obvious near-term advantages that the moon would offer as a site for an astronomical observatory. There are some other advantages that would accrue in the longer term. Once a lunar base is established and we learn how people can best work on the lunar surface, then the proximity of people and support facilities on the moon will be extremely important. The people would be able to maintain and modify the astronomical instruments as dictated by operational requirements. They would also be able to perform oversight functions of any automated equipment on the lunar surface used to control the operation of the observatory. Another long-term advantage would be the use of lunar materials to construct the observatory. There is no question that this will ultimately be the major advantage of conducting astronomical observations from the moon rather than from earth orbit or somewhere else in space.

These are all formidable advantages. It might be worthwhile to dwell for just a few minutes on what a lunar astronomical observatory would look like. It would be a facility with the capability of looking at the entire spectrum of radiations emitted by stellar and galactic objects. These would range from low-frequency radio waves down to perhaps 30 meters or the few megahertz region, up to gamma rays with quantum energies of many billions of electron volts. The observation of very energetic cosmic-ray primaries would also be possible. In the case of radio astronomy, there would also be the large Arecibo-type antennas mounted in lunar craters that I have already discussed. Of equal importance would be the very extensive phased arrays that would be spread over large areas to do low-frequency radio interferometry.

In the optical and the infrared domain, the construction and maintenance of very large mirrors, perhaps up to 100 meters or more in diameter, can be contemplated. These would eventually be built with local materials, since aluminum and silicon are plentiful on the moon. The construction and control of the mountings would be easier because of the relatively low level of gravity. In the ultraviolet, x-ray, and gamma-ray astronomy regions, there are also extremely important possibilities. Very large area detectors, for example, are possible since real estate, once again, is no problem. These, again, would be made mostly of local materials.

Finally, there is the fascinating question of neutrino astronomy. Currently, neutrino astronomy on Earth is conducted from very deep mine shafts. Detectors are placed far underground in order to avoid the high level of cosmic-ray background that exists on the surface. The limiting feature of such detectors is the natural radioactive background radiation from materials in the Earth's crust. There are some who argue that such background, that is, the background due to natural radioactivity in the lunar crust or mantle, would be smaller than it is on Earth. Placing neutrino detectors at the bottom of these holes on the moon might therefore be a major advantage. I should hasten to add that the calculations to determine whether this suggestion is really practical have not been performed.

Perhaps the most important feature of such an observatory is that all of the instruments covering the entire frequency spectrum could be pointed at the same object and, therefore, conduct what I like to call "broadband" astronomy. About ten years ago, I made the suggestion that an astronomical payload for the space shuttle should be designed for the purpose of looking at various objects in quite different wavelength regions simultaneously.[11] I know from personal experience that broadband astronomy would be a most powerful investigative tool. In 1967, while looking at the x-ray star SCO-XR1, we decided that it would be valuable to observe, simultaneously, the x-ray flux and the visible light emitted by the star. In due course, an arrangement was made so that the large telescope at the Cerro Tololo Observatory was focused on the star and used to measure the fluctuations in the visible spectrum. At the same time, sounding rockets were poised at the launch site at Kauai in the Hawaiian Islands, and the rockets were launched to observe x rays at a time when it was deemed interesting, based on the observation of the visual light curve. The results of this experiment were significant in unraveling the mystery of the energy source in this particular star. This experience taught me the value of making synergistic measurements of this type. If I had to guess right now, I would say that it is the ability of the lunar observatory to conduct broadband astronomy that will eventually be its most important feature, and that will lead to the most important scientific contributions of such an installation.[12]

What would we learn if we placed an observatory on the lunar surface? What are the major mysteries? I know that in answering such questions everyone can make his or her own list. Let me give you mine. First, I believe that the kind of broadband observation of various objects that I have been discussing would lead to the unraveling of the mysteries of quasars. There is no doubt that this is one of the critically important problems in modern astronomy, and my belief is that this would be one of the first important results. The nature of the energy source in a quasar is probably related somehow to high-energy particle physics as well, and therefore one would bridge the gap between these two most important branches of fundamental knowledge. A second important result from a lunar astronomical observatory would be the ability to observe in much greater detail the three-degree Kelvin background radiation. My guess is that if this could be done with really great precision, then it would be possible to make much better quantitative calculations than can be made today of what actually happened in the first few moments of the creation. In a sense, such measurements would be the rewriting of the Book of Genesis. The next important result would be the discovery of huge numbers of dark stars. Since most of the stars in the Hertzsprung/Russell sequence simply die a quiet death, these will continue to exist. My guess is that observations made from the lunar observatory would give us the first accurate account of these dark stars and perhaps resolve the riddle of the "missing matter" in the universe. The detection and tracing of a supernova explosion, using broadband astronomy, could have extremely valuable consequences in understanding these objects. No one has yet made a really good light curve of a supernova in several different regions of the electromagnetic spectrum. There is very good reason to believe that a spectral light curve is very sensitive to the model that modern theories propose for supernova explosions. Thus such measurements should be an excellent discriminator between these models.

Finally, many of you know that I have a favorite subject of my own. I believe that perhaps the most important single category of measurements that can be made from the lunar observatory will deal with the observation of planetary formation and the search for other planets like the Earth in our galaxy. Astronomy to find planets moving around very distant stars is much more easily done from the lunar surface than from Earth orbit or from the ground. Once the frequency occurrence of planetary systems in the galaxy is established by observation, then the question of searching for extraterrestrial life can be placed on a much firmer scientific basis.

I have provided you now with my list of what I believe should be done at the lunar astronomical observatory that I am proposing. These observations are both exciting and fundamentally important from the viewpoint of increasing human knowledge in critical areas. There is no better way, in my view, that our country can demonstrate its leadership in international space science than by initiating the establishment of an international astronomical observatory on the moon.

REFERENCES

1. Emily R. Morey and David J. Baylink, "Inhibition of Bone Formation During Space Flight," *Science*, **201**, 1138 (22 September 1978).
2. H. Sandler, V. P. Krotov, J. Hines, V. S. Magadev, B. A. Benjamin, A. M. Badekeva, B. M. Halpryn, H. L. Stone, and V. S. Krilov, "Cardiovascular Results From a Rhesus Monkey

Flown Aboard the Cosmos 1514 Spacecraft," Aviat. Space and Environ. Med. **58**(6), 529 (June 1987).

3. Gerry Neugebauer, "The Infrared Astronomical Satellite," Proc. Am. Philos. Soc. **130**(2), 155 (1986).
4. Hans Mark, "The Planetary Report," Vol. IV, No. 4, 6 (July/August 1984).
5. David Griggs, Louis Ridenour, Francis Clauser *et al.*, "Preliminary Design of an Experimental World-Circling Spaceship," Douglas Aircraft Company, Santa Monica, California, Report No. SM-11827, Contract W33-038 ac-14105 (2 May 1946).
6. Wernher von Braun, Fred L. Whipple, Joseph Kaplan, Heinz Haber, and Willy Ley, "Man Will Conquer Space Soon," Collier's Magazine (22 March 1952).
7. Hans Mark, *The Space Station—A Personal Journey* (Duke University Press, Durham, 1987), pp. 195–196.
8. Sally K. Ride, "Leadership and America's Future in Space," NASA Document (August 1987).
9. Harlan J. Smith, "Astronomy From the Moon," Sky and Telescope **74**, 27 (July 1987).
10. Dietrick E. Thomsen, "Man in the Moon," Sci. News **129**, 154 (8 March 1986).
11. Hans Mark, "Space Shuttle—A Personal View," J. Vac. Sci. Technol. **14**, 1234 (1977).
12. G. Chodil, Hans Mark, R. Rodrigues, F. D. Seward, C. D. Swift, Isaac Turiel, W. A. Hiltner, George Wallerstein, and E. J. Mannery, "Simultaneous Observations of the Optical and X-Ray Spectra of SCO XR-1," Astrophys. J. **154**, 645 (1968).

"Physics at the Edge of
the Earth"/ JOSEPH P. ALLEN

I thank you for the opportunity to speak to the International Union of Pure and Applied Physics and the Corporate Associates of the American Institute of Physics.

It is particularly appropriate that I voice my gratitude and appreciation for the invitation because I, in many ways, am an outsider to this audience of physicists. For example, I am in no way now a practicing physicist. By this, I mean one who studies, teaches, and carries out research on the behavior of matter or one who applies to advanced technical projects the new insights gleaned from such research; and more than 20 years have passed since I was a serious student of physics. Clearly, I am not a practicing physicist.

On the other hand, I am not totally without qualification to share with you today some thoughts on the roots of high technology. To begin, I have always been fascinated by the behavior of the world we perceive, a fascination honed by the study of physics through the undergraduate and graduate levels. Secondly, in the last several years I have, in fact, practiced physics as a physician would practice medicine. By this I mean "practice" in the literal sense—to go over and over a task in a mechanical way until that task becomes intuitive, second nature, and achievable without great mental effort. I will give an example of practicing physics shortly. But, for the moment, let me return to that time I was a university student.

Thirty years ago today, plus or minus 24 hours, I sat in undergraduate Physics 101 at DePauw University and read for the first time in my life of the discoveries and the remarkable concepts of Copernicus, Kepler, and Newton. At that time many of the concepts of these extraordinary thinkers were dramatically demonstrated by the satellite Sputnik as it orbited around our planet Earth. In itself, a satellite was not unique of course, since even in those years we had long been quite accustomed to the Moon. But Sputnik was an artificial satellite, homemade by Russian physicists, engineers, and technicians who were, understandably, quite proud. The rest of us were appropriately impressed at the achievement of its successful launch. A short time later, as I was finishing my

JOSEPH P. ALLEN *obtained a Ph.D. in physics at Yale University and then went into the United States space program. After retiring from being an astronaut he became Executive Vice-President of Space Industries, Inc., in Webster, Texas.*

graduate studies of physics at Yale, Yuri Gagarin and then John Glenn orbited the Earth in spaceships—artificial satellites large enough to accommodate a person. Suddenly our understanding of the classical laws of nature, and the application of these laws, made it possible for us to travel outside the Earth's atmosphere at speeds far beyond normal human experience. I want to say "at speeds beyond human imagination," but of course a form of imagination led to those very concepts initially envisioned by Copernicus, Kepler, and Newton. So clearly, the spaceship velocity of about five miles per second was not beyond human imagination.

Nevertheless, the speed needed to keep homemade satellites in orbit about planet Earth (about 18 000 miles per hour in everyday terms) was, and still is, bold by human standards. Even so, over the last 30 years the combined knowledge, skill, and audacity of thousands of scientists, applied scientists, engineers, and technicians have made possible the construction and orbital flight of many such satellites. What is also remarkable is that scientists as well as pilots, politicians, and others, have been privileged to make space journeys aboard them. Thanks in large part to my interest and education in physics, I have had the good fortune to travel into space on two occasions, have made altogether over 200 orbits of the earth and have reentered the atmosphere twice—landing aboard *Columbia* in California and aboard *Discovery* in Florida. In addition, I have experienced the feeling of floating, untethered, out from the open hatch of an orbiting space shuttle, and have traveled as a satellite myself along one-quarter of a full Earth orbit.

As students of physics we all have carried out *gedanken* experiments, picturing ourselves in our mind's eye as, for example, moving with one reference frame at near-light speed with respect to a second reference frame, then observing meter sticks, clocks, and masses in the two reference frames. The spaceships that orbit Earth do not travel at rates approaching the speed of light, but I assure you that to be a passenger aboard one, as far as classical physics is concerned, is to experience a *gedanken* experiment in real life. To live in the perpetual free fall of orbital flight is to be confronted with example after example of simple classical physics in wonderful and extraordinary detail.

My purpose today is to share with you some of these examples of physics as experienced in orbit—or, as expressed in my title, to reflect on physics at the edge of the Earth. I will use photographs taken from various flights to illustrate the points I wish to make. In that regard, it occurs to me that if Newton or Kepler were somehow to know of our orbiting the Earth, exactly as their equations so elegantly predicted but in artificial satellites, they would not be particularly surprised. Yet they would surely be amazed to see for themselves the views of life aboard these ships that we have captured with the modern invention called the camera. The photography that I use today was taken with quite ordinary cameras and represents to a large degree what your own eyes would see were you to make such a journey.

We are approaching the 500th anniversary of the historic voyage of Christopher Columbus. Although Columbus and his crew were contemporaries of Copernicus, whom I mentioned earlier, the convictions of Copernicus were unknown, or at least unconvincing, to the crew in that they feared, we are told, sailing to the edge of the Earth and then falling beyond. Interestingly enough, each space journey begins precisely that way. A spaceship, modern sailors inside, is propelled by rockets up through the atmosphere and into the vacuum of space with enough velocity to perpetually fall beyond. But there are some dramatic differences as well. The journey from sea level to orbit for a space shuttle, for example, takes 8 minutes, 50 seconds; and contrary to Columbus's direction, space launches are to the east to take advantage of the eastward rotation of the Earth. For the space shuttle the linear outbound acceleration is about $3g$'s during the launch phase. Of course, at the instant of engine shutdown the acceleration experienced within the spacecraft goes to zero to first order, and the spaceship is then in the silent vacuum of space well beyond the edge of the Earth. As the ship coasts along its orbital path, everything within the ship floats and all human feeling of up or down mysteriously vanishes.

Everything floats—pencils, clothes, cameras, relaxed limbs, and bodies. The appearance of any solid object, of course, does not change, but, because of the floating, ordinary objects can be used for amusing demonstrations. The stability of a book-shaped object spinning around its long axis or its short axis is an easy example. But when gently spun around its intermediate axis, the object rotates, swaps ends, rotates, swaps ends, and so on as though magically demonstrating Euler angles and Euler's equations for an animated videotape on physics instruction.

In contrast to solid objects, the behavior of liquids in microgravity is so different from that of liquids on Earth that the mere sight of water, orange juice, or lemonade out of its proper container borders on comedy. For example, water squeezed from its drink container remains on the end of the container's straw as a perfectly motionless, but glistening globe. When you disturb the straw slightly by shaking or blowing on it, the globe quivers, tiny waves of motion circling its circumference. If disturbed more sharply, the globe shivers, vibrates, and oscillates—veritable Legendre functions of deformation spreading as tidal waves around the sphere (Fig. 1). Excited by a final puff of air, the globe can suddenly fission into two smaller spheres that drift apart, still trembling. If the straw is pulled away from the water, the liberated globe of course floats freely in the crew cabin until air currents move it to a wall, ceiling or floor. Under these conditions it acts very well as a spherical lens (Fig. 2). On touching any relatively flat surface, the liquid changes shape instantly from a sphere to a hemisphere—looking now like weak Jello dessert molded in a round mixing bowl and dumped onto a serving plate. The water will stay there indefinitely unless mopped with an absorbent towel or drunk by a thirsty crewman.

One can become more innovative in physics demonstrations by adding solid objects into the fluid (iron filings, or "BB's" for example) or by introducing gas, such as air bubbles, into the fluid (Fig. 3). I will not go into more detail here but rather will leave these demonstrations as *gedanken* experiments for the students of microgravity who, I am confident, clearly understand that the processes of

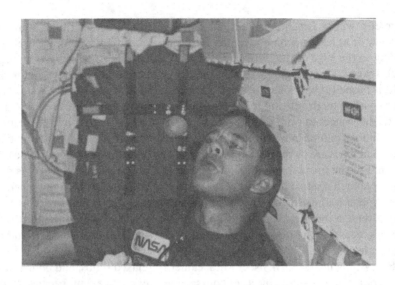

FIGURE 1   Astronaut generates oscillations in a sphere of liquid.

sedimentation, buoyancy, and convection are all driven by the gravity we enjoy right now.

Let me return to the example I mentioned earlier of "practicing physics". Consider the satellite (Fig. 4) being deployed from the space shuttle *Columbia*. It is a typical communications satellite—cylindrical, covered on the outside with solar cells with which it generates its needed power as it moves in geosynch-

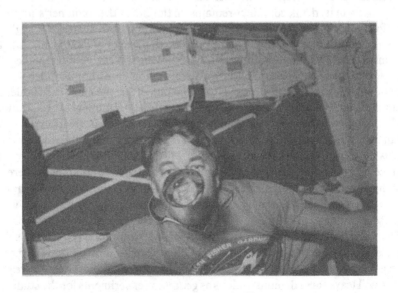

FIGURE 2   A globule of liquid forming an inverted image.

FIGURE 3   A spheroid of liquid containing a volume distribution of bubbles.

FIGURE 4   A communications satellite being launched from the cargo bay of a spacecraft.

ronous orbit about 22 000 miles above the equator. Such satellites in space are spin stabilized, that is to say they are their own gyroscopes. As this photograph was taken, the satellite was spinning at 30 rpm and, of course, will continue at this approximate spin rate for its full lifetime of ten years or so. In 1984 two such satellites, properly deployed from the space shuttle *Challenger* and in perfectly good working order, unfortunately later failed to reach their intended destination of geosynchronous orbit because of auxiliary rocket malfunctions.

In the spring of 1984, five astronauts (Hauck, Walker, Fisher, Gardner, and I) were given the assignment of recapturing the lost satellites, if at all possible, and returning them to Earth for refitting and for reuse. The task would clearly be challenging since the satellites were not small (each weighed about 1000 pounds on the Earth's surface), were spinning at an impressive speed and were not equipped with grapple points, handholds, or attachment fittings of any sort for just such a contingency. However, the errant devices *were* in orbits easily reached by a space shuttle. Moreover, their spins could be slowed by ground commands, and we, as salvage crew members, had for our use the manned maneuvering unit (the MMU—see Fig. 5) that could be attached to the back of a suited astronaut and would, in theory, enable the astronaut at least to maneuver up to the satellite. If we could approach the satellite without causing it to wobble or tumble, we could probably seize it through the nozzle of its spent rocket—an opening conveniently aligned through the satellite's center of rotation.

The function of the manned maneuvering unit is to an astronaut what the function of a dory (a small boat) is to a sailor. Each can carry a passenger a short distance out from the mother ship, can move around in the local area and, if handled properly, can return to the ship. The dory, of course, is controlled by the sailor using oars to move the vessel over the two dimensions of the water's surface. The maneuvering unit, on the other hand, is controlled by a space-suited astronaut using two hand controllers connected to thrusters which move the unit across the three dimensions of space. I do not need to point out that the maneuvering unit is unconstrained by a surface of any kind, nor is it tethered, and thus it may move in all six degrees of freedom, three linear and three rotational. Furthermore, by Newton's first law, any motion imparted along or around any one axis continues unchanged until the astronaut–MMU "cluster" is acted upon by still another force.

Although called at times an "overstuffed rocket chair" because of its appearance, the maneuvering unit is actually a very simple device using compressed gas, typically nitrogen, meted out in bursts through any of 24 finger-sized thrusters mounted on the corners of the back pack. To pilot it, we think of ourselves as wearing a cube-shaped back pack with three perpendicular thrusters at each of the eight corners of the cube. The thrusters are ganged to work in sets of 12. [Any one set of 12 (two thrusters at each of six corners) can provide adequate control to return the astronaut to the mother ship if the other set of 12 should fail.] The astronaut controls the linear motion along the body axes by moving the left-hand controller along those same axes, and the rotational mo-

FIGURE 5 An astronaut in a manned maneuvering unit (MMU).

tion around the body axes by twisting, tilting, and rocking the right-hand controller in the direction of the motion to be induced. The hand controllers are coupled electronically to solenoid valves which in turn allot the bursts of compressed gas to the appropriate thrusters. To fly the maneuvering unit in the frictionless, undamped, and silent isolation of space is an extraordinary experience. This is especially the case since there are complications in implementing what I have just described which will, I think, intrigue you. I can testify that these complications indeed do challenge the astronaut piloting the maneuvering unit, particularly when in search of a wayward and potentially skittish satellite. More important to this discussion, I wish to stress that to fly the maneuvering unit, in simulation or in space, is literally to "practice physics", and in this example you are simultaneously the experiment and the experimenter. Let me be more specific.

You will recognize immediately that the center of thrust of the 24 thrusters (the 12 couples) does not correspond exactly with the center of mass of the "cluster" to be maneuvered, i.e., the MMU, the suited astronaut and the attached auxiliary equipment. Of course, this cluster itself is not exactly a rigid body in that the astronaut is not tightly bound by the spacesuit but, rather, floats within this inflated cocoon of air. Some astronauts do qualify as ponderous bodies perhaps, but, in the strict physics sense of the word, certainly not as rigid bodies.

Because of the offset just mentioned, each linear motion commanded by the pilot has an admixture of rotational drift that is added to the resulting linear motion, and for each rotational motion commanded an admixture of linear drift is added to the resulting rotational motion. Put in everyday terms, the MMU "skids" during its translations and "drifts" during its rotations. This skidding and drifting must be imagined as occurring in all three dimensions which is, of course, more complicated than, for example, the simple skidding of a car.

A second perturbation, or complication, in stalking a spinning satellite follows immediately from the undeniable facts of orbital mechanics. The mathematics describing the motions of each satellite—the communication satellite and the astronaut satellite—is straightforward enough, and the position of one object with respect to the other is easy to visualize and predict by, for example, a physics student watching the chase scene from a vantage point above the North Pole. But the relative motion of the two, the change of position of astronaut with respect to satellite, is not necessarily intuitive to the astronaut at the time.

Please consider the following example: The satellite and the astronaut are in identical circular orbits. The astronaut is 20 feet behind the satellite, feet pointed to the Earth's center, attitude inertially stable, with the lance (the capture probe) perfectly aligned with the rocket nozzle of the satellite. (Assume for the moment that the MMU thrusters are not being used.) Then, one-half orbit later (approximately 45 min), the situation will have reversed. The astronaut, head now pointed to the Earth's center, still trails the satellite in the orbital sense, but the lance is pointed away from the satellite which, from the astronaut's perspective, is now behind him. Thus, on the time scale of minutes the astronaut has made a half circle around the satellite, staying always 20 feet from it—a change of relative position due only to the orbital mechanics, *not* due to motion induced by the MMU.

Now add to this simple example of two objects in identical orbits the further complication of the astronaut moving above or below the target satellite and then off to one side (out of the satellite's orbital plane). Finally, recognize that one's field of view from within a space suit is restricted, so that it is difficult to judge simultaneously one's position with respect to the satellite, and the direction to the center of the Earth—information which in theory could enable anticipation of the "Kepler-induced windage" of the situation.

A third complication presents itself, not from physics directly, but rather from the way we humans perceive our surroundings. To attempt to explain this fully would require a long lecture on the physiology and psychology of seeing.

An easy description, however, does apply to this situation. In attempting to line oneself up with a large satellite of simple geometric shape, starkly illuminated by direct sunlight, against the velvet black background of deep space, it is very difficult to estimate distances. In fact, it is difficult even to distinguish, for instance, a translation of your position to the right from a rotation of your attitude to the right. In both cases the object against which you are judging moves to the left in your field of vision. Yet, it is important to distinguish the difference since you are commanding, separately, translation and rotation.

The fourth complication is the most challenging, and in my view the most amusing. The communication satellite spins, and its stability (its resistance to possible wobble induced by plumes of compressed gas from the approaching astronaut and MMU) is a direct function of its spin rate. The suited crewman could, in theory, set up an equal spin rate around the axis of the capture probe jutting out in front of him. Initial contact with the satellite would then be both elegant and the least demanding on the mechanisms to be mated, since sudden torques imposed at impact would be eliminated. It was exactly this spinning capture procedure we planned, and we began to practice this technique in ground-based simulations several months before the mission was to take place (Fig. 6).

I suspect that several, if not all, of you now anticipate the results. The seemingly straightforward human task of flying a conceptually simple maneuvering unit in Earth orbit is complicated by the cross-coupling of translation–rotation controls, by the easily predicted but perplexing effect of orbital mechanics on the position of one satellite with respect to another, and by the limitations of human vision in judging $\mathbf{R}$, $\dot{\mathbf{R}}$, and $\ddot{\mathbf{R}}$, where $\mathbf{R}$ is the vector from astronaut to

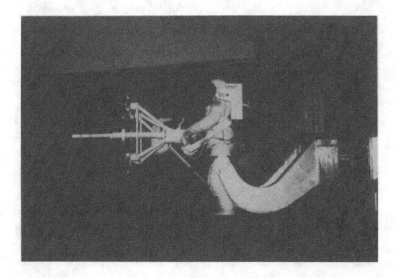

FIGURE 6   Astronaut with "lance" in training to recapture a communications satellite.

satellite. Each complication is slight, acts slowly and of itself is not insurmountable. But to these three compounded factors we attempted to add a fourth, that is to fly the approach to the target satellite with both astronaut and satellite in a rotating reference frame. We quickly learned from simulation that, even with rotation rates as low as 3 rpm and the approach rate as small as three feet per second, the Coriolis force (the $-2m\,\omega\times\mathbf{R}$ term in the force equation) and the centrifugal force [the $-m\,\omega\times(\omega\times\mathbf{R})$ term] impose a flying task on the human pilots that is confusing in the extreme.

Several test simulations showed that the gas pressure available in the MMU often was depleted before the capture task was accomplished, and the MMU pilot was mentally exhausted by the process. This "practice of physics" clearly indicated to us well before the mission itself that we must fly, *inertially* fixed, onto the satellite and simply attempt the capture with equipment sturdy enough to withstand the imposed torques of the impact.

Figure 7 shows that our revised flying procedures worked as planned. In November 1984, two satellites (one of them being the astronaut + MMU) were joined on a modern jousting field some 200 miles beyond the edge of the Earth,

FIGURE 7   A successful rendezvous with a communications satellite.

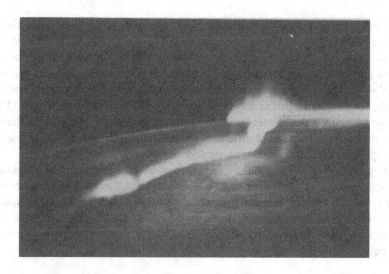

FIGURE 8   The northern lights seen from above.

FIGURE 9   Ion glow around the contours of a spacecraft.

and both astronaut–satellite collisions were on target and perfectly inelastic, as we had hoped.

I would enjoy digressing from my subject of observing classical physics in action and include several examples of quantum phenomena—plasma physics mostly—which are spectacular from the vantage point of Earth orbit. For example, the northern lights as observed from above (Fig. 8) or the ion glow (Fig. 9) that surrounds the orbiter itself on occasion, or perhaps the plasma sheath that spreads out from the shuttle as we start the energy management task of reentry. But these are subjects for another lecture.

I hope you have enjoyed these excerpts from an amateur physicist's logbook of a space voyage. I can assure you that both basic and applied physics are not only essential, but can be great fun when associated with voyages into space. Moreover, a physics education travels very well even over great distances, and an appreciation of physics enhances beautifully these already astounding journeys.

The Scanning Tunneling Microscope: A New Era of Science and Microtechnology/ JOSEPH E. DEMUTH

Size is one of the many physical limits that hinders man in exploring his natural world. Various telescopes have allowed us to reach out into the vastness of space. The manned space program has also allowed man to escape his natural environment, cover vast distances, and directly explore the moon or probe other planets in more detail than ever before. Likewise, on the opposite end of the size spectrum, man has developed a variety of microscopes to allow him to explore the nature and structure of matter. Included among these is the scanning tunneling microscope (STM) which represents an unusual departure from all previous methods of studying atomic structure. Namely, it allows man to *directly* probe the micro-cosmos at the atomic level. No less surprising is the relative simplicity of this method, which refreshingly contrasts with the growing tendency towards bigger, more complex scientific instruments. Invented only six years ago, scanning tunneling microscopy has captivated our imagination and promises to be a major scientific tool for the future—impacting virtually all areas of science where surfaces are important. It may equally well mark the beginning of a new scientific era on the near atomic scale.

In this paper, I describe this new method and its potential to provide novel information about atomic-scale phenomena. Various historical perspectives,[1-3] several topical reviews,[3-5] and a recent review article by Hansma and Tersoff[6] should provide further information and more specific references about work in this area. Here I have chosen to provide examples of what one individual feels to be the most physically significant and illustrative of this method. Also described are modifications and variations of the STM which allow it to operate in aqueous solution or to measure atomic-scale forces. Such microscopes are being used to understand a wide range of phenomena. This includes for example, the structure, bonding, epitaxy, diffusion, chemical reactions, and even friction, on a wide variety of surfaces in vacuum, air, or liquid environments. Overall, the most important feature of the STM is its ability to provide a *direct* link to the

JOSEPH E. DEMUTH *works in the field of surface physics. Since obtaining a Ph.D. in the School of Applied and Engineering Physics at Cornell University, he has conducted his research at the IBM Thomas J. Watson Research Center, Yorktown Heights, New York.*

quantum-mechanical world where one can observe, interact, or even modify matter *at the atomic level*! This theme will be the main focus of my paper.

A METHOD FOR SURFACE TOPOGRAPHY

The Scanning Tunneling Microscope was invented and developed by Gert Binnig and Heine Rohrer at the IBM Research Laboratory in Zurich, Switzerland.[1] Figure 1 schematically illustrates the STM. The enlarged view on the right shows the "tip" positioned about 5 Å above the surface. It is well known that, at this distance, electrons can tunnel between the tip and the sample, and if one applies a voltage difference between them, anywhere from 1 mV to 2 V, a tunneling current flows. This current is monitored as the tip, positioned on a piezoelectric scanning device, is raster-scanned over the surface. A feedback mechanism compares the tunnel current to a fixed value and produces an offset voltage proportional to the difference. This feedback voltage is then applied to the z-piezo device so as to maintain a constant tunneling current during the x,y scanning. By monitoring the voltage, and thus the relative position of the z-piezo device, one can plot out the contour of the tip as it is raster-scanned along the surface. One of the first atomically resolved scans is shown in Fig. 2 and readily displays atomic steps and terraces on a Au(100) surface. Here, repetitive line scans of the surface are made in the x direction and successively displaced in the y direction to produce topographic images. In this mode of operation, the tip can "follow" large variations in the surface topography, the scan rate being limited by the bandwidth of the feedback circuit (usually ~ 3 kHz) and the mechanical response of the scanning assembly. Another mode of operation involves fast scanning where the tip scans a nearly flat plane and variations in the tunnel current are recorded.[3,6] The higher bandwidth of the current channel has allowed video-rate imaging[3,6] as well as real-time spectroscopy[5] described later. As will be discussed, the large amount of information obtained from the STM makes computer control of the raster-scanning, data acquisition and image display almost essential.

This high sensitivity and resolution in the z direction, now typically ~ 0.1 Å,

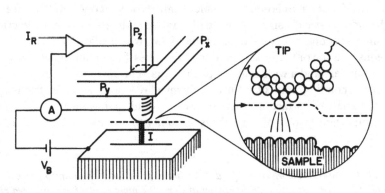

FIGURE 1   Schematic diagram of the Scanning Tunneling Microscope.

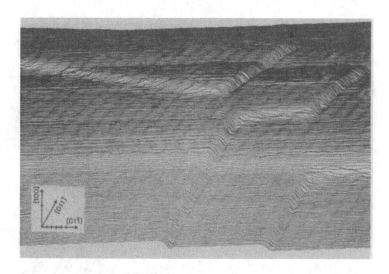

FIGURE 2   STM topography of a clean Au(100) surface. Each unit of the scale corresponds to 5 Å. [Courtesy of G. Binnig, H. Rohrer, Ch. Gerber, and E. Stoll, Surf. Sci. **144**, 321 (1984)].

is expected from the exponential dependence of the tunneling current on the separation of the sample and tip. A 1-Å change in sample-tip separation produces an order of magnitude change in tunneling current, with typical tunneling currents being ~ 1 nA.

The high lateral resolution of about 5 Å was initially very surprising, but, as theoretical considerations later showed,[6,7] it is reasonable for a single atom or a few atoms on the end of the probe tip. In practice, STM probes are mechanically ground or electrochemically etched and end up with an overall tip radius of 100–1000 Å. The presence of a variety of different asperities, together with the exponential dependence of tunneling, appears to allow a self-selection of the asperity closest to the surface for tunneling.

The most important technical achievement of the STM is the mechanical precision, control, and stability at these small distances. All previous tunneling was performed between materials separated by a thin mechanical layer—usually an oxide. The topografiner, a precursor to Binnig's scanning instrument, was limited to 1000-Å resolution by vibrations and stability problems.[8] Binnig and Rohrer succeeded in vibrationally isolating their new microscope from the outside world, as well as developing a method for positioning the sample from relatively large distances to within ~ 5 Å of the sample *without* inadvertently crashing the sample into the tip. To achieve these properties, they developed several vibration isolation systems and a piezoelectric walker called the "louse".[1] Since then, several different types of sample positioners—electromagnetic or mechanical positioners using levers, pivots or differential have been developed and are in use. A picture of several generations of STM's developed by Binnig and Rohrer is shown in Fig. 3. Over a period of only five years, they built four different microscopes, simplifying and evolving their design to a ver-

FIGURE 3   Four generations of Scanning Tunneling Microscopes. The first microscope (a) employed superconducting levitation of the microscope, the second (b) a simple spring suspension system, the third (c) a more complex spring suspension system, and finally (d) simple viton dampers and springs in a compact design. All were designed for use in vacuum. (Courtesy of G. Binnig and H. Rohrer.)

sion that fits into the palm of one's hand—rightfully called the "pocket-STM". Several STM's have been made as small as 2 cm³ for low-temperature studies.

Many STM's in use today are designed to fit into a vacuum chamber where samples can be processed and characterized, and can remain intact for long periods of time. Inert samples, such as graphite or gold, are stable in air where they can be studied but sometimes reveal problems associated with contamination layers on the tip. Nevertheless, STM's have operated in air on a wide variety of materials and reveal useful structural information.[9] Atomic resolution can also be achieved in other environments such as liquid helium, water, or electrolytic solutions.[6] The STM studies of well-defined surfaces in vacuum have provided fundamental new information about the physics of the technique as well as about atomic structure and bonding at surfaces.

Semiconductor surfaces have long attracted interest because of the presence of surface states well known to occur since the development of the transistor. The broken covalent bonds of the semiconductor surface are energetically unfa-

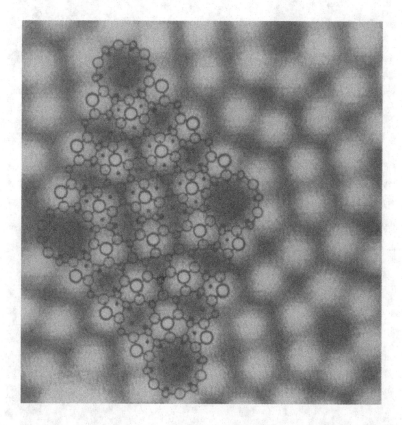

FIGURE 4   STM image of the Si(111) 7 × 7 surface with a superimposed model of the now widely established structure. This is unprocessed data taken every 0.25 Å over a scan area of 60 Å × 60 Å. The grey scale here corresponds to 1 Å.

vorable and try to recombine. Due to their strongly directional bonding, this leads to a restructuring of the last few surface layers. One such case is the Si(111) 7×7 surface where a 7×7 reconstruction occurs upon annealing above ~900 °C to produce a periodic arrangement of atoms having a unit cell seven times larger than the primitive "ideal" unit cell of the (111) surface. A grey scale STM image of this surface is shown in Fig. 4. Superimposed on this image is one unit cell of the now widely established DAS (dimer-adatom-stacking fault) model.[4]

Binnig's early STM work provided key information to understand this complex reconstruction. Namely, he first identified the 12 protrusions per unit cell

FIGURE 5   Large area scan of a Si(111) 7×7 surface. U. Koehler, J. E. Demuth, R. J. Hamers IBM Thomas J. Watson Research Center, Yorktown Heights.

as adatom structures (i.e., three-fold-coordinated Si atoms in the top layers).[2] Later STM work directly verified the existence of a stacking fault in one-half of the unit cell from the location of three-fold coordinated atoms in the second layer as identified from their surface states.[5] The Si dimer pairs which terminate the silicon dangling bonds along the sides of the unit cell cannot be directly seen. Much of the detailed atomic structure is not observed in the STM image, since the electron density sampled in tunneling is rather smoothly distributed about these atoms.

In contrast to this "small area" image, one can also obtain large-area scans of well-prepared Si(111) 7×7 surfaces as shown in Fig. 5. Here the scan area is $\sim 800 \times 1000$ Å and displays a rather low density ($\sim 10^{11} - 10^{12}/cm^2$) of structural defects. Also found here is a surface grain boundary caused by the initial (random) nucleation of two 7×7 structures at different locations on the surface. Even though the 7×7 structure is rather complex, it is the most thermally stable of the silicon surfaces found.

Coarser surface topographic features can also be observed with the STM, as shown in the large area perspective view shown in Fig. 6. Here an atomically flat Si(100) surface was heated to 1150 °C and rapidly cooled. Several flat regions are found amongst a mostly irregular "landscape" where some atomic-scale facets and terrace steps can be observed. Many of the higher hillsides here probably reflect the shape of the tip and not just the true topology of the sample. Such uncertainties in the shape of the tip complicate the interpretation of such large-scale topographs. Except for surfaces that are very inert and/or cleaved in ultra-high vacuum, most crystal surfaces when first examined with the STM

FIGURE 6   Perspective view of a thermally roughened Si(100) surface. The area corresponds to 1000 Å × 1000 Å with a maximum height of 75 Å.

have also looked rough. Most workers have found that the flat, ideal-periodic crystalline surfaces emerge only after arduous, careful fine-tuning of the preparation conditions—using the STM as a monitor. Even then substantial defects frequently exist.

The ability to obtain topographic information on this scale also stimulated the study of complex biological systems.[1] Initially, the PHI 29 virus and DNA strands were examined on metallic substrates, but these first results lacked the fine details that could be detected on metal and semiconductor surfaces. More recent studies of DNA have used a metallic coating to provide a good conduction path for tunneling. As an example, Fig. 7 shows a topograph of a C-Pt overcoated, Rec-A protein-decorated DNA strand, obtained by Travaglini and co-workers in the IBM Research Lab in Zurich.[10] The novel feature of this STM picture is that one can directly observe the right-handed helix repeating every 90–100 Å, as well as the Rec-A monomers on the top portion of each helical

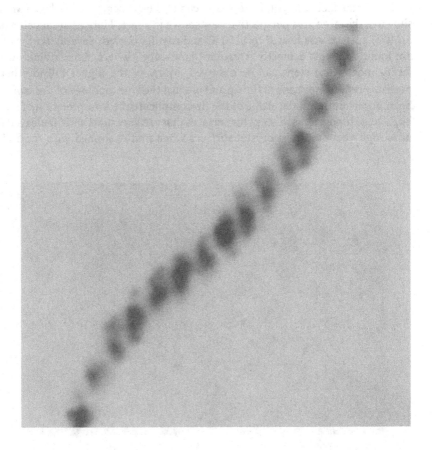

FIGURE 7   STM image of a Rec-A DNA strand. The spacings between the turns of the helix are ~95 Å. (Courtesy of G. Travaglini, Ref. 10).

turn. X-ray and TEM studies of such systems show only the projected density of these structures, and not a topology. In many biological systems, the specific surface topology and location of molecular groups along these surfaces provide keys to understanding how certain molecules bond or chemically react. Related active sites on other catalytic surfaces may also be observed and identified some day with the STM.

SPATIALLY RESOLVED SPECTROSCOPY

The atomic-scale features seen with the STM actually reflect the charge density contours at the surfaces.[6] In some cases, these may not even reflect the positions of the atoms! In order to understand these effects, one must consider the nature of the electronic states at the surface (and tip!) and the tunneling process. N. Lang has utilized an exact calculation using real wave functions and has extended the Bardeen-tunneling Hamiltonian formalism to calculate various tunneling properties.[7] In Fig. 8(a), the calculated trajectory that maintains a constant current when a Na atom on one surface (the tip) is scanned against a variety of atoms on another surface is shown. The resulting trajectory for the small bias condition reflects not just the size of the atom, but a path which closely follows the contour of constant charge density for states at the Fermi level. For helium, the states near the Fermi level are suppressed by the Pauli repulsion of the surface charge density, and this leads to a "negative" topograph. Increasing the bias voltage also allows tunneling from a variety of surface states between the Fermi levels of the tip and the sample. As shown in Fig. 8(b) and (c), this produces a tunneling amplitude whose magnitude can strongly depend on the density of states of the tip and the surface atoms involved in tunneling. For negative bias, one tunnels from the filled states of the sample to the empty states of the tip (and vice-versa for a positive bias).

The role of these density-of-state effects on STM images is seen in many STM experiments. As an example, we show, in Fig. 9, different topographic scans of

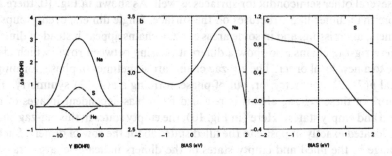

FIGURE 8   The role of electronic structure in determining (a) the apparent lateral size and height of a topographic image, and (b,c) the apparent height as a function of bias. (Courtesy of N. Lang, Ref. 7.)

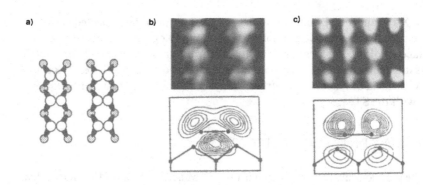

FIGURE 9   Bias dependent STM images of Si(100) 2×1 and theory. The dimer rows are separated by 7.7 Å, individual dimers in each row by 3.85 Å. The filled states (b) are observed at − 2 V and the empty states at (c) at + 2 V. The corresponding calculated charge densities for one dimer are shown below, courtesy of J. Ihm, M. L. Cohen, and D. J. Chadi, Phys. Rev. B **21**, 4592 (1980).

an ideal region of the Si(100) 2×1 reconstructed surface. This surface reconstructs very simply by using the one "dangling" silicon bond on each surface atom to form Si dimer pairs. The STM image for tunneling out of the filled surface states (b) shows elongated structures, expected of such dimer pairs, whereas tunneling into the empty states (c) shows different features. The relative alignment of these filled and empty states shown here has been determined from a simultaneous imaging technique described later. The calculated charge density contours for the valence and conduction band states for the reconstructed dimer are shown below each image and readily account for the observed images. Interestingly, the node in the empty surface state shown in panel (c) produces the depression running through the center of the dimers. As seen here, the dimers are most readily apparent from the STM image employing tunneling from their filled states.

Larger-area scans of Si(100) reveal a more complex behavior characteristic of several other semiconductor surfaces as well. As shown in Fig. 10, there are more than dimer-like structures on the surface. Single dimers, even groups of dimers, are missing, and in some areas zig-zag chains appear instead of dimers. These zig-zag chains also show a different phasing between rows, which gives rise to a new local order. The zig-zag chains are sometimes in phase, forming a local $p(2×2)$ symmetry, or, out of phase, forming a $c(4×2)$ symmetry. The nature of these zig-zag chains is revealed from bias-dependent images of the filled and empty states.[5] Here (in Fig. 10), the empty states in this zig-zag chain are located exactly opposite to the filled states about the row of dimers. Such a change in the filled and empty states of the dimers indicates charge transfer between the atoms in the dimer. Here the two electrons in the dimer bond form an ionic configuration, placing both electrons in one dangling bond leaving the other dangling bond empty. A buckling or asymmetric dimer pair is expected to arise from such charge transfer. As shown in Fig. 10, this charge transfer, and

FIGURE 10   40 Å × 80 Å region of a clean Si(100) surface. The grey scale range is 0.8 Å.

the formation of asymmetric dimers, appears to originate near defects and to "heal" out along the dimer row. It is unclear whether the observed "symmetric dimers" actually ever exist, since these could be time-averaged asymmetric dimers. Such charge transfer between atoms has also been used by Feenstra and Stroscio to determine the location of the anion and cation atoms of the GaAs(110) surface.[11]

While these changes in images reflect different state densities, one would like to perform a spatially resolved spectroscopy to study surfaces in general. The capability of doing spectroscopy using tunneling is well established from tunneling experiments using fixed oxide barriers, and was one of the motivations for Binnig's initial vacuum tunneling experiments.[1] A schematic of such tunneling current versus voltage, or I–V spectroscopy with the STM, is displayed in Fig. 11. Here, the tunneling current increases more rapidly for larger bias whenever additional state density occurs. The onsets from these density-of-state contributions can be observed more readily by taking dI/dV. Because of the strong change in conductance with voltage and tunneling distance, it turns out that dI/dV normalized by I/V is closely related to the surface density of states.[7] Similar density-of-state effects associated with the tip can also occur, and have to be considered. Luckily, clean tungsten tips in ultra-high vacuum seem to have a flat structureless density of states at the distances required for tunneling, as deduced from tunneling spectroscopy on a wide range of systems.

Performing I–V measurements at well-defined atomic positions at the surface is also problematical with the STM, primarily because of thermal drift and the limited lifetime of the highest-resolution tips. To overcome these problems, a multiplexing method for operating the STM has been developed to permit simultaneous topographic and I–V information to be obtained.[5] Here, the sample and tip bias is not maintained constant but is modulated at ~2.5 kHz with a sawtooth-shaped signal. In order to stabilize the feedback and plot the topography, a synchronized sample-and-hold circuit measures the tunneling current at one particular bias selected for the feedback control. The tunneling currents

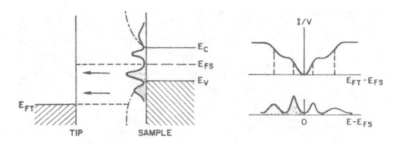

FIGURE 11   Spectroscopy using the STM.

at other biases are displayed in real time on an oscilloscope as $I-V$ spectra, even while scanning a topograph. This real-time display is particularly convenient as it provides a type of chemical fingerprint for different structures or atoms on the surface. These $I-V$ spectra are usually recorded during the experiment for each pixel of the topograph, and later plotted and analyzed in more detail. The spatially dependent tunneling current can also be used for spectroscopic imaging. This has permitted the determination of the atomic locations of different surface states on several semiconductor surfaces.

Figure 12 shows simultaneously acquired topographic and spectroscopic images of the two occupied surface states on the Si(111) 7×7 that are well known to exist from conventional surface spectroscopies. As seen in the $I-V$ spectra and shown in these spectroscopic images, the first state occurs from E_F to -0.6 eV and resides on the 12 threefold-coordinated Si atoms (adatoms) on the top layer of the unit cell (see Fig. 4). The second state from -0.6 eV to -1.0 eV is localized on the six threefold coordinated atoms in the second silicon layer as well as on the one threefold coordinated atom in the corner hole. From these states, one can identify a charge transfer that reduces the number of partly occupied dangling bonds on the adatoms and contributes to stabilizing this surface. Here, the seven low-lying states on the second layer's threefold coordinated atoms are occupied and have gained seven electrons. The remaining 12

FIGURE 12   Simultaneously acquired topographic (a) and spectroscopic images (b,c) of Si(111) 7×7. The filled surface states between (b) 0 to -0.6 eV and (c) -0.6 to -1.0 eV are shown.

adatoms must lose seven electrons total, thereby eliminating seven electrons and the higher kinetic energy associated with these high-lying dangling bond states. The five remaining electrons in the adatom dangling bonds produce a metallic band which seems distributed differently on the various adatoms. Such electron counting at the atomic level provides a simple way to understand some of the surface electronic interactions that drive reconstruction and surface bonding.

Such spectroscopic imaging also shows important charge redistribution near imperfections and defects. In Fig. 13, a topograph (a) and simultaneously acquired current images nearer the Fermi level (b–c) and above it (d) are shown. Here two dimer defects, which can be considered as microscopic potential wells along the surface, show subtle differences in their topographs and very different spectroscopic images. These dimer defects differ in terms of the energies at which they trap charge as well as in the spatial location of this charge. Interestingly, for either defect the electrons seem to pile up near the *edges* of these quantum boxes!

This type of local spectroscopy of semiconductor surfaces and their defects can be invaluable for understanding the origin of their micro- and macro-electrical properties. The useful spectroscopic energy range accessible with vacuum tunneling is restricted to about ± 4 V due to the limited stability of the tip as well as the presence of tip-sample standing waves at higher energies.[4] While this type of spectroscopy does not permit true elemental analysis, this energy range is ideal for probing the valence electronic states which are generally believed important for transport as well as for bonding and chemical reactions. Inelastic tunneling spectroscopy, again well known from fixed-thickness, oxide tunnel junction experiments, offers the potential for chemically specific identification, and is being actively pursued in the STM community.

One additional consequence of the existence of such localized electronic states, and of one's ability to probe them with the STM, is the possibility of inducing specific chemical transformations (i.e., desorption or decomposition) at surfaces. In fact, for several systems, including GaAs and Ge, the tunneling current must be reduced by a factor of ten to prevent the sample from being damaged.

FIGURE 13 Simultaneous topographic (a) and spectroscopic images (b–d) of surface defects on Si(100). The topography is at -2 V, and spectroscopic (current) images at -1.2, -0.8, and $+0.8$ eV, respectively.

EPITAXY

The ability of the STM to probe atomic level details makes it ideal for understanding the initial epitaxy and growth of materials on surfaces. It is typical of almost all STM work to date that surprising features are found whenever one examines a new system with atomic resolution. Figure 14 shows the initial formation of a stable Ag two-dimensional island formed on a cleaved graphite substrate in UHV by Clarke and coworkers at the Lawrence Berkeley Laboratory at the University of California.[12] What is very curious here is that Ag forms a rectangular structure which is not directly related to the fcc structure of the bulk crystal. This rectangular lattice also shows a quasi-incommensurate structure. In several other regions, large distortions in this lattice occur, which suggests Ag–C interactions that are stronger than the Ag–Ag interactions! Several fundamental issues concerning the relative energetics of different microstructures have the possibility of being resolved with the STM.

Many metals readily form large regions of well-ordered two-dimensional arrays on Si surfaces but contain specific types of defects. Figure 15 shows an STM image for nearly a third of a monolayer of Al atoms on Si(111) which ideally forms a $\sqrt{3} \times \sqrt{3}$ ordered array on top of this surface. The point defects seen here disappear for higher Al coverages and can be understood via tunneling spectroscopy performed in a 1 Å^2 area over them. As shown in the lower portion of Fig. 15, these point defects have a markedly different electronic structure. The filled state density of this defect near E_F is characteristic of an occupied dangling bond state as found, for example, on the 7×7 surface. Such dangling bonds would be expected from the substitution of Si for Al. This, together with the dependence of the defect concentration on Al coverage, identifies these as substitutional defects. Interestingly, the energetics here are such that point defects are favored, instead of a phase separation and the formation of perfect $\sqrt{3} \times \sqrt{3}$ Al islands.

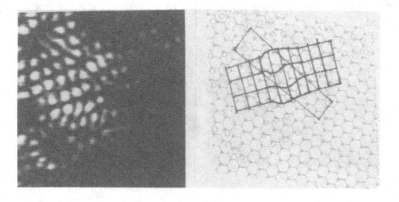

FIGURE 14   STM image of Ag 2-D clusters on graphite (right) and a drift-corrected model of the locations of Ag atoms on the underlying graphite surface. (Courtesy of J. Clarke, Ref. 12.)

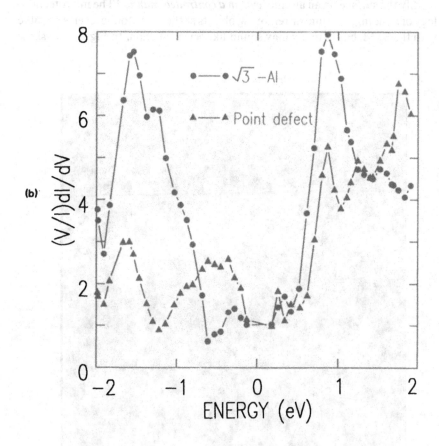

FIGURE 15 (a) STM image of point defects in a $\sqrt{3} \times \sqrt{3}$ Al adsorbed layer on Si(111) and (b) their local spectroscopy.

For more chemically reactive systems where compound formation occurs, the situation becomes even more complex. Figure 16 shows the strong inhomogeneities arising for a Pd monolayer deposited and annealed on Si(111). Regions of the clean 7×7 reconstructed surface persist (lower left), a small lattice mismatched structure nucleates at steps on the surface (top), a different strongly mismatched ordered phase occurs (lower right) as well as random clumps. It is clear that such systems will be a real challenge to understand!

SURFACE MODIFICATION

The STM is not only a passive probe but, as mentioned earlier, can be used to modify the surface on an atomic level *in a controlled manner!* The microtechnology of building, moving, or removing objects at the near atomic level is possible with the STM, but to many may sound like science fiction. In Fig. 17, we show

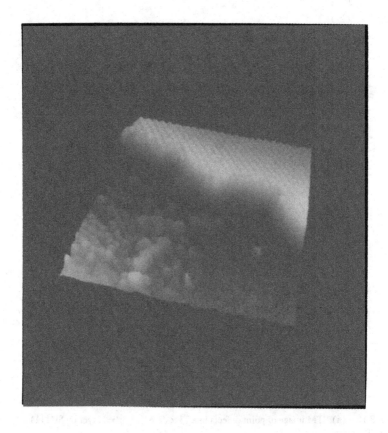

FIGURE 16   STM image of the inhomogeneous metallization of Si(111) after deposition of one monolayer of Pd and annealing to 350 °C. The scan area is 150 Å × 150 Å.

FIGURE 17   STM images of Ge(100) before and after modification. (Courtesy of R. Becker and J. Golovchenko.)

two sequential STM images of a Ge(100) surface taken by Becker and co-workers at the AT&T Bell Laboratories.[13] In between the scans to obtain these images, the bias between the tip and sample was momentarily increased, and produced the new clump of material on the surface. While the nature of this deposit is unknown, the capability for localized deposition is clear.

Control of such deposition, removal, or rearrangement of material on surfaces is now being developed. Here, one of the technical difficulties is separating the STM control process, i.e., the tip tracking and observation, from the interaction or modification process. Recently, Gimzewski and Moeller at the IBM Research Laboratory in Zurich[14] have used the sample-and-hold method described earlier for operating the STM feedback control, so as to precisely control the tip-sample contact. Such controlled contact to the surface allows one to better understand the micromechanics of contact. Here the tip-sample distance is continuously stabilized by the feedback circuit and at a controlled location the tip is extended into the surface. As shown in Fig. 18, the tunnel current increases, and a discontinuity indicates the initiation of contact. After contact, the surface topography has changed, leaving the bump shown. Not surprisingly, the cleanliness of the tip determines whether a depression or mound is left behind. Using this control system, Gimzewski and Moeller have written various patterns on surfaces over areas as small as 100 Å square. One might ultimately consider such use of the STM for genetic engineering, provided that the "pa-

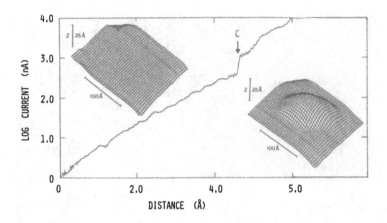

FIGURE 18   Controlled sample-tip contacting and the resulting surface modification for a metallic tip. (Courtesy of J. Gimzewski, Ref. 14).

tients" could be maintained alive during their operation! One solution to this is to perform STM in a liquid environment!

STM IN AQUEOUS SOLUTION

Although pure water is not a good conductor, most solutions are, or become so from residual contamination. The current through this additional conduction path becomes substantially larger than the tunneling current, thereby short-circuiting the much-needed feedback control system. Hansma and Sonnenfeld at the University of California at Santa Barbara have, however, demonstrated that atomic resolution is possible in solution[15] as shown in Fig. 19. Here, an oxide-coated tip prevents a large "leakage" current so that a large fraction of

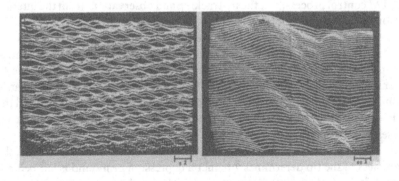

FIGURE 19   STM images of (a) graphite and (b) a Ag-film under water. [Courtesy of P. K. Hansma, Ref. 15).

the conduction occurs via tunneling at the end of the tip where the oxide is purposely ruptured. They have observed the evolution of electroplating as well as corrosion with nanometer resolution. Returning to the biological systems, these "patients" can now live but don't necessarily survive the 10^8 A/cm^2 current densities normally used in the STM. Thermal and electrical conduction by the sample is also a major concern for the STM.

THE ATOMIC FORCE MICROSCOPE

Binnig also conceived the idea of exploiting the small forces between the STM tip and the sample to control the feedback system, and he subsequently developed the atomic force microscope (AFM).[16] Here as shown in Fig. 20, a sharp tip on a very sensitive lever interacts with the surface, becoming attracted or repelled. A tunneling tip or sensitive laser interferometer measures these deflections and/or the changes in the lever resonance frequency. In most cases, AFM utilizes the repulsive interactions between the tip and sample. Although atomic resolution has been achieved on graphite surfaces, it is my opinion that this method will be most widely exploited to study larger-scale features on real surfaces under normal laboratory conditions. Wickramasinghe and co-workers at the IBM Research Laboratory in Yorktown Heights have developed such an instrument[17] and have demonstrated its novel capabilities as shown in Fig. 21. Here, grooves were reactive-ion-etched into a silicon wafer which then becomes oxidized in air. The different microstructures on top of and in these grooves associated with the etching method are clearly resolved over a wide range of scales, even though the surface is insulating. What looks like noise in the grooves at low magnification is real structure as seen in the grey scale image (center) taken at the start of a groove. At the highest resolution, features as small as 50 Å high are easily seen. Such force microscopy also offers a means to study tribological phenomena. Recently, McClelland and co-workers at the IBM Research

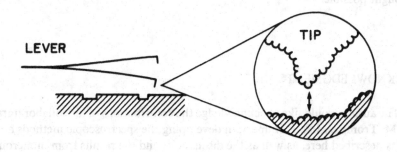

FIGURE 20 Schematic of the Atomic Force Microscope. An STM or laser interferometer is used to measure the displacements of the probe tip as it is scanned along the surface.

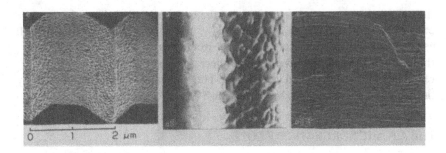

FIGURE 21   AFM images of grooves etched in Si for several increased magnifications. The highest magnification scans show the details of the groove substructure which is only 50 Å high. (Courtesy of K. Wickramasinghe, Ref. 17).

Laboratory in Almaden, CA, have looked at the lateral forces in an AFM and can resolve frictional forces and stick–slide phenomena with atomic resolution.[18]

In summary, the STM and its related cousins represent a new direction in the way science can be done—directly monitoring and even interacting with matter at the atomic level. In any technology involving miniaturization, or in any science investigating novel physical or chemical effects arising at the quantum limit, this capability represents a major leap forward. In fact, these micro-capabilities may just mark the beginning of a new technological revolution. Surprisingly, the principles of the STM are simple—it is the vibration isolation, control, and precision at these size scales that has been conquered. There still remain numerous aspects of this technology and science to be pursued, from building better tips and understanding their properties, to exploring applications in a wide range of scientific and engineering fields. It has been our fortunate experience that most new directions or experiments with the STM reveal surprises and, more often than not, novel phenomena never observed before or even thought possible!

ACKNOWLEDGMENT

The author would like to acknowledge the contributions of his collaborators, R. M. Tromp and R. J. Hamers, in developing the spectroscopic methods and ideas described here, as well as the discussions and the results from numerous colleagues outside and within the IBM Research Laboratories worldwide. Some of the author's own work has been partially supported by the Office of Naval Research.

REFERENCES

1. G. Binnig and H. Rohrer, Helv. Phys. Acta **55**, 726 (1982).
2. G. Binnig and H. Rohrer, Sci. Am. **253**, 50 (August 1985).
3. C. Quate, Phys. Today **39**(8), 26 (1986).
4. J. A. Golovchenko, Science **232**, 48 (1985).
5. R. M. Tromp, R. J. Hamers, and J. E. Demuth, Science **234**, 304 (1985).
6. P. K. Hansma and J. Tersoff, J. Appl. Phys. **61**, 121 (1987).
7. N. Lang, Phys. Rev. Lett. **56**, 1164 (1986); **58**, 45 (1987); Phys. Rev. B **34**, 5947 (1986).
8. R. Young, J. Ward, and F. Scire, Rev. Sci. Instrum. **47**, 1303 (1976).
9. A. M. Baró, R. Miranda, and J. L. Carrascoa, IBM J. of Res. and Dev. **30**, 380 (1986).
10. M. Amrein, A. Stasiak, H. Gross, E. Stoll, and G. Travaglini, Science **240**, 514 (1988).
11. R. M. Feenstra, J. A. Stroscio, J. Tersoff, and A. P. Fein, Phys. Rev. Lett. **58**, 1192 (1987).
12. E. Ganz, K. Sattler, and J. Clarke, Phys. Rev. Lett. **60**, 1856 (1988).
13. R. S. Becker, J. A. Golovchenko, and B. S. Swartzentruber, Nature **325**, 419 (1987).
14. J. K. Gimzewski and R. Moeller, Phys. Rev. B **36**, 1284 (1987).
15. R. Sonnenfeld and P. K. Hansma, Science **232**, 211 (1986).
16. G. Binnig, C. F. Quate, and Ch. Gerber, Phys. Rev. Lett. **56**, 930 (1986).
17. Y. Martin, C. C. Williams, and H. K. Wickramasinghe, J. Appl. Phys. **61**, 4723 (1987).
18. C. M. Mate, G. M. McClelland, R. Erlandsson, and S. Chiang, Phys. Rev. Lett. **59**, 1942 (1987).

Artificially Structured Materials/ A. Y. CHO

INTRODUCTION

Recent developments in crystal growth methods such as molecular beam epitaxy (MBE) and metal–organic chemical vapor deposition (MOCVD) allow us to artificially structure new materials on an atomic scale. These structures may have electrical or optical properties that cannot be obtained in bulk crystals. There has been a dramatic increase in the study of layered structures during the past decade which has led to the discovery of many unexpected physical phenomena and opened a completely new branch of device physics. Since the advanced crystal growth techniques can tailor the compositions and doping profiles of the material to atomic scales, it pushes the frontier of devices to the ultimate imagination of device physicists and engineers. It is likely that for the next century the new generation of devices will rely heavily on artificially structured materials.

This article will be limited to a discussion of recent developments in the area of semiconductor thin epitaxial films which may have technological impact.

DEPOSITION OF ATOMS LAYER BY LAYER

Most semiconductor devices are made of single-crystal materials. Some devices require complicated compositional and doping profiles to achieve high performance. In order to artificially structure the material to atomic scale, we need a technique for monitoring the deposition that can resolve one atomic layer. Since MBE is conducted in a vacuum, reflection high-energy electron diffraction (RHEED) has become a useful technique for monitoring the structure of layers grown.[1] Besides the information that can be gained about the surface structure or the presence of steps and facets, RHEED has been recently used to count the deposition of atoms layer by layer.[2] Upon the initiation of

ALFRED Y. CHO *has worked since 1968 with AT&T Bell Laboratories, Murray Hill, New Jersey, where he is Director of the Materials Processing Research Laboratory. His field of interest is microwave and optoelectronic devices, and particularly the related thin-film technology and molecular-beam epitaxy.*

MBE growth, the intensity of the diffracted electron beams shows an oscillatory behavior which is directly related to the initiation and completion of each monolayer.

Imagine that the growth by MBE proceeds with two-dimensional nucleation one atomic layer at a time. The diffracted electron intensity is at the highest value when the surface is atomically smooth (i.e., completion of an atomic layer) and at the lowest value when the surface least resembles a complete layer (i.e., one-half atomic layer). Figure 1 shows the intensity of diffracted electrons as a function of time for the growth of three monolayers of GaAs followed by three monolayers of AlAs and then twenty layers of GaAs.[3] If one wants to change the composition at the exact completion of a monolayer, one must execute the change at the instant when the diffracted electron intensity is at its maximum. This becomes more critical for the preparation of very thin layers of alternating compositions as, for instance, a quantum well superlattice with only two or three atomic layers for each well.

LOW-DIMENSIONAL STRUCTURES

The precise control in the preparation of ultra-thin layers allows us to study carrier confinement in two-dimensional systems.[4] The discovery of fractional quantum Hall effect with such structures is described in the paper by Stormer in this symposium. Modulation-doped $Al_x Ga_{1-x} As/GaAs$ structures with electron mobilities of 5×10^6 cm^2/V sec at a two-dimensional electron areal density of 1.6×10^{11} cm^{-2} have been achieved.[5] The structures were grown with an As$_4$ beam sublimed from a solid arsenic rod and Ga and Al beams evaporated from high-purity pellets in a conventional MBE system. These high-electron-mobil-

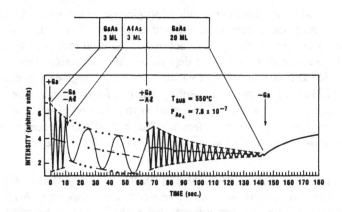

REFLECTION HIGH ENERGY ELECTRON DIFFRACTION (RHEED)
INTENSITY OSCILLATION FOR PHASE-LOCKED EPITAXY

FIGURE 1   The intensity of diffracted electrons as a function of time during MBE.

ity structures may be used to fabricate high-speed devices.[6] Fujitsu has recently announced a 16-bit multiplier featuring the world's fastest multiplication time of 4.1 nsec.

It is expected that quantum confinement to one and zero degrees of freedom (quantum wires and quantum boxes) will lead to further new discoveries of physical phenomena. Lateral quantum confinement has been achieved by using gallium ion beams to disorder around a tungsten masked area.[7] For GaAs/AlGaAs multilayers, the interdiffusion of Al in GaAs after rapid thermal annealing is much larger in the disordered region than in the masked region. One can therefore prepare high-quality laterally confined structures such as quantum wires and quantum boxes.

A more interesting system to study is InGaAs and InP. As the structure gets smaller and smaller, the surface-to-volume ratio becomes larger. Therefore, the surface properties become important. The recombination velocity of electrons and holes for InP is about a hundred times smaller than that of GaAs and therefore high luminescence efficiency from the InGaAs/InP structures is expected. Recently, gas source molecular beam epitaxy (GSMBE, MOMBE,

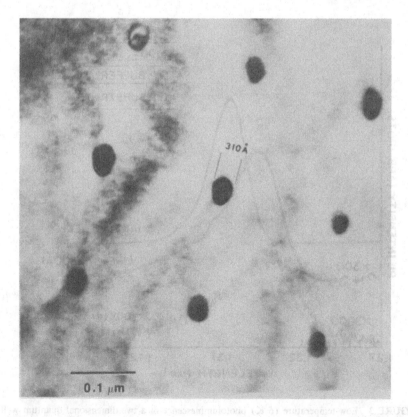

FIGURE 2 Transmission electron micrograph of InGaAs/InP quantum boxes with an average box diameter of 300 Å.

CBE, or VCE) has been used successfully to grow high-quality phosphorus compounds.[8,9] Figure 2 shows a transmission electron micrograph of the In-GaAs/InP quantum boxes with an average box diameter of 300 Å. The photoluminescence spectra, as illustrated in Fig. 3, show a shift of 8 to 14 meV, which is as expected for this degree of lateral confinement.

"BAND STRUCTURE ENGINEERING"

Molecular-beam epitaxy can prepare hetero-epitaxial structures consisting of alternating ultra-thin layers of two different materials (superlattice), graded compositions (variable bandgap) and spike-doping (tunable band-edge discontinuities). These structures may have unique electronic and optical properties, which will not only result in many discoveries of new physical phenomena but also may be applied to novel devices. Engineers take advantage of the capability

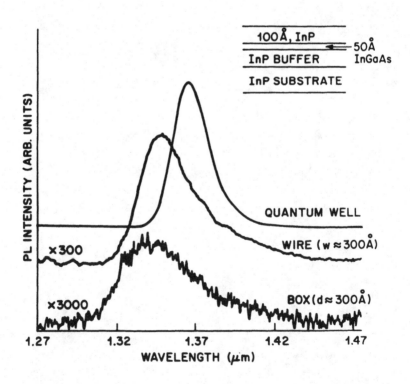

FIGURE 3   Low-temperature (6 K) photoluminescence of a two-dimensional quantum well (50 Å) sample (upper curve) and the artificial low-dimensional structures (lower curves). Inset shows a schematic diagram of the As-grown quantum well structure.

to independently tune the transport properties of electrons and holes, using the built-in electric fields in a graded bandgap material, and the freedom to modify the conduction and valence band discontinuities in a given heterojunction, to prepare the most efficient devices. For instance, a heterojunction bipolar transistor should have a larger bandgap in the emitter than in the base in order to suppress hole injection of a heavily doped base. However, the spike at the heterojunction interface impedes the high emitter injection efficiency. A graded bandgap emitter has improved the current gain by 100% compared with that of an abrupt emitter-base junction.[10] Tunable barrier heights and band discontinuities obtained by doping interface dipoles (Fig. 4) have improved the photo collection efficiency of a photodetector by 40–50%.[11] Using a sheet charge (Fig. 5) of a 30-Å-thick highly doped layer to modify the energy diagram of a heterostructure, a new optoelectronic switch has been demonstrated.[12] Quantum well structures have been used in lasers, photodetectors, and

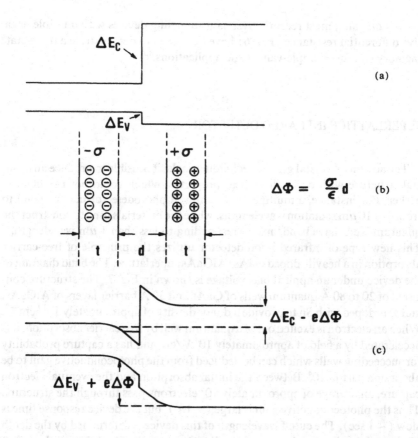

FIGURE 4   (a) Band diagram of an intrinsic heterojunction. (b) Schematics of doping interface dipole. σ is the sheet charge density and $\Delta\phi$ the dipole potential difference. (c) Band diagram of an intrinsic heterojunction with doping interface dipole.

ENERGY BAND DIAGRAM OF THE PHOTONIC SWITCH

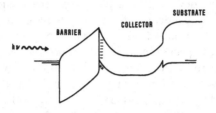

FIGURE 5   Energy-band diagram of a photonic switch where the optical gate is formed by a sheet charge.

transistors; and, most recently, resonant tunneling devices with multiple negative differential resistance (Fig. 6) have been used to demonstrate a three-state memory cell for multiple-value logic applications.[13]

SUPERLATTICE INFRARED DETECTOR

The advanced crystal growth techniques make it possible to produce an artificial structure with desirable optical properties which could not be achieved earlier. For instance, a multiple quantum well photodetector may respond to from 3 to 10 μm radiation wavelengths, while the materials used to construct the quantum wells have bandgaps corresponding to less than 1 μm wavelength.[14] This new type of infrared photodetector utilizes the principle of free-carrier absorption in a heavily doped GaAs/AlGaAs superlattice. The band diagram of the device under an applied bias voltage is shown in Fig. 7. The structure consisted of 20 to 80 Å quantum wells of GaAs and 30 Å barrier layers of AlGaAs, and was doped with Sn to provide a donor density of approximately 10^{19} cm^{-3}. When an electron is excited out of a quantum well by free carrier absorption, it is accelerated by a field of approximately 10^4 V/cm and has a capture probability for succeeding wells which can be deduced from the photoconductive gain to be about one part in 10^4. Between the initial absorption and the eventual electron capture, an average of approximately 10^4 electrons pass through the structure. Thus the photoconductive gain is large ($\sim 10^4$), but the device response time is slow (~ 1 sec). The cutoff wavelength of this device is determined by the depth of the conduction-band quantum wells. One can therefore design the cutoff wavelength, which may allow the fabrication of devices operating at several bands of near- and mid-infrared.[14]

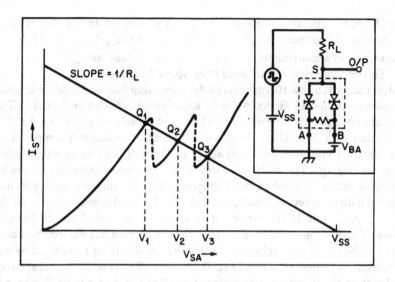

FIGURE 6 Schematic of the current–voltage characteristic and diagram of the three-state memory cell utilizing the negative resistance of the resonant tunneling device. The intersections of the load line with the positive-slope portions of the *I–V* curve represent the stable states of the circuit.

Another class of superlattice photodetector is based on carrier tunneling. The effective-mass filtering photodetector[15] utilizes the large difference between the tunneling rates of electrons and holes which gives rise to a photoconductivity effect of quantum-mechanical origin. The photoconductive gain is given by the ratio of the electron lifetime (time before being captured) to the electron transit

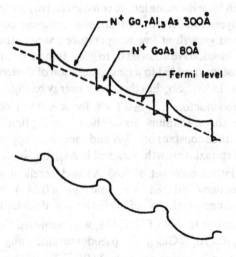

FIGURE 7 The band diagram of a multiple quantum-well photodetector under an applied bias voltage.

time ($L/\mu_e F$). The structure was grown on an n^+ InP substrate and consists of 100 periods of undoped $Al_{0.48}In_{0.52}As$ (35 Å) and $Ga_{0.47}In_{0.53}As$ (35 Å). A current gain of approximately 2×10^4 is obtained at bias voltages as low as 20 mV. The response time is estimated to be about 10^{-3} sec.[15]

Most recently, a new $10\,\mu$m infrared detector using intersub-band absorption in resonant tunneling GaAs/AlGaAs superlattice was demonstrated.[16] The structure consisted of 50 periods of 65 Å GaAs wells (doped $n = 1.4 \times 10^{18}$ cm^{-3}) and 95 Å $Al_{0.25}Ga_{0.75}As$ barriers. These thicknesses and compositions were chosen to produce only two states in the well with an energy spacing corresponding approximately to $10\,\mu$m. The operation of this detector can be illustrated in the conduction band diagram (Fig. 8). Infrared light which is resonant with the intersub-band transition (E_2-E_1) excites an electron from the doped ground state to the excited state where it can tunnel out of the well through the thin top of the barrier. This photogenerated hot electron will contribute to the current (travel a mean free path L) before being recaptured by one of the wells. A responsivity of 0.52 A/W at $\lambda = 10.8\,\mu$m and estimated speed of ~ 30 psec were achieved.[16] The detectivity D^* of this device is 10^{-9}, which is comparatively inferior to HgCdTe (2×10^{-10} cm Hz$^{1/2}$ W$^{-1/2}$) because of the excess value of dark current.

STRAINED-LAYER EPITAXY

Most of the single-crystal thin films are epitaxial layers consisting of materials having lattice constants the same as those of the substrate. The limited pairs of compounds which have the same lattice constants have restricted the freedom of design of multilayered structures. Advances in epitaxial techniques, such as using nonequilibrium growth at low temperatures, superlattice buffer layers, and graded compositions, have allowed us to grow thin heterostructures having different lattice constants and led to a new generation of devices which had not been realized before. $InAs_{1-x}Sb_x$ has the lowest energy bandgap among the III–V compound semiconductors ($E_g \approx 0.1$ eV for $x = 0.65$) corresponding to $\lambda \approx 12\,\mu$m where the minimum atmospheric absorption is present. $InAs_{0.35}Sb_{0.65}$ has a lattice constant of 6.3 Å and there is no appropriate substrate for lattice-matched epitaxial growth. Strained $InAs_{1-x}Sb_x$ layers were grown on GaAs substrates (lattice constant of 5.65 Å) and excellent long-wavelength photodetectors were demonstrated. A responsivity of 0.84 A/W at $\lambda = 5.4\,\mu$m with internal quantum efficiency of 47% and a detectivity of 3×10^{-10} cm Hz$^{1/2}$ W$^{-1/2}$, similar to that of HgCdTe, were achieved.[17]

An $In_{0.15}Ga_{0.85}As/Al_{0.15}Ga_{0.85}As$ pseudomorphic single-quantum-well modulation-doped transistor was grown by MBE.[18] The higher electron saturation velocity in $In_{0.15}Ga_{0.85}As$, compared to that of GaAs, provides the large saturation current and transconductance needed to drive capacitances rapidly

10 µm QUANTUM WELL DETECTOR

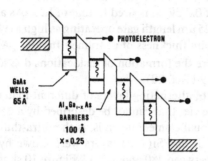

FIGURE 8 Photoconductive current produced by absorption of intersub-band radiation followed by tunneling out of the quantum well.

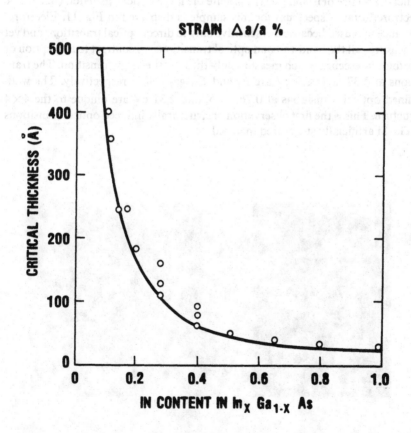

FIGURE 9 Experimental data and calculated value (solid line) of maximum $In_x Ga_{1-x}As$ layer thickness on a GaAs substrate before the formation of misfit dislocations.

in integrated circuits. Furthermore, this compound (with about 1% lattice mismatch to GaAs) has high electron mobility and large energy minima difference $\Delta E = E_\Gamma - E_L$, resulting in higher value of overshoot velocity. GaAs has a relatively small ΔE of 0.3 eV compared to that of InGaAs alloy of about 0.6 eV. MODFETs with 0.25 μm length gate operating as high as 98 GHz have been reported.[19] The maximum thickness of a strained $In_x Ga_{1-x}$As layer on GaAs that can be grown before the formation of dislocations due to the lattice mismatch is shown in Fig. 9 (Ref. 20).

Another illustration of alternating layers of different materials, which may have entirely new properties that cannot be described by a combination of the properties of the individual elemental solids, is the ultra-thin Ge–Si superlattice.[21] The alternating Ge and Si(100) layers were grown by MBE with a Si substrate temperature between 480 and 530 °C. Figure 10 shows a high-resolution transmission micrograph of four periods of the 4×4 (4 monolayers of Ge and 4 monolayers of Si) ordered structure in the $\langle 110 \rangle$ cross section where each dot represents a pair of atom columns. The uniform layers of the superlattice indicate absence of island growth despite the large lattice mismatch (4%). The electroreflectance spectrum for this sample is displayed in Fig. 11. Electroreflectance was used because it is sensitive only to direct optical transitions and yet it measures all the band-to-band optical transitions, in contrast to absorption or photoluminescence which measure only the lowest energy transition. The transitions at 3.37 and 3.13 eV are E_1 and E_0' gaps of Si respectively. The well-defined optical transitions at 0.76, 1.25, and 2.31 eV are unique to the 4×4 structure. This is the first observation of structurally induced optical transitions in Ge–Si artificially structured material.

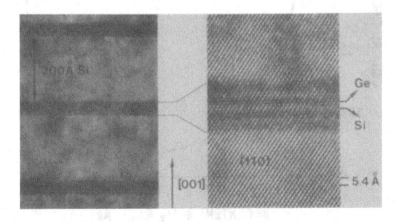

FIGURE 10   High-resolution electron micrographs of four periods of a 4×4 ordered Ge–Si superlattice.

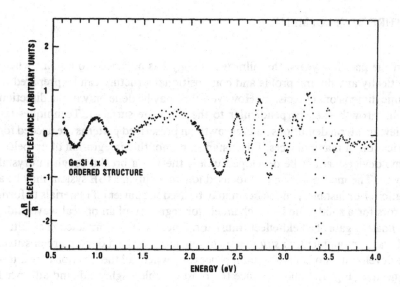

FIGURE 11 Electroreflectance spectrum of the Ge–Si (4×4) superlattice on Si substrate. The transitions at 0.76, 1.25, and 2.31 eV are due to the artificially structured (4×4) layers.

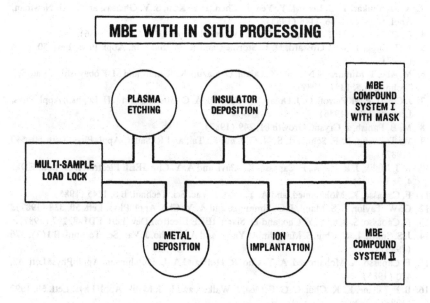

FIGURE 12 Conceptual arrangement of an *in situ* processing MBE system.

FUTURE DEVELOPMENT

In the past few years, thin-film technology has advanced to a point where practically any doping profile and compositional structure can be prepared to atomic dimensional precision. However, this may be done only in the direction of film growth (i.e., perpendicular to the substrate surface). Techniques for achieving lateral definitions and regrowth on prescribed patterns are needed for fabrication of many devices. For instance, monolithic integration of optoelectronic devices cannot be developed unless there is a breakthrough in crystal growth. The incorporation of a focused ion beam in an MBE system may be a solution. For instance, an ion beam may be used to sputter off materials to form mirrors for a solid-state laser, channels for regrowth of an optical waveguide, and floating gate of a field-effect transistor. Since MBE is conducted in an ultrahigh vacuum ambient, *in situ* metallization and ion beam doping are possible. The development of an all-vacuum processing where all the functions are interconnected (Fig. 12) may produce the most reliable, high yield, and advanced devices.

REFERENCES

1. A. Y. Cho, J. Appl. Phys. **42**, 2074 (1971).
2. J. J. Harris, B. A. Joyce, and P. J. Dobson, Surf. Sci. **103**, L90 (1981).
3. A. Madhukar, T. C. Lee, M. Y. Yen, P. Chen, J. Y. Kim, S. V. Ghaisas, and P. G. Newman, Appl. Phys. Lett. **46**, 1148 (1985).
4. D. C. Tsui and H. L. Stormer, IEEE J. Quant. Electron. **QE-22**, 1711 (1986).
5. J. H. English, A. C. Gossard, H. L. Stormer, and K. W. Baldwin, Appl. Phys. Lett. **50**, 1826 (1987).
6. M. Abe, T. Mimura, S. Notomi, K. Odani, O. Kazuo, K. Kondo, and M. Kobayashi, J. Vac. Sci. Technol. A **5**, 1387 (1987).
7. J. Cibert, P. M. Petroff, G. J. Dolan, S. J. Pearton, A. C. Gossard, and J. H. English, Appl. Phys. Lett. **49**, 1275 (1986).
8. M. B. Panish, J. Crystal Growth **81**, 249 (1987).
9. W. T. Tsang, E. F. Schubert, S. N. G. Chu, K. Tai, and R. Sauer, Appl. Phys. Lett. **50**, 540 (1987).
10. R. J. Malik, J. R. Hayes, F. Capasso, K. Alavi, and A. Y. Cho, IEEE Electron Dev. Lett. **EDL-4**, 383 (1983).
11. F. Capasso, K. Mohammed, and A. Y. Cho, J. Vac. Sci. Technol. B **3**, 1245 (1985).
12. G. W. Taylor, F. S. Mand, J. G. Simmons, and A. Y. Cho, Appl. Phys. Lett. **50**, 338 (1987).
13. F. Capasso, S. Sen, A. Y. Cho, and D. Sivco, IEEE Electron Dev. Lett. **EDL-8**, 297 (1987).
14. J. S. Smith, L. C. Chiu, S. Margalit, A. Yariv, and A. Y. Cho, J. Vac. Sci. Technol. B **1**(2), 376 (1983).
15. F. Capasso, K. Mohammed, A. Y. Cho, R. Hull, and A. L. Hutchinson, Appl. Phys. Lett. **47**, 420 (1985).
16. B. F. Levine, K. K. Choi, C. G. Bethea, J. Walker, and R. J. Malik, Appl. Phys. Lett. **50**, 1092 (1987).
17. C. G. Bethea, M. Y. Yen, B. F. Levine, K. K. Choi, and A. Y. Cho, Appl. Phys. Lett. **51**, 1431 (1987).

18. T. Henderson, M. I. Aksum, C. K. Peng, H. Morkoc, P. C. Chao, P. M. Smith, K. H. G. Duh, and L. F. Lester, IEEE Electron Dev. Lett. **EDL-7**, 649 (1986).
19. A. Fischer-Colbrie, J. N. Miller, S. S. Laderman, S. J. Rosner, and R. Hull, Eighth MBE Workshop (U.C.L.A., Los Angeles, CA, Sept. 9–11, 1987), WP-2, p. 17.
20. T. G. Anderson, Z. G. Chen, V. D. Kulakovskii, A. Uddin, and J. T. Vallin, Appl. Phys. Lett. **51**, 752 (1987).
21. J. Bevk, A. Ourmazd, L. C. Feldman, T. P. Pearsall, J. M. Bonar, B. A. Davidson, and J. P. Mannaerts, Appl. Phys. Lett. **50**, 760 (1987).

Phases and Phase Transitions in Less Than Three Dimensions/ MICHAEL E. FISHER

SUMMARY

The lecture aimed to present a broad overview of phase transitions in less than three dimensions by following various theoretical threads reaching back four decades[1] through to the present day. Our understanding of matter, relating delicate observations to subtle theory, has increased dramatically over this period. It has culminated recently in the application of conformally covariant field theory to continuous phase transitions and multicriticality in two dimensions.[2-4]

A typical bulk phase diagram in the pressure–temperature or (p,T) plane, as sketched in Fig. 1, exhibits a gas–liquid critical point at (p_c, T_c) but also contains various critical lines, $T_c(p)$, separating, e.g., a ferromagnetic from a paramagnetic crystalline phase. Furthermore, a critical line may terminate at a *tricritical* point beyond which the transition becomes first order.[5] The shape of the coexistence curve, or of the plot of spontaneous order, vs. T is characterized by a power law, $|t|^\beta$ with $t = (T - T_c)/T_c$, as $T \to T_c$. Classical theory (due to van der Waals, Landau, and others) predicts a parabolic shape, meaning $\beta = \frac{1}{2}$, but is generally valid *only* for spatial dimensionality $d \geqslant 4$. Normal bulk $(d = 3)$ dimensionality is reflected in β values close to 0.33. By contrast, Onsager's profound study[6] in 1944–49 of *two*-dimensional lattice gases or Ising-model magnets yielded[7] $\beta = \frac{1}{8}$. It also predicted specific heats diverging as $|t|^{-\alpha}$ with $\alpha = 0$, corresponding to a logarithmic law, $\log |t|$.

To study two-dimensional phase transitions in the real world and test the calculations of Onsager and others one may, first, study bulk but layered materi-

MICHAEL E. FISHER *has spanned the fields of chemistry, mathematics, and theoretical physics. After a number of years on the faculty of King's College, London, he became Horace Mann Professor of Physics, Chemistry and Mathematics at Cornell University. He is now Wilson H. Elkins Professor at the Institute for Physical Science and Technology, University of Maryland, College Park, Maryland.*

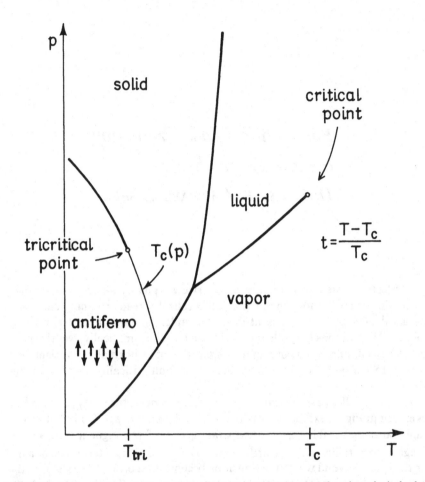

FIGURE 1 Schematic phase diagram in the pressure-temperature plane for a hypothetical substance which, in its solid phase, exhibits antiferromagnetism, as indicated by the spin arrows: there is an associated critical line, $T_c(p)$, ending in a tricritical point at T_{tri} beyond which the magnetic transition becomes first order.

als in which the relevant interplane interactions are weak or cancel in leading order. Thus experiments by Birgeneau, Guggenheim, and Shirane[8] in 1969 on K_2NiF_4 and other magnetic perovskite crystals (with structures like the recently discovered "high-T_c" superconductors) yield $\beta = 0.13 \pm 0.02$ in nice verification of the prediction $\beta = \frac{1}{8}$. Bulk ($d = 3$)-dimensional specific heat peaks are characterized by $\alpha \simeq 0.11$ (for Ising-model-like systems); however, 1983 measurements by Belanger, Jaccarino, and co-workers on the layered antiferromagnet Rb_2CoF_2 yield a behavior[9] closely matching the logarithmic ($\alpha = 0$) specific heat divergence predicted for $d = 2$ Ising systems.

On going beyond the thermodynamics of bulk systems one comes first to *interfaces*, as between a crystal and its vapor or melt. The surface of a crystal may change its structure as a function of temperature, undergoing a sharp "surface reconstruction" transition. For example, a gold (110) face exhibits a so-called (2×1) to (1×1) transition near $T_c = 650$ K. Because of the two-fold character of the ordered state, an Ising transition is predicted; electron scattering experiments in 1985 by Campuzano *et al.*[10] have, indeed, found $\beta = 0.13 \pm 0.02$. (The experiments also check other detailed predictions for scattering from $d = 2$ Ising systems.) Another type of surface transition is seen in vanadium. Bulk vanadium is a paramagnet but the (100) surface becomes ferromagnetic at $T_c = 475.1$ K; owing to magnetic anisotropy the transition should be of Ising type. Using an ingenious deuteron electron-capture spectroscopy, Rau[11] has recently observed the spontaneous order and concluded $\beta \simeq 0.128 \pm 0.020$.

Intermediate between purely two-dimensional and bulk transitions are roughening and wetting transitions in which an interface extends significantly into the third dimension, interacts with the container walls and reorganizes the shapes of the bulk crystalline or liquid phases. Striking examples of these phenomena have been observed in experiments on helium-four on its melting curve.[12-14] Thus a wall-wetting transition of the (0001) face has been discovered recently at 1.03 K (although this appears to be associated with a few parts per million helium-three impurities[14d]). Roughening transitions of the (0001), $(1\bar{1}00)$ and $(1\bar{1}01)$ faces of He^4 crystals are found at $T_R \simeq 1.28, 0.9$, and 0.36 K, respectively.[12-14] Below T_R the face in question is atomically smooth; above T_R it becomes rough and fluctuates over dimensions diverging logarithmically with the crystal size. The Kosterlitz–Thouless theory[15] of XY-like systems in $d = 2$ dimensions and the Pokrovsky–Talapov[16] analysis of domain wall melting describe many details of the temperature and crystal shape dependence.[17] Recent synchrotron radiation scattering observations[18] of the roughening transition at $T_R \simeq 450\,°C$ on the (110) face of crystalline silver confirm further theoretical predictions.

Truly two-dimensional systems are found in studying adsorption from a bulk vapor onto the walls of a container. For a crystalline wall or 'substrate' in the form of a graphite surface—Union Carbide's UCAR-ZYX exfoliated graphite has been the foundation for many landmark experiments—one finds a wide range of surface phase diagrams. Some closely mirror bulk phase diagrams.[19] At submonolayer coverages one has a close-to-ideal lattice gas: precise specific heat measurements of methane adsorbed on graphite by Kim and Chan[20] have shown that the corresponding two-dimensional liquid-vapor coexistence curve is described by $\beta = 0.127 \pm 0.02$, a beautiful verification of the original Onsager lattice-gas prediction nearly forty years later! At higher coverages, layering transitions occur[21] and the corresponding behavior around the layer critical points, $T_c(n), n = 1, 2, \ldots$, also obeys two-dimensional Ising laws. In addition, there are "solid" phases which can be either *commensurate* or *incommensurate* with the substrate lattice.[22] In contrast to standard bulk behavior, two-dimensional melting may occur via continuous transitions; furthermore, *tricritical* points appear to arise on the melting lines.[23]

Helium, hydrogen, and other small molecules exhibit a so-called $\sqrt{3} \times \sqrt{3}$ commensurate solid phase when adsorbed on graphite. Ordered phases of this type have a *three*-fold character. On this basis Alexander[24] suggested that these systems should provide realizations of the *Potts three-state-model*. Landau's group-theoretic rules for the melting of such a system predict a first-order transition; Straley and Fisher,[25] however, argued that in *two* dimensions the transition was, in fact, continuous with non-Ising critical exponents α, β, etc. Early theoretical estimates for α were erratic: 0.05 to 0.26 or larger.[25,26] But in 1977 Bretz[26] made precise high-resolution specific heat measurements of He^4 on graphite: his conclusion was $\alpha = 0.36 \pm 0.04$. This exponent value was vindicated by later theory: R. J. Baxter[27] proved that the transition was indeed continuous, and further theoretical work by Baxter, den Nijs, Nienhuis, and others[28] led to the rational value $\alpha = \frac{1}{3}$ (and $\beta = \frac{1}{9}$, etc.). Experiments on other systems[29] confirm the new Potts values. One learns that fluctuations play an enhanced role in two-dimensional physics.

Why are the critical exponents for two-dimensional systems rational fractions? The first point is that critical phenomena obey *scaling laws* as $T \to T_c$. As a result various exponents are rationally related to one another. Then, as stressed by Kadanoff, one can characterize critical behavior in terms of basic exponents, ω_1 and ω_2, defined purely *at criticality* via the decay of the order–order and energy–energy correlation functions with distance r according to $1/r^{2\omega_1}$ and $1/r^{2\omega_2}$.[1,4] One finds $(\omega_1, \omega_2) = (\frac{1}{8}, 1)$ for Ising and $(\frac{2}{15}, \frac{4}{3})$ for Potts three-state systems, respectively. In 1970 Polyakov[30] suggested that scaling invariance should, at criticality, be associated with *conformal covariance*. In general dimensions this means the correlation functions should transform in an appropriate way, controlled by the ω_i, under the inversion transformation $\mathbf{r}' = 1/\mathbf{r} \equiv \mathbf{r}/|\mathbf{r}|^2$ (or its infinitesimal analogs).[2,4] One surprising consequence, appreciated only more recently,[31,2,4] is that, for example, the $1/z^{\omega_1}$ enhancement of the order at criticality near a planar boundary *wall* should directly yield the behavior inside *or* outside a large but *finite* sphere.

More far-reaching, however, was the realization that in $d = 2$ dimensions the full conformal group encompasses much more than the inversion $1/\mathbf{r}$. Indeed, if one writes $z = x + iy$ for a point (x,y), then any analytic mapping $z' = w(z)$ provides a conformal transformation. This fact had already been exploited in the field theory of strings.[3] A relativistic one-dimensional string traces out a two-dimensional world sheet or surface and the corresponding massless field theory has the same formal structure as the statistical mechanics of a two-dimensional (continuum) system at criticality.[2,3] In a seminal paper in 1983, Belavin, Polyakov, and Zamolodchikov[32] (BPZ) presented a classification of all such conformally covariant two-dimensional field theories and showed that the critical exponents, ω_i, should derive from a set of eigenvalues, $h_{jk}(c)$ and $\bar{h}_{jk}(c)$, related to the corresponding commutation algebra of the generators of the conformal transformations: this so-called Virasoro algebra[2,3] contains a free parameter, the *central* or *conformal charge*, c. For the special values $c = 1 - 6/m(m + 1)$ with rational $m \equiv r/s \geqslant 2$ and $j = 1, 2, \ldots, (r - 1)$ and

$k = 1, 2, \ldots, (r + s - 1)$, BPZ showed that only a *finite* number of basic field operators (or 'order parameters') arose. Furthermore, the corresponding correlation functions satisfied explicit linear differential equations. They proposed that such "magical" field theories should describe two-dimensional systems at criticality. Indeed, they showed that for $m = 3$ (or $r = 3$, $s = 1$) the table of eigenvalues h_{jk} and $\bar{h}_{jk}$ generated precisely the Ising exponents $(\omega_1, \omega_2) = (\frac{1}{8}, 1)$ via $\omega = h + \bar{h}$.

The physical principle underlying the magical values of c was exposed by Friedan, Qiu, and Shenker[33] who proved that the requirement of *unitarity* imposed on the field theory restricted m to the integers $m = 3, 4, 5, \ldots$; unitarity should characterize many (although not all) physical systems near criticality. Indeed, the value $m = 5$ was found to provide a description of the three-state Potts model with $(\omega_1, \omega_2) = (\frac{2}{15}, \frac{4}{5})$.[34] From the table of h_{jk} for $m = 4$ and 6 one finds *tri*critical exponents for Ising and three-states Potts tricriticality, respectively. Further values of m can also be identified with higher order multicritical points.[35]

For larger values of m the eigenvalue table allows construction of many ω_i. Some of these are needed since, for example, Ising tricriticality actually demands four independent ω_i rather than just two as for ordinary criticality. This can be seen most directly by looking at tricriticality in bulk fluid mixtures, such as benzene, ethanol, and water with ammonium sulphate:[36] at atmospheric pressure one can depart from the tricritical point by varying T or any one of three independent mole fractions specifying the overall chemical composition. However, the superfluity of the remaining potential exponents, ω_i, was an embarrassing puzzle. Very recent progress by Cardy[37] and by Itzykson and Zuber[38] resolves this puzzle by considering criticality on a large skew torus. Invoking modular invariance—which says that different unit cells in the corresponding space lattice must yield the same physical description—restricts the allowed combinations of h_{jk} and $\bar{h}_{jk}$. All the required basic 'critical operators' or order parameters are provided for in the well understood examples, and complete sets are predicted for the higher-order cases. The challenge is back in the experimental court: many further experiments are needed to test theory and its ramifications.

APOLOGIA AND CAVEAT

This lecture surveyed much ground but was in no sense a scholarly review. Many fascinating and important topics, which could well have been discussed, were necessarily omitted under pressure of time. The experimental and theoretical examples chosen were intended to illustrate rather than to paint a complete and balanced picture. Likewise, the references listed should provide an entry to the literature but are unsystematic and far from exhaustive. Apologies are offered to authors whose research or writings have been arbitrarily left unmentioned in the process of sampling from the rich material that makes up the tapestry of modern condensed matter physics.

ACKNOWLEDGMENTS

I am grateful to many colleagues for assistance in preparing and absorbing the material on which this lecture was based. The ongoing support of the National Science Foundation, principally through the Condensed Matter Theory Program is much appreciated.

REFERENCES

1. See, e.g., M. E. Fisher, Repts. Prog. Phys. **30**, 615 (1967).
2. J. L. Cardy in *Phase Transitions and Critical Phenomena*, edited by C. Domb and J. L. Lebowitz, Vol. 11 (Academic, London, 1987), pp. 55–126.
3. D. Friedan , Z. Qiu, and S. H. Shenker, J. Magn. Mag. Matls. **54–57**, 649 (1986).
4. For some informal reviews see M. E. Fisher (a) J. Appl. Phys. **57**, 3265 (1985); (b) *Spatial Symmetries and Critical Phenomena* in Proc. Fifth Philip Morris Science Symp., October 1985 (Philip Morris, New York, 1986); (c) *The States of Matter—a theoretical perspective* in Proc. R. A. Welch Conf. XXIII, *Modern Structural Methods* (Welch Foundation, Houston, 1986).
5. See, e.g., M. E. Fisher, AIP Conf. Proc. No. 24, *Magnetism and Magnetic Materials*, 1974 (AIP, New York, 1975), pp. 272–280 and Ref. 4(c).
6. L. Onsager, Phys. Rev. **65**, 117 (1944); Nuovo Cim. (Suppl.) **6**, 261 (1949). (See also Ref. 1.)
7. The first derivation published in the literature was presented by C. N. Yang, Phys. Rev. **85**, 808 (1952).
8. R. J. Birgeneau, H. J. Guggenheim, and G. Shirane, Phys. Rev. Lett. **22**, 720 (1969); Phys. Rev. B **1**, 2211 (1970).
9. P. Nordblad, D. P. Belanger, A. R. King, V. Jaccarino, and H. Ikeda, Phys. Rev. B **28**, 278 (1983).
10. J. C. Campuzano, M. S. Foster, G. Jennings, R. F. Willis, and W. Unertl, Phys. Rev. Lett. **54**, 2684 (1985).
11. C. Rau, G. Xing, C. Liu, and M. Robert, Phys. Rev. Lett. (to be published).
12. See, e.g., H. J. Maris and A. F. Andreev in *Physics Today*, Feb. 1987 (AIP, New York, 1987), p. 25.
13. S. Balibar and B. Castaing , J. Phys. Lett. **41**, L329 (1980); P. E. Wolf, S. Balibar, and F. Gallet, Phys. Rev. Lett. **51**, 1366 (1983).
14. (a) J. E. Avron, L. S. Balfour, C. G. Kuper, J. Landau, S. G. Lipson, and L. S. Schulman, Phys. Rev. Lett. **45**, 814 (1980); (b) Y. Carmi, S. G. Lipson, and E. Polturak, Phys. Rev. B **36**, 1894 (1987); (c) S. G. Lipson, Contemp. Phys. **28**, 117 (1987); (d) S. G. Lipson (private communication).
15. J. M. Kosterlitz and D. J. Thouless, J. Phys. C **6**, 118 (1973).
16. V. L. Pokrovsky and A. L. Talapov, Zh. Eksp. Teor. Fiz. **78**, 269 (1980) [Sov. Phys. JETP **51**, 134 (1980)]; see also J. Villain in *Ordering in Strongly Fluctuating Condensed Matter Systems*, edited by T. Riste (Plenum, New York, 1980), pp. 221–260.
17. C. Rottman and M. Wortis, Phys. Repts. **103**, 59 (1984).
18. G. A. Held, J. L. Jordan–Sweet, P. M. Horn, A. Mak, and R. J. Birgeneau, Phys. Rev. Lett. **59**, 2075 (1987).
19. See, e.g., J. G. Dash in *Physics Today*, Dec. 1985 (AIP, New York, 1985), p. 26.
20. H. K. Kim and M. H. W. Chan, Phys. Rev. Lett. **53**, 170 (1984).
21. Q. M. Zhang, Y. P. Feng, H. K. Kim, and M. H. W. Chan, Phys. Rev. Lett. **57**, 1456 (1986); see also M. Drir and G. B. Hess, Phys. Rev. B **33**, 4758 (1986).
22. R. J. Birgeneau and P. M. Horn, Science **232**, 329 (1986).
23. K. D. Miner, Jr., M. H. W. Chan, and A. D. Migone, Phys. Rev. Lett. **51**, 1465 (1983).
24. S. Alexander, Phys. Lett. **54A**, 353 (1975).

25. J. P. Straley and M. E. Fisher, J. Phys. A **6**, 1310 (1973).

26. M. Bretz, Phys. Rev. Lett. **38**, 501 (1977).

27. R. J. Baxter, J. Phys. C **6**, L445 (1973); J. Phys. A **13**, L61 (1980); see also *Exactly Solved Models in Statistical Mechanics* (Academic, London, 1982).

28. M. P. M. den Nijs, J. Phys. A **12**, 1857 (1979); B. Nienhuis, E. K. Riedel, and M. Schick, Phys. Rev. B **23**, 6055 (1981).

29. For example, F. A. B. Chaves, M. E. B. P. Cortez, R. E. Rapp, and E. Lerner, Surf. Sci. **150**, 80 (1985) studied para-H_2 on graphite finding $\alpha = 0.37 \pm 0.03$.

30. A. M. Polyakov, Pis'ma Zh. Eksp. Teor. Fiz. **12**, 538 (1970) [Sov. Phys. JETP Lett. **12**, 381 (1970)].

31. T. W. Burkhardt and E. Eisenriegler, J. Phys. A **18**, L83 (1985).

32. A. A. Belavin, A. M. Polyakov, and A. B. Zamolodchikov, J. Stat. Phys. **34**, 763 (1984); Nucl. Phys. B **241**, 333 (1984).

33. D. Friedan, Z. Qiu, and S. H. Shenker, Phys. Rev. Lett. **52**, 1575 (1984).

34. Vl. S. Dotsenko, J. Stat. Phys. **34**, 781 (1984); Nucl. Phys. B **235**, 54 (1984).

35. D. A. Huse, Phys. Rev. B **30**, 3908 (1984) showed that a sequence of exactly soluble models discovered by G. F. Andrews, R. J. Baxter, and P. J. Forrester, J. Stat. Phys. **35**, 193 (1984) provided realizations of the unitary conformal field theories for all $m = 3, 4, \ldots$.

36. J. C. Lang, Jr. and B. Widom, Physica **81A**, 190 (1975).

37. J. L. Cardy, Nucl. Phys. B **270**, 186 (1986).

38. C. Itzykson and J.-B. Zuber, Nucl. Phys. B **275**, 580 (1986).

The Fractional Quantum Hall
Effect/ HORST L. STORMER

The fractional quantum Hall effect (FQHE),[1] is the manifestation of a new, highly correlated, many-particle ground state[2] that forms in a two-dimensional electron system at low temperatures and in high magnetic fields. It is an example of the new physics that has grown out of the tremendous recent advances in semiconductor material science,[3] which has provided us with high-quality, lower-dimensional carrier systems.[4] The novel electronic state exposes itself in transport experiments through quantization of the Hall resistance to an exact rational fraction of h/e^2, and concomitantly vanishing longitudinal resistivity. Its relevant energy scale is only a few degrees kelvin. The quantization is a consequence of the spontaneous formation of an energy gap separating the condensed ground state from its rather elusive quasiparticle excitations. According to the prevailing theory, the quasiparticles carry charges which are an exact rational fraction of the elementary charge e. The multitude of experimentally observed rational quantum numbers[5-12] 1/3, 2/3, 1/5, 2/5, 2/7, 3/7, 4/9, 5/11, etc. correspond to an equivalent string of different condensates which are believed to reflect a hierarchical order[13-16] of exclusively odd-denominator fractional states with weaker daughter states developing from more robust parental ones. Each sequence is expected to be terminated by the formation of a yet undetected quantum solid removing the characteristic transport features. The theoretical understanding of the novel quantum liquids which underlie the FQHE has predominantly emerged from an ingenious many-particle wave function[2] strongly supported by numerous few-particle simulations.[17-22]

Theory has now constructed a complex model for ideal two-dimensional electron systems in the presence of high magnetic fields and makes definitive, often fascinating predictions. Experiments have successively uncovered odd-denominator fractional states reaching presently to 7/13. Coupled with further progress in material preparation techniques, leading to yet "more ideal" electron systems,[23] experiments are now able to make contact with theory and test some of the calculable quantities. The application of new experimental tools to the

HORST L. STORMER *obtained his doctorate in Germany, but then moved to the United States, where he has worked at the AT&T Bell Laboratories, Murray Hill, New Jersey, since 1978. His research field is solid-state physics, especially the phenomena associated with high magnetic fields.*

FQHE, such as optics,[24,25] microwaves, and phonon techniques promises the direct observation of such parameters as the gap energy and possibly even some of the more elusive quantities in the future.

While theory and experiment in the FQHE appear to be converging, there remains considerable room for challenging surprises. The very recent discovery of the first even-denominator fraction at the FQHE represents such a totally unexpected event.[26] Even-denominator fractions are unaccounted for by the theoretical construct which links odd denominators directly with the Pauli exclusion principle for fermions. It remains to be seen whether this finding requires revision or merely expansion of the existing theoretical model. In any case, the new discovery is bound to further enrich the FQHE.

This paper provides a concise overview of the FQHE.[27] It focuses on the experimental aspects and states, but does not expand on the theoretical advances.[28]

TWO-DIMENSIONAL SYSTEMS

The fractional quantum Hall effect was first observed in the two-dimensional (2D) electron system of a modulation-doped GaAs-(AlGa)As heterostructure. These materials are a result of two decades of intense material research in molecular beam epitaxy. They have provided exceedingly smooth semiconductor–semiconductor interfaces with unprecedentedly low disorder.[29] In many respects these structures resemble the Si-MOSFET (Metal-Oxide-Semiconductor Field-Effect Transistor), the device traditionally employed in 2D physics. Compared to quantum-confined heterostructures, the Si-MOSFET is more versatile but cannot match the exceedingly low scattering rates of the heterosystems.

In the GaAs-(AlGa)As system, electrons are quantum-mechanically bound to the hetero-interface by a selective doping technique (Fig. 1). As in all semiconductors, the presence of the periodic potential alters the electron mass to become an effective mass m^*, different from the free-electron mass m_0 ($m^* \approx 0.07\, m_0$ in GaAs). Their motion in the direction z perpendicular to the interface is quantized and the available states are discrete with a typical separation of ~ 10 meV. At low temperatures, when carriers occupy only the ground state of this discrete set, all degrees of freedom in the z direction are frozen out while the x–y motion remains unimpeded. These quantum-mechanically bound electron systems at the interface between GaAs and (AlGa)As are presently the most truthful realization of the theoretical concept of an ideal 2D metal. At present electron mobilities as high as $\mu = e\tau/m^* = 5 \times 10^6$ cm^2/V sec, equivalent to scattering times τ as long as $\tau = 2 \times 10^{-10}$ sec, have been achieved[23] at low temperatures where scattering by phonons has ceased (Fig. 2). Assuming completely randomizing scattering events, such a low scattering rate translates into an elastic mean free path of ~ 20 μm, a truly macroscopic distance. Such

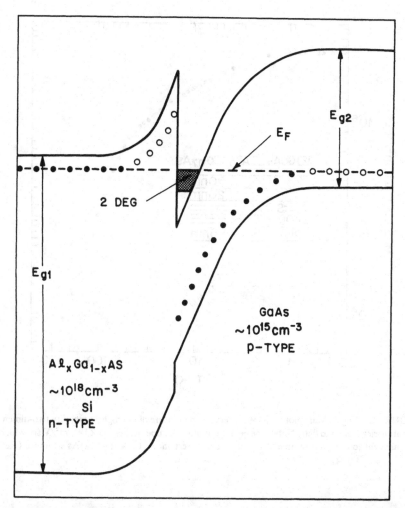

FIGURE 1   Two-dimensional electron system in a modulation-doped GaAs-(AlGa)As hetero-junction. Electrons from the Si impurities in the (AlGa)As transfer to the GaAs side of the interface and become quantum-mechanically confined. E_F is the Fermi energy. The motion in the plane of the interface is free.

extremely low-disorder materials are a prerequisite for the observation of carrier correlation effects such as the FQHE with energy scales of only a few degrees kelvin.

TWO-DIMENSIONAL CARRIERS IN THE PRESENCE OF HIGH MAGNETIC FIELD

The hallmark of the FQHE is the quantization of the Hall resistance to $(h/e^2)/f$, where f is an exact rational fraction, and concomitantly vanishing resistivity. Of course, quantization of the Hall resistance per se is not a novelty.

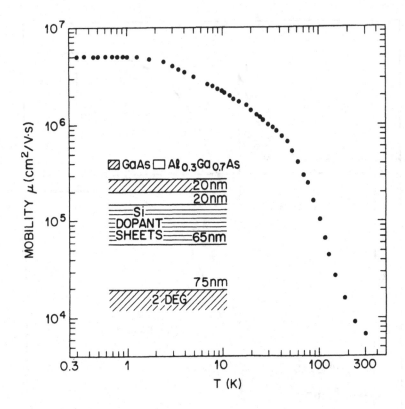

FIGURE 2 Logarithmic plot of mobility versus temperature of the highest mobility two-dimensional carrier system to date. Notice the peak value at low temperatures of 5×10^6 cm^2/V sec which is equivalent to a scattering time of 2×10^{-10} sec. Insert shows GaAs-(AlGa)As structure (Ref. 23).

The integral quantum Hall effect (IQHE) has convincingly demonstrated the existence of such quantization[31,32] and the accuracy to which it reflects h/e^2 has been tested[33] to a few parts in 10^8. The remarkable attributes of the FQHE are the values of its quantum numbers which are exact rational fractions. However, the IQHE represents an important prior step, since it has led to a new understanding of 2D transport in high magnetic fields, and a deeper appreciation of the important role played by random potentials.

Classically, transport in an ideal two-dimensional system of areal carrier density n with a perpendicular magnetic field **B** is easily visualized by recalling the motion of a charged particle in crossed fields **E** and **B**.[34] Contrary to intuition, the carriers drift uniformly in the direction *normal* to **E** and **B** with a velocity $v = E/B$ providing a current density $j = neE/B$ in this direction. The lack of carrier motion along the **E** field results in a vanishing diagonal component σ_{xx} of the two-dimensional conductivity tensor $\hat{\sigma}$ while the off-diagonal component becomes $\sigma_{xy} = ne/B$. Following the tensor relationship between conductivity and resistivity this leads to

$$\rho_{xx} = \frac{\sigma_{xx}}{\sigma_{xx}^2 + \sigma_{xy}^2} = 0 ,$$

$$\rho_{xy} = \frac{\sigma_{xy}}{\sigma_{xx}^2 + \sigma_{xy}^2} = B/ne.$$

Thus an ideal system in the absence of disorder has vanishing, longitudinal resistivity ρ_{xx} and a Hall resistance which is strictly proportional to B. In the presence of disorder the carriers are scattered, the drift motion develops a component parallel to $\mathbf{E}$, and dissipation sets in. Consequently, the transport coefficients deviate from their ideal values.

The quantum-mechanical treatment is rather more complex, largely due to the unusual quantization condition imposed by the magnetic field.[30] Quantization of carrier orbits in a magnetic field is equivalent to the quantum-mechanical problem of the harmonic oscillator, except that the force constant is replaced by the field strength. Under its influence, the continuum of states splits into a sequence of discrete Landau and spin levels at energies given by $E_N = (N + \frac{1}{2})\hbar\omega_c \pm g^*\mu_B B$ separated by wide forbidden gaps (Fig. 3a). $\omega_c = eB/m^*$ is the cyclotron frequency and g^* is the effective g factor. Independently of any material parameters, the degeneracy of each of these magnetic levels is $d = B/\Phi_0$ with the flux quantum $\Phi_0 = h/e = 4.14 \times 10^{-7}$G cm^2. For constant carrier density n there exists a sequence of specific field values B_i at which an exact multiple i of magnetic levels are occupied and the Fermi energy resides in the gap between levels i and $(i + 1)$. Since the areal density at these singular field positions is given by $n = id$, the off-diagonal conductivity acquires exactly the value $\sigma_{xy} = ie^2/h$. Concomitantly, the vanishing density of states at the Fermi level leads inevitably to $\sigma_{xx} = 0$. Thus 2D magneto-transport in an idealized quantum-mechanical system is expected to assume the values $\rho_{xx} = 0$ and $\rho_{xy} = h/ie^2$ at a sequence of singular values of B when exactly i magnetic levels are completely filled. However, these "quantized" values of ρ_{xy} are not distinct on the linear ρ_{xy} versus B trace and are only indirectly defined by the positions of vanishing ρ_{xx}. Formation of quantized plateaus at $\rho_{xy} = h/ie^2$ cannot be expected from such a free-electron model alone.

THE INTEGRAL QUANTUM HALL EFFECT

Constancy of ρ_{xy} over wide stretches of magnetic field quantized to $\rho_{xy} = h/ie^2$, $i = 1, 2, 3 \ldots$ and concomitantly vanishing resistivity ρ_{xx} are the characteristics of the IQHE. This was first observed in a Si-MOSFET.[31] With the advent of quantum-confined heterostructures, the IQHE is now being observed in a multitude of two-dimensional structures.[35-40] A smaller effective

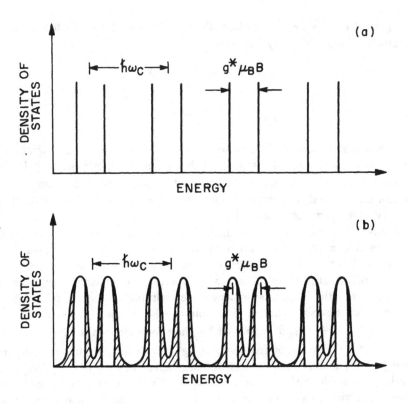

FIGURE 3 Energy spectrum of a two-dimensional electron system in a high magnetic field B. The spectrum consists of discrete Landau levels separated by the cyclotron energy $\hbar\omega_c$ and spin splitting $g^*\mu_B B$ (a) in the absence of disorder, (b) in the presence of disorder. States in the shaded parts of the spectrum are localized.

mass m^* and therefore less stringent requirements on magnetic field strengths, and a much reduced interfacial disorder, make heterostructures presently the preferred system for the study of the IQHE (Fig. 4). GaAs-(AlGa)As systems also dominate the high-precision work[33] on the IQHE which is trying to determine the fine structure constant $\alpha = e^2/\hbar c$ to higher accuracy.

It is now well established that plateau formation, quantization, and vanishing resistivity in the IQHE can be understood in terms of the free-electron model developed above and the omnipresence of disorder which gives rise to macroscopically localized electronic states.[41-47] Disorder broadens the heretofore sharp magnetic levels (Fig. 3b). The spatial extent of states located in the wings of the distribution shrinks with increasing energetic distance of the state from the center of the magnetic level. For any given device geometry, a finite fraction of states no longer extends to the boundaries and ceases to participate in carrier transport across the specimen. The distinct features of the IQHE arise when the Fermi energy passes through these localized states, changing the occupation of

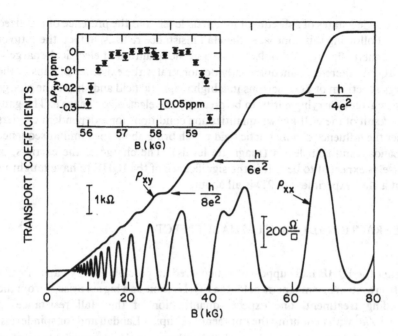

FIGURE 4 Resistivity, ρ_{xx}, and Hall resistance, ρ_{xy}, versus magnetic field B of a low-mobility GaAs-(AlGa)As heterojunction at 1.2 K. Inset shows precision data on the $i = 4$ plateau (Ref. 33).

orbits that are unimportant for the macroscopic transport properties. The current is carried by electrons in the extended states of the filled magnetic levels which are free from scattering by virtue of their wide separation in energy from any empty states. Consequently, ρ_{xx} vanishes and the current flow is dissipationless over these wide regions of magnetic field where E_F combs through the localized states. Constancy of ρ_{xy} ensues necessarily since only the unaffected extended states determine the transport parameters.

Precise quantization of the Hall resistance to h/ie^2, in spite of the fact that a large fraction of carriers does not participate in transport, is less obvious. It is a consequence of the spatial separation of localized and extended states. Although a large section of a given specimen may not contribute to the macroscopic transport, there remain sufficiently wide regions of extended states joining the external current and voltage probes at its circumference. When the Fermi level resides in the localized regime of a magnetic level, the regions of extended states are indeed filled to an exact multiple i of the degeneracy d (i.e., $n = id$). The localized regions act as a buffer ensuring the quantization. Since the electrical probes sample only the network of extended states, this local density quantization is reflected in the macroscopic transport coefficients as $\rho_{xy} = B/ne = B/ide = h/ie^2$.

Laughlin has provided an elegant *gedanken* experiment[42] proving the quantization of ρ_{xy}. His argument requires the existence of an energy gap separating the extended states of subsequent magnetic levels and the presence of localized states. Following this approach, the Hall resistance reflects, in fact, the ratio of the magnetic flux quantum $\Phi_0 = h/e$ and the quantum of electrical charge e. The IQHE, therefore, unequivocally demonstrates the existence of gaps in the energy spectrum of 2D electrons in a high magnetic field and shows the charge of the current-carrying entity to be precisely the electronic charge e. The gaps are a result of the well-known quantization conditions for independent carriers under the influence of a magnetic field which bring about the familiar sequence of Landau and spin levels (magnetic levels). The charge of the carriers, of course, is expected to be e. It is the significance of the IQHE to have taught us what a Hall experiment in 2D is all about.

THE FRACTIONAL QUANTUM HALL EFFECT

After the IQHE had supposedly uncovered all possible energy gaps of a 2D electron system in a magnetic field, the FQHE came as a big surprise.[1] From the preceding treatment, one expects quantization of the Hall resistance to $\rho_{xy} = h/ie^2$ with i counting the number of occupied Landau and/or spin levels. Nonintegral values of i are not associated with gaps in the energy spectrum of an independent particle picture and are, therefore, inexplicable within the framework of the IQHE. Yet, exactly such quantization of $\rho_{xy} = h/fe^2$ with f being precise rational fractions, and concomitantly vanishing ρ_{xx}, was observed in a high-mobility specimen of GaAs-(AlGa)As heterostructure. Figure 5 shows its first appearance. Below ~ 50 kG, the IQHE is well developed. Plateaus in ρ_{xy} quantized to $\rho_{xy} = h/ie^2$, $i = 1, 2, 3$, and 4 are present in the low-temperature traces accompanied by equivalent stretches of vanishing resistivity in ρ_{xx}. The novel features appear around 150 kG. At this magnetic field the degeneracy of the lowest level, d, has outgrown the carrier concentration n by a factor of three. Consequently, it is filled only to $\nu = 1/3$ of its capacity, and the Fermi energy resides within the level and not in a single-particle gap. In spite of this fact, a plateau develops in ρ_{xy} in the vicinity of $(h/e^2)/(1/3)$ and the resistivity ρ_{xx} shows a deep minimum. The accuracy of the quantum number $f = 1/3$ was independently established[6] at 90 mK to one part in 10^5 and ρ_{xx} was found to be less than 0.1 $\Omega/\square$, both limited by instrumental resolution.

As is evident from Fig. 5, the IQHE and the FQHE are phenomenologically very similar. Quantization of the Hall effect and vanishing resistivity is common to both. The difference is found in the associated quantum numbers, being of integral or rational value respectively. The appearance of fractional quantum numbers and their inexplicability in terms of an independent-particle picture already suggests the manifestation of a new physical phenomenon. These early experiments have stimulated a large body of new work[5-12] and many experiments have since been carried out on higher-mobility samples at lower temperature and higher magnetic field, and on other 2D systems. Fractional quantiza-

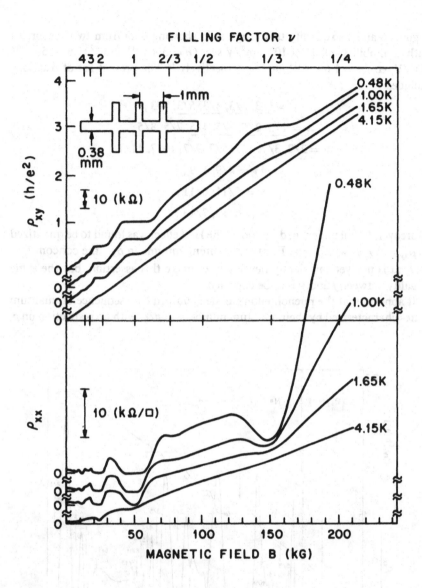

FIGURE 5 First observation of the FQHE in a GaAs-(AlGa)As heterostructure with an areal carrier density of $n = 1.23 \times 10^{11}$ cm^{-2} and a mobility of $\mu = 90\,000$ cm^2/V sec. The Hall resistance ρ_{xy} assumes a plateau at fractional filling $\nu = 1/3$ quantized to $\rho_{xy} = (h/e^2)/(1/3)$ indicating a rational quantum number $f = 1/3$ (Ref. 1).

tion of the Hall effect was found in all high-mobility materials for electrons and holes alike. More importantly, it has been established that the phenomenon is much more general and not limited to a quantum number of $f = 1/3$ at fractional filling of $\nu = 1/3$. To date, a whole sequence of rational quantum numbers f apparent in the vicinity of fractional filling $\nu \sim f$ has been uncovered.

Figures 6 and 7 exemplify this progress in showing data from two specimens with a mobility of 1.3×10^6 cm^2/V sec ($n = 3 \times 10^{11}$ cm^{-2}) and 5×10^6 cm^2/V sec ($n = 1.6 \times 10^{11}$ cm^{-2}) respectively.[12] The present list of fractions includes:

$$v = 1/3,\ 2/3,\ 4/3,\ 5/3,\ 7/3\ ,$$
$$v = 1/5,\ 2/5,\ 3/5,\ 4/5,\ 7/5,\ 8/5\ ,$$
$$v = 2/7,\ 3/7,\ 4/7,\ 5/7,\ 9/7,\ 10/7,\ 11/7\ ,$$
$$v = 4/9,\ 5/9\ ,$$
$$v = 5/11,\ 6/11\ ,$$
$$v = 6/13,\ 7/13\ .$$

Moreover, for all underlined fractions, the Hall effect was found to be quantized to $\rho_{xy} = h/fe^2$ with $f = v$. For the remaining minima in ρ_{xx}, the concomitant plateau is not yet sufficiently developed to make this assertion, but the same equality between f and v is to be expected.

It appears that the phenomenon can be identified by a sequence of quantum states characterized by their quantum number $f = p/q$, with no restriction on p,

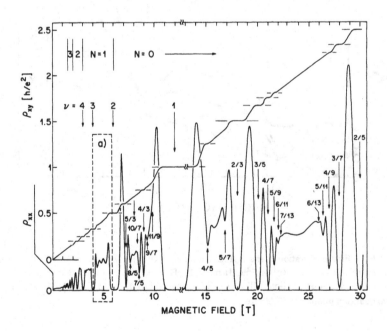

FIGURE 6   Present high-field, low-temperature ($T \sim 0.1$ K) data on the FQHE taken from a high-mobility ($\mu \sim 1.3 \times 10^6$ cm^2/V sec) specimen. The familiar IQHE appears at filling factor $v = 1, 2, 3...$ All fractional numbers are a result of the FQHE. Fractions as high as 7/13 are now being observed. Section a is reproduced in Fig. 9 (Ref. 26).

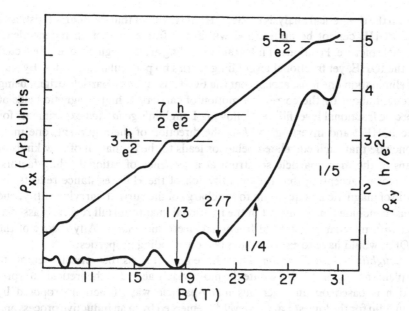

FIGURE 7 The FQHE in a very high-mobility (5×10^6 cm^2/V sec), low-density (1.6×10^{11} cm^{-2}) specimen. The spectrum extends to very low filling factor and demonstrates the existence of a primitive state at fractional filling of $\nu = 1/5$ (Ref. 12).

whereas q is always odd. Experimental research of the past years has elucidated several properties of these quantum states:

1. They occur at fractional filling $\nu \sim p/q$ of Landau levels, as deduced from their position on the B axis.

2. They are associated with a quantum number $f = \nu = p/q$, as determined from the concomitant quantization of the Hall resistance.

3. They are sensitive to disorder. Only low-disorder specimens, as characterized by their carrier mobility, show the FQHE. Progressive improvement in mobility leads to the appearance of more quantum states, generally with higher denominators. Low-mobility samples do not show a FQHE.

4. The FQHE is a low-temperature phenomenon with a characteristic energy scale of only a few degrees kelvin. With increasing temperatures its transport features disappear.

5. There is a tendency for quantum states with higher denominators to show weaker transport features, to require the highest mobility materials and to vanish first when raising the temperature. Such a progression is particularly evident around half filling.

6. Higher magnetic fields promote the observation of the FQHE. At lower fields the FQHE features disappear.

7. In the well-developed quantum states, the resistivity ρ_{xx} vanishes, indicating current flow without dissipation.

From the theoretical analysis of the IQHE it is evident that the characteristics of the FQHE cannot be understood within the framework of an independent-particle picture. Positions of integral level filling are distinguished and reflected in the IQHE, yet fractional level filling carries no particular attributes. By way of elimination one must accept that the FQHE is of many-particle origin, being brought about by the correlated motion of carriers in high magnetic fields at specific fractional level filling. Following Laughlin's *gedanken* experiments for the IQHE[42] and inverting, *ad hoc*, the direction of the argument, one must conclude that such correlated behavior leads to the formation of new kinds of gaps in the many-particle spectrum at a sequence of rational filling factors. Moreover, accepting that the quantization of the Hall resistance reflects the ratio of magnetic flux quantum to the charge of the current-carrying entity, one must conclude that transport in the fractional quantum Hall regime is associated with peculiar particles of fractional electronic charge. Any theory of the FQHE would have to explain these two outstanding properties:

Laughlin's Wave Function. There have been various theoretical attempts to explain the FQHE.[2,48-50] To date the most widely accepted theoretical interpretation is based on an ingenious many-particle wave function proposed by Laughlin for the lowest Landau level.[2] It emerged from an inductive process and is a bold generalization from two- and three-particle correlation in a high magnetic field to the behavior of many such particles. The wave function turns out to be exact for short range interactions and still a very good approximation for the case of coulombic interaction. This is corroborated by sophisticated numerical few-particle calculations[17-22] that find close agreement between the numerically determined pair-correlation function and the one derived from Laughlin's analytical scheme. Laughlin's wave function has been highly successful in explaining the most distinct implications of the FQHE: the existence of energy gaps and of quasiparticles having fractional charge. In its most compact notation the 2D coordinates of all N electrons are expressed as complex numbers $z_1, z_2 \ldots z_n$ in units of the Landau radius $l_0 = \sqrt{\hbar/eB}$.

It is a product over all complex "pair-distances" $(z_i - z_j)$ raised to the mth power. The integer m is identified with a filling factor $v = 1/m$.

$$\psi_m(z_1, z_2, \ldots, z_N) = C \prod_{i<j}^{N} (z_i - z_j)^m \exp\left(-\tfrac{1}{4} \sum_k^N |z_k|^2 \right).$$

The many-electron state described by Laughlin's wave function has the following properties:

1. It is a stable state only at primitive filling factors $v = 1/m$.

2. A case can also be made for $v = 1 - (1/m)$ by assuming electron hole symmetry within each magnetic level.[51]

3. Its pair-correlation function indicates that it is a novel quantum fluid rather than an electron solid for all $m \lesssim 7$.

4. Its elementary excitations are separated from the groundstate by a finite energy gap which increases as $B^{1/2}$ and decreases with increasing m.

5. These quasiparticle excitations are fractionally charged with $e^* = e/m$ and, being neither fermions nor bosons, obey fractional statistics.

6. The quantum fluid has no low-lying excitations. Hence, it flows without dissipation at $T = 0$.

7. For $m \gtrsim 7$ the quantum fluid is expected to crystallize to form an electron solid, still not observed to date.

The characteristic features of the FQHE develop in the following way. At fractional magnetic level occupancy $\nu = 1/m$, the 2D electrons are in the highly correlated quantum fluid state ψ_m which can carry currents without dissipation at $T = 0$. A filling factor $\nu = 1 - (1/m)$ can be treated equivalently, regarding it as a completely occupied, and therefore homogeneously charged, level filled to $\nu = 1/m$ with holes. As the filling factor slightly exceeds $\nu = 1/m$, e.g., by variation of the external magnetic field, some of the electrons are expelled from the quantum liquid in order to maintain its stable configuration. Those additional electrons are promoted above the energy gap and transform into quasiparticles of charge $e^* = e/m$. The fluid state accommodates filling factors slightly below $\nu = 1/m$ by the creation of a few quasiholes, also of fractional charge. The formation of Hall plateaus in the vicinity of primitive fractional filling $\nu = 1/m$ is then caused by localization of these additional quasiparticles. Quasiparticles can be localized by residual potential fluctuations either individually or by forming an electron lattice, which becomes pinned as a whole. In this way the transport coefficient remains constant in the vicinity of $\nu = 1/m$ and $\nu = 1 - (1/m)$. The fractional value of the Hall quantum number is a direct consequence of the fractional quasiparticle charge. The occurrence of the FQHE for filling factor $2 < \nu < 1$, which represents the upper spin of the lowest Landau level, can also be interpreted within this scheme since the orbital wavefunctions of these carriers are identical and the completely filled lower level only provides a uniform background.

Hierarchy. Laughlin's wave function provides an explanation for a FQHE only at primitive fractional occupation $\nu = 1/m$ and $\nu = 1 - (1/m)$. Many pronounced quantum states such as $\nu = 2/5, 3/5, 3/7, 4/7 \ldots$ are not included. Interpretation of these higher-order fractional states required the introduction of a hierarchical model.[13-16] It was recognized that the charged quasiparticles above the primitive rational ground state can once again execute correlated motion. When the filling factor deviates sufficiently from $\nu = 1/m$ so that the ensuing quasiparticles can overcome localization, they are able to form a condensed state among themselves. This state again can be described by Laughlin's wave function where now the coordinates represent the positions of fractionally charged quasiparticles originating from the primitive fractional states. The argument can be repeated *ad infinitum*, if not limited by the formation of the yet unobserved quantum crystal.[52] In this way, the quasiparticles of each new groundstate give rise to a new generation of condensed state. The resulting quantum fluids can be ordered in a hierarchical scheme whose fractional filling factors derive from a continued fraction[13]

$$v = \cfrac{1}{p + \cfrac{\alpha_1}{p_1 + \cfrac{\alpha_2}{p_2 + \cdots}}}$$

where p is odd, the p_i's are even, and $\alpha_i = 0, \pm 1$. This model covers all odd-denominator rational numbers. The hierarchy creates order in the quantum fluids. Any given higher-order state derives uniquely from a sequence of lower-order states which can be traced back to one particular primitive state. Hence, the observation of a higher-order state is expected to require first the observation of the whole sequence of parental states. As an example, Fig. 8 shows the beginning of the hierarchy derived from the 1/3 and 2/3 states. The sequences of states progressing along the edge of the graph and converging toward $v = 1/2$ are beautifully reflected in Fig. 6. So far, in the lowest magnetic level no excep-

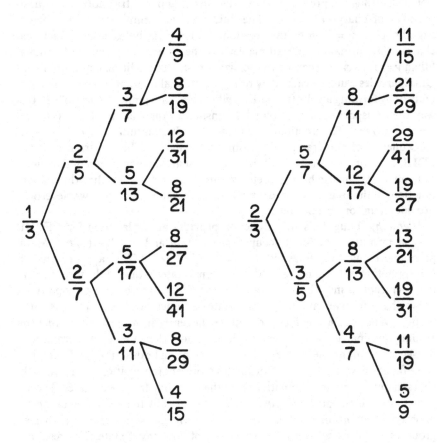

FIGURE 8   The first few higher-order fractional states derived from the primitive 1/3 and 2/3 states.

tion to the hierarchical order of quantum fluids has been found. The most recent reports on the hierarchy in the upper spin state of the lowest Landau level $1 < \nu < 2$ may have detected such an exception,[53] but more data are needed to substantiate these findings.

While the hierarchical model provides an explanation for nonprimitive quantum states and orders their successive appearance, it does not provide an equivalent order for their relative strength. Since higher-order states develop from the quasiparticles of their lower-order parental states, one must expect them to be less robust and show weaker transport features than their parents. This is supported by experiment in the sequences converging towards $\nu = 1/2$. However, in comparing two quantum states in different lines of descent, the hierarchy does not give a clue as to their relative strength. Here Halperin has provided an approximate scheme[14] which entails that the size of the energy gap, i.e., the strength with which a given fractional state $\nu = p/q$ appears in the FQHE, is proportional to $q^{-5/2}B^{1/2}$, where B is the magnetic field. So far this approximate rule could be neither confirmed nor refuted. At present, disorder effects prevent a reliable determination of the energy gaps associated with higher-order fractional states.

EVEN-DENOMINATOR FRACTIONS

The observation of exclusively odd denominators in the FQHE always represented a surprising experimental fact. With the advent of Laughlin's wave function there appears to exist a fundamental reason for the existence of exclusively odd denominators since, within this model, only they ensure antisymmetry of the wave function under electron exchange. From Laughlin's wave function the odd-denominator rule propagates to the hierarchical model which comprises all odd-denominator rational fractions. However, there exists no string of arguments excluding even-denominator quantum numbers in the FQHE on the basis of a general physical symmetry. There may well exist wave functions different from Laughlin's which describe such even-denominator states. Even within Laughlin's model, even denominators are acceptable, although the carriers would have to be bosons rather than fermions. In fact, Halperin[50] and Yoshioka[54] have considered such cases in which electrons are paired to form bosons.

Experimentally, there had been no compelling evidence for even-denominator fractions in the FQHE, although shallow minima in ρ_{xx} at $\nu = 1/2$ and $\nu = 3/4$ had been noted early on.[5,9] These structures were never associated with any inflection in the Hall resistance; ρ_{xy} followed the straight line of the classical Hall effect. The lack of plateaus in ρ_{xy} quantized to the appropriate value makes the association with even-denominator quantization rather unlikely. The broad depression in ρ_{xx} may rather be the consequence of an accumulation[14] of odd-denominator fractional states of the hierarchy. At present there exists no experimental evidence for an even-denominator FQHE in the lowest Landau level $(\nu < 2)$.

Recently, research on the FQHE has been extended to higher Landau levels. Here the most prominent, although weak, minima may be associated with even-denominator fractional filling[55] of $\nu = 9/4$, $5/2$ and $11/4$. The lack of even-denominator quantization in the Hall effect, and a considerable shift of the position of the minima with temperature, renders such an assertion problematic. However, the very latest experimental data[26] on the FQHE in a higher Landau level do provide convincing evidence for an even-denominator fractional state (Fig. 9). At temperatures as low as 25 mK, a plateau is forming in ρ_{xy} at a magnetic field position corresponding to $\nu = 5/2$. The plateau is centered at $\rho_{xy} = (h/e^2)/(5/2)$ to within 0.5%, intersecting the classical Hall line. Simultaneously a deep minimum develops in ρ_{xx}. While not yet fully developed, these features emerge in a manner analogous to conventional odd-denominator fractional states. Taken together these data provide striking evidence for an even-denominator FQHE.

At present there exists no theoretical model for an even-denominator state. So far there has not appeared to be a particularly pressing need to search for analytical expressions that describe such even states. However, it is surprising that the extensive numerical schemes have not uncovered a state at half-filling. Yet, most of these calculations have focused on the lowest Landau level, where even-denominator quantization indeed remains unobserved. The application of numerical schemes to higher Landau levels, whose electronic wave functions differ from those of the ground Landau level, has started only recently and one has to await further results. One obviously may speculate on the origin of even-denominator states. Theory has generally adopted the infinite-field approach, neglecting mixing between Landau levels and spin coupling. Effects of Landau level mixing on the Laughlin ground state have been treated perturbatively. Spin-related modifications of the ground state have been investigated,[56] although not in the context of an even denominator state. It is possible that such inter-level mixing can bring about even denominators. In particular, electron pairs with antiparallel spin configurations come to mind. Such pairs can be regarded as bosons, and application of Laughlin's wave function would require even exponents m resulting in even denominator quantum states. The observation of the 5/2 state at rather low magnetic field, where the Zeeman energy $g^*\mu_B B$ ($g^* = $ effective g factor ≈ 0.4 for GaAs) is small and possibly comparable with typical condensation energies, gives some credibility to speculations of this kind.

It certainly is now evident that the FQHE is not limited to odd-denominator fractions. If the FQHE at odd denominators is any guide, one must expect to find more and possibly different even denominators in the future. It remains to be seen whether a common description can be found or whether, after all, one is dealing with two rather different groundstates.

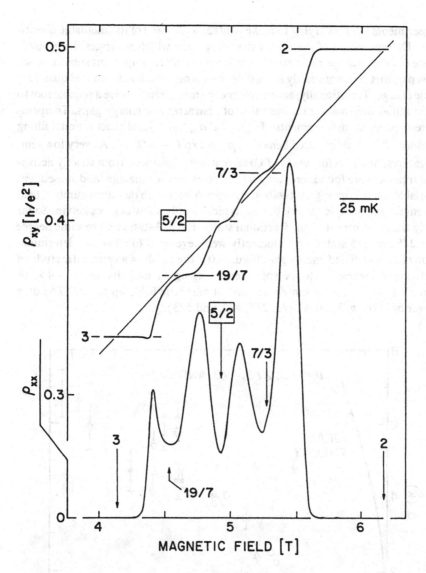

FIGURE 9   The first observation of an even-denominator quantum number in the FQHE [Part (a) of Fig. 6]. At temperatures below 0.1 K a plateau develops in the Hall resistance quantized to $\rho_{xy} = (h/e^2)/(5/2)$ while concomitantly a minimum develops at $\nu \sim 5/2$ level filling (Ref. 26).

ENERGY GAPS

After its initial discovery, experiments on the FQHE focused predominantly on establishing the quantization to rational quantum numbers and determining its qualitative features such as the expanding sequences of odd-denominator fractions. With the emergence of a theoretical description of the ground state, experiments are now trying to make contact with some of its calculable quantities. Determination of the size of the energy gap which separates the ground-state from its quasiparticle excitations is a particularly important subject. Also, this parameter is more easily accessible than others, such as fractional quasiparticle charge. Traditionally, activation energy measurements are a reliable tool to gain initial information on the value of characteristic energy gaps. Temperature-dependent measurements of ρ_{xx} (and σ_{xx}) at several exact rational filling factors[57-62] reveal activated behavior, $\rho_{xx} \propto \exp(-\Delta/2kT)$. At very low temperatures, and also for weaker FQHE features, deviations from strictly activated transport are found, probably related to carrier localization and subsequent variable range hopping. Activation energies Δ vary with quantum number, field strength, and sample quality but are typically only a few degrees kelvin. So far only the most robust of the fractional states, the $p/3$ states and to some degree the 2/5 and 3/5 states, are sufficiently well developed to allow the determination of an associated energy gap. Figure 10 is the result of a systematic study of activation energies[59] in six specimens ranging in mobility from $\sim 4 \times 10^2$ cm^2/V sec to $\sim 1 \times 10^6$ cm^2/V sec and in magnetic fields up to $30T$. The data cover only the $p/3$ states (1/3, 2/3, 4/3, and 5/3).

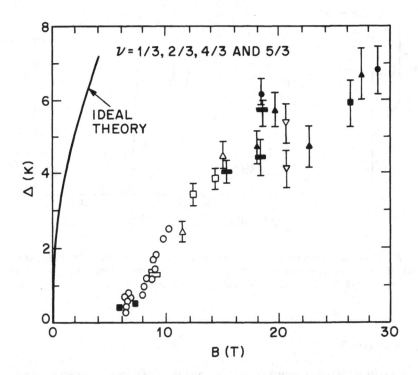

FIGURE 10   The result of a systematic study of activation energies of the $p/3$ state in FQHE on six different specimens of comparable mobility. The data reflect the characteristic scale of the energy gaps in the FQHE. Energy gaps vanish below ~ 6 T and increase with increasing magnetic field. The solid line represents the theoretical result for an ideal 2D system (Ref. 59).

Identifying activation energies with quasiparticle gap energies, the following can be stated for these samples:

1. Within the accuracy of the experiment all $p/3$ states have similar gaps.
2. The gaps have a finite threshold ($\sim 6T$).
3. Gap energies increase with increasing field strength.

The similarity among the gap energies of $p/3$ states with different numerators is gratifying since it reflects approximate electron-hole symmetry within each magnetic level (1/3, 2/3, 4/3, 5/3) and orthogonality of the spin states. The theoretical dependence of the energy gap is $\Delta = Ce^2/\epsilon l_0$, where ϵ is the dielectric constant of GaAs and $l_0 = \sqrt{\hbar/eB}$ is the Landau radius. The value of C has now converged to $C \sim 0.10$ from several theories on the energy gap of the $p/3$ states.[17-22] The full line in Fig. 10 indicates the theoretical B dependence of the energy gap $\Delta = 0.1\ e^2/\epsilon l_0$. The discrepancies with experiments are considerable. They are attributed to the fact that the theory applies only to an infinitely thin two-dimensional system at exceedingly high magnetic field and in the absence of disorder. With the adoption of a few adjustable parameters and treatment of some of the complicating factors in an approximate way, theory has been able to account for the experimental dependence.[63,64] While the general reduction in gap energy is largely a result of Landau level mixing[65] and nonzero wave function extent,[65-67] the finite threshold is caused by residual disorder that reduces the gap by a roughly field-independent constant amount. A theoretical assessment of the effects of disorder is particularly demanding especially since only a single parameter, the carrier mobility at $B = 0$, is accessible to characterize it. Experiments, therefore, aim to further reduce disorder and achieve a more meaningful comparison with theory.

A recent very high-mobility (5×10^6 cm^2/V sec) specimen[23] shows energy gaps which considerably exceed the gap values of earlier measurements (Fig. 11). Moreover, the extent of the electronic wave function parallel to the magnetic field can be estimated. The full line in Fig. 11 represents the result of a theoretical few-particle calculation using such a finite layer thickness.[66] Landau level mixing remains excluded but is considered of secondary importance in the higher field range. Evidently, the agreement between theory and experiment is much improved compared to the earlier data of Fig. 10. The remaining discrepancies, quite confidently, must be attributed to disorder and, to a lesser extent, to Landau level mixing.

As sample quality further improves and disorder is being suppressed, theory can be subjected to a more and more critical inspection. Reduction of disorder is also a necessity for the reliable determination of higher-order energy gaps in order to scrutinize the hierarchical model of the odd-denominator FQHE.

QUASIPARTICLES

Quasiparticles charged to an exact rational fraction of the elemental electronic charge probably represent the most extraordinary aspect of the theory for

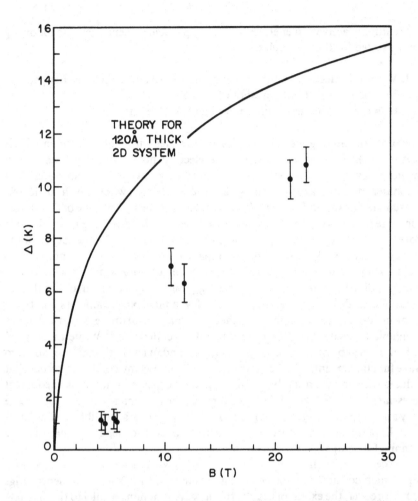

FIGURE 11　Activation energies on the $p/3$ states of the FQHE from the highest mobility specimen presently available ($\mu = 5 \times 10^6$ cm^2/V sec). The associated energy gaps considerably exceed those of Fig. 10. Solid line represents a theoretical calculation for this sample including finite thickness of $32_0 = 120$ Å for the 2D system (Ref. 12).

the FQHE. Apart from their unusual charge state they are also predicted to have a unique dispersion relation exhibiting a roton minimum at finite wavevector.[18,68] This terminology is used due to the phenomenological analogies between the dispersion of the quasiparticles and the dispersion relation of phonons in superfluid He. Experimental determination of the quasiparticle charge is a formidable task. Of course, supposedly the quantum number determined in a quantized Hall experiment already reflects the charge of the current-carrying objects. Yet, one wonders whether there is not more reality to those quasiparticles and whether they might not show up in a more traditional experiment.

Measurements of low-temperature shot noise caused by the statistical fluctuations in the number of carriers may reveal the fractional charge of the quasiparticle. However, in the FQHE at low temperatures the conductivity approaches $\sigma_{xx} \rightarrow 0$ and the associated Johnson noise becomes prohibitively large. Possibilities of determining the quasiparticle charge via the Aharanov–Bohm effect are presently being discussed. It remains unclear whether in a suitably designed experiment the Aharanov–Bohm oscillation due to the interference of 1/3-quasiparticle wave functions would differ from those of normal electrons by a factor of 3 in frequency. It remains even more unclear how to design a suitable experiment.

Most recently Clark *et al.* have made an intriguing observation.[53] Extrapolation of the temperature-dependent conductivity of the $p/3$ states to $T \rightarrow \infty$ seems to converge towards a universal value. This extrapolation was achieved by means of the standard Arrhenius plot used to infer the gap energies. The extrapolated value is found to be in the range 0.013–0.015 $e^2/\hbar$. Comparing this value with the minimum conductivity 0.12 $e^2/\hbar$ earlier proposed for traditional two-dimensional electron systems, one finds a discrepancy of a factor close to $\sim 1/9 = (1/3)^2$. With some courage one might associate this discrepancy with the presence of 1/3 fractionally charged quasiparticles since they are the current-carrying objects rather than real electrons. Certainly, there are many conundrums. For one, theory of the past years has shown that the concept of minimum conductivity does not apply to 2D and therefore the value 0.12 $e^2/\hbar$ already is questionable. Moreover, the theory was not considered to apply in the presence of a magnetic field. However, this is an appealing thought that ought to be pursued.

ELECTRON SOLID

At very low filling factor a quantum solid rather than Laughlin's quantum liquid is found theoretically to be the ground state.[52] The transition occurs at $v \sim 1/7$. The solid state is gapless, loses its incompressibility, and its flow becomes dissipative. Moreover, the quantum solid accommodates any carrier density and consequently does not show preference for a given filling factor. In a way similar to that by which the quantum solid of electrons terminates the sequence of primitive fractional Laughlin states at $v \sim 1/7$, a quantum solid of quasiparticles is expected to terminate the hierarchy when their number required to form the next higher-level state is below a critical value. Carrier transport by the quantum solid is probably pinned—i.e., a few impurities or other disorder can impede the motion of the complete phase and a finite electric field is required to overcome the localization. Nonlinear current/voltage characteristics will probably ensue.

At present this quantum solid has not been observed. In transport experiments its existence is very difficult to differentiate from Anderson localization of individual carriers, and its observation will probably have to await further progress in the reduction of disorder in 2D systems.

OTHER EXPERIMENTS

Tunneling. Low-temperature tunneling experiments have traditionally provided a wealth of information on electronic energy spectra in solid state physics. Their impact on superconductivity, in particular the determination of the associated energy gap and its fine-structure, is well established. The FQHE with its wide range of field- and quantum-number-dependent energy gaps represents a subject ideally suited for the application of this powerful tool. In practice, however, and possibly even in principle, there exist a few severe difficulties. The conductivity of the specimen is an important parameter in tunneling experiments. For tunneling of carriers into a general point of a 2D system, the conductivity σ_{xx} governs subsequent transport of the carriers within the sample. In the FQHE at $T \rightarrow 0$, σ_{xx} vanishes and therefore the 2D system represents an infinite series resistance. This situation becomes equivalent to an attempt to tunnel into an insulator. Tunneling into the edge of a 2D system may avoid such complication. However, apart from the technical difficulties of performing such an experiment, the gap structure at the edge might well differ from the internal one. Moreover, 2D systems with their extremely low carrier concentration are easily electrostatically distorted by weak external potentials.

In spite of these inherent difficulties of practical feasibility, as well as of principle, tunneling experiments have been attempted.[69,70] A considerable reduction in current associated with the presence of a FQHE has been found in several cases. The relative contributions of tunneling inhibition, due to the formation of a gap, and of conductivity loss, due to a vanishing conductivity of the specimen, are still being debated. In any case, the method has not yet allowed a determination of the value of the energy gap.

One would hope that, in spite of the inherent complication, this very powerful experimental tool can be made useful for future studies on the FQHE.

Optical Experiments. Optical measurements with photon energies of ~ 1 eV would seem to be unsuited for the study of carrier correlation effect on an energy scale of just a few degrees kelvin at temperatures as low as a few hundred millidegree kelvin. However, two different groups have recently succeeded in performing such experiments and reported remarkable results.[24,25] The optical experiments consisted of extremely low-power luminescence measurements in which holes are created by optical means in the valence band of the semiconductor. The energetic condition of carriers in the 2D system is sampled by monitoring the radiation emerging from the subsequent recombination of electrons and holes.

Very surprisingly, one of the experiments[24] was performed on a Si-MOSFET on the $\nu = 2 + 1/3$ state at a relatively high temperature and in a rather low magnetic field where transport data showed only very weak structure in ρ_{xx}. In spite of these considerable shortcomings, the recombination radiation in the vicinity of $\nu = 2 + 1/3$ seems to shift substantially on lowering the temperatures from $T \approx 4$ K to $T \approx 1.6$ K, while a similar shift is absent at other filling factors. The authors interpret their result as observation of the energy gap separating the quasiparticles from the many-particle ground state and claim to be able to determine separately the quasihole and quasielectron excitation gaps. Indeed, the extracted values fall within a factor of three of the theoretical ones. These data are presently being passionately debated due to the apparent conflict between strong optical effects and the lack of structure in transport. Moreover, a direct association of the luminescence line shift with the quasiparticle energy gap is not necessarily warranted due to the presence of electron–hole Coulomb interactions which presently cannot be quantified.

This fact was pointed out by a second group[25] that succeeded with optical experiments on the FQHE. Their specimen was of the GaAs-(AlGa)As type, slightly modified to be suitable for an optical probe. With a rather ingenious optical arrangement, the experiment could be conducted at 0.4 K where ρ_{xx} exhibits a deep minimum at $\nu = 2/3$ and $25T$. The low-temperature optical spectra in a wide region around $\nu = 2/3$ reveal a splitting which disappears, as temperature is raised, on a scale characteristic for the FQHE.

These new optical data are very encouraging, although a satisfactory interpretation of their origin has not yet been achieved. There is much hope that an improvement in the quality of specimens that are suitable for optical experiments will allow for higher resolution measurements which may provide further clues to the energy spectrum of 2D electrons in the presence of a magnetic field and strong carrier correlation.

SUMMARY

The discovery of the FQHE has prompted a thorough re-evaluation of carrier correlation in 2D systems in a very high magnetic field. Experiment has successively unraveled a surprising amount of reproducible fine-structure where formerly one had expected uniformity. Theory has now developed an extensive model portraying an intriguing scenario full of fascinating phenomena, such as hierarchically ordered quantum liquids, rational quantum numbers, fractionally charged quasiparticles, quantum solid of electrons and of quasiparticles, fractional statistics and dispersion relations with roton character. With the continuing progress in material science, the generation of increasingly higher magnetic fields and the application of novel experimental tools combined with improved refrigeration schemes to reach successively lower temperatures, one will be able to further explore this fascinating environment and scrutinize our understanding of carrier correlation phenomena in high magnetic fields.

A new challenge to the theoretical model has just been provided by the discovery of the first even-denominator fractional state. One can be confident that there is more to come.

ACKNOWLEDGMENTS

Most of the work presented in this chapter results from a very enjoyable collaboration with my colleagues, D. C. Tsui, J. P. Eisenstein, A. M. Chang, G. S. Boebinger, and R. L. Willett. The invaluable high-quality MBE material was provided by A. C. Gossard, J. H. English, W. Wiegmann, J. C. M. Hwang, A. Y. Cho, and G. Weimann. K. W. Baldwin provided excellent technical support. I would also like to thank R. C. Dynes for helpful advice on the manuscript. A large fraction of the experiments took place at the Francis Bitter National Magnet Lab., Cambridge, Massachusetts.

REFERENCES

1. D. C. Tsui, H. L. Stormer, and A. C. Gossard, Phys. Rev. Lett. **48**, 1559 (1982).
2. R. B. Laughlin, Phys. Rev. Lett. **50**, 1395 (1983).
3. A. Y. Cho, "Artificially Structured Materials," pp. 163–175 in this volume.
4. H. L. Stormer, R. Dingle, A. C. Gossard, W. Wiegmann, and M. D. Sturge, Solid State Commun. **29**, 705 (1979).
5. H. L. Stormer, A. Chang, D. C. Tsui, J. C. M. Hwang, and A. C. Gossard, Phys. Rev. Lett. **50**, 1953 (1983).
6. D. C. Tsui, H. L. Stormer, J. C. M. Hwang, J. S. Brooks, and M. J. Naughton, Phys. Rev. B **28**, 2274 (1983).
7. E. E. Mendez, M. Heiblum, L. L. Chang, and L. Esaki, Phys. Rev. B **28**, 4886 (1983).
8. A. M. Chang, P. Berglund, D. C. Tsui, H. L. Stormer, and J. C. M. Hwang, Phys. Rev. Lett. **53**, 997 (1984).
9. G. Ebert, K. von Klitzing, J. C. Moom, G. Remenyi, C. Probst, G. Weimann, and W. Schlapp, J. Phys. C **17**, L775 (1984).
10. G. S. Boebinger, A. M. Chang, H. L. Stormer, and D. C. Tsui, Phys. Rev. B **32**, 4268 (1985).
11. R. J. Nicholas, R. G. Clark, A. Usher, C. T. Foxon, and J. J. Harris, Solid State Commun. **60**, 183 (1986).
12. R. Willett, H. L. Stormer, D. C. Tsui, A. C. Gossard, J. H. English, and K. Baldwin, Surf. Sci. **196**, 257 (1988).
13. F. D. M. Haldane, Phys. Rev. Lett. **51**, 605 (1983).
14. B. I. Halperin, Phys. Rev. Lett. **52**, 1583 (1984).
15. R. B. Laughlin, Surf. Sci. **142**, 163 (1984).
16. A. H. MacDonald, G. C. Aers, and M. W. C. Dharma-Wardana, Phys. Rev. B **31**, 5529 (1985).
17. D. Yoshioka, B. I. Halperin, and P. A. Lee, Phys. Rev. Lett. **50**, 1219 (1983).
18. F. D. M. Haldane and E. H. Rezayi, Phys. Rev. Lett. **54**, 237 (1985).
19. R. Morf and B. I. Halperin, Phys. Rev. B **33**, 2221 (1986).
20. W. P. Su, Phys. Rev. B **32**, 2617 (1985).
21. G. Fano, F. Ortolani, and E. Colombo, Phys. Rev. B **34**, 2670 (1986).

22. A. H. MacDonald and G. C. Aers, Phys. Rev. B **34**, 2906 (1986).
23. J. H. English, A. C. Gossard, H. L. Stormer, and K. Baldwin, Appl. Phys. Lett. **50**, 1826 (1987).
24. I. V. Kukushkin and V. B. Timofeev, JETP Lett. **44**, 228 (1986).
25. B. B. Goldberg, D. Heiman, A. Pinczuk, C. W. Tu, A. C. Gossard, and J. H. English, Surf. Sci. **196**, 209 (1988).
26. R. Willett, J. P. Eisenstein, H. L. Stormer, D. C. Tsui, A. C. Gossard, and J. H. English, Phys. Rev. Lett. **59**, 1776 (1987).
27. D. C. Tsui and H. L. Stormer, IEEE QE-**22**, 1711 (1986).
28. A recent monograph contains excellent review articles on the FQHE, "The Quantum Hall Effect," in *Graduate Text in Contemporary Physics*, edited by R. E. Prange and S. M. Girvin (Springer, New York, 1987).
29. H. L. Stormer, Surf. Sci. **132**, 519 (1983).
30. T. Ando, A. B. Fowler, and F. Stern, Rev. Mod. Phys. **54**, 437 (1983).
31. K. von Klitzing, G. Dorda, and M. Pepper, Phys. Rev. Lett. **45**, 494 (1980).
32. K. von Klitzing, Festkorperprobleme XXI, *Advances in Solid State Physics*, edited by J. Treusch (Braunschweig, Germany, 1981).
33. M. E. Cage, R. F. Dziuba, and B. F. Field, IEEE Trans. Instrum. Meas. IM-**34**, 301 (1985).
34. H. L. Stormer and D. C. Tsui, Science **220**, 1241 (1983).
35. D. C. Tsui and A. C. Gossard, Appl. Phys. Lett. **38**, 550 (1981).
36. Y. Guldner, J. P. Hirtz, J. P. Vieren, P. Voisin, M. Voos, and M. Razeghi, J. Phys. **43**, L613 (1982).
37. M. Inoue and S. Nakajima, Solid State Commun. **50**, 1023 (1984).
38. E. E. Mendez, L. L. Chang, C. A. Chang, L. Alexander, and L. Esaki, Surf. Sci. **142**, 215 (1984).
39. W. P. Kirk, D. S. Kobiela, R. A. Schiebel, and M. A. Reed, *Proc. 18th Intl. Conf. Phys. Semicon. Stockholm, 1986*, edited by O. Engstrom (World Scientific, Singapore, 1987), p. 497.
40. H. L. Stormer, Z. Schlesinger, A. M. Chang, D. C. Tsui, A. C. Gossard, and W. Wiegmann, Phys. Rev. Lett. **51**, 126 (1983).
41. R. E. Prange, Phys. Rev. B **23**, 4802 (1981).
42. R. B. Laughlin, Phys. Rev. B **23**, 5623 (1981).
43. B. I. Halperin, Phys. Rev. B **25**, 2185 (1982).
44. H. Aoki and T. Ando, Solid State Commun. **38**, 1079 (1981).
45. P. Streda, J. Phys. C **15**, L717 (1982).
46. S. Luryi and R. F. Kazarinov, Phys. Rev. B **27**, 1386 (1983).
47. S. Trugman, Phys. Rev. B **27**, pp. 7539 (1983).
48. S. Kivelson, C. Kallin, D. Avrovas, and J. R. Schrieffer, Phys. Rev. Lett. **56**, 873 (1986).
49. S. T. Chui, T. M. Hakim, and K. B. Ma, Phys. Rev. B **33**, 7110 (1986).
50. B. I. Halperin, Helv. Phys. Acta **56**, 75 (1983).
51. M. Girvin, Phys. Rev. B **29**, 6012 (1984).
52. P. K. Lam and S. M. Girvin, Phys. Rev. B **30**, 473 (1984).
53. R. G. Clark, J. R. Mallett, A. Usher, A. M. Suckling, R. J. Nicholas, S. R. Haynes, Y. Journaux, J. J. Harris, and C. T. Foxon, Surf. Sci. **196**, 219 (1988).
54. D. Yoshioka, Phys. Rev. B **29**, 6833 (1984).
55. R. G. Clark, R. J. Nicholas, and A. Usher, Surf. Sci. **170**, 141 (1986).
56. T. Chakraborty and F. C. Zhang, Phys. Rev. B **29**, 7032 (1984).
57. H. L. Stormer, D. C. Tsui, A. C. Gossard, and J. C. M. Hwang, Physica **117B**, **118B**, 688 (1983).
58. A. M. Chang, M. A. Paalanen, D. C. Tsui, H. L. Stormer, and J. C. M. Hwang, Phys. Rev. B **28**, 6133 (1983).
59. G. S. Boebinger, A. M. Chang, H. L. Stormer, and D. C. Tsui, Phys. Rev. Lett. **55**, 1606 (1985).
60. S. Kawaji, J. Wakabayashi, Y. Yoshino, and H. Sakaki, J. Phys. Soc. Jpn. **53**, 1915 (1984).
61. J. Wakabayashi, S. Kawaji, J. Yoshino, and H. Sakaki, J. Phys. Soc. Jpn. **55**, 1319 (1986).
62. I. V. Kukushkin and V. B. Timofeev, Surf. Sci. **170**, 148 (1986).
63. A. H. MacDonald, K. L. Liu, S. M. Girvin and P. M. Platzman, Phys. Rev. B **33**, 4014 (1985).

64. A. Gold, Phys. Rev. B **33**, 5959 (1986).
65. D. Yoshioka, J. Phys. Soc. Jpn. **55**, 885 (1986).
66. F. C. Zhang and S. Das Sarma, Phys. Rev. B **33**, 2903 (1986).
67. A. H. MacDonald and G. C. Aers, Phys. Rev. B **29**, 5976 (1984).
68. S. M. Girvin, A. H. MacDonald, and P. M. Platzman, Phys. Rev. Lett. **54**, 581 (1985).
69. T. W. Hickmott, Phys. Rev. Lett. **57**, 751 (1986).
70. T. P. Smith, W. I. Wang, and P. J. Stiles, Phys. Rev. B **34**, 2995 (1986).

Modern High-Temperature Superconductivity/ CHING WU CHU

I. INTRODUCTION

Ever since the discovery of superconductivity in 1911, its unusual scientific challenge and great technological potential have been recognized. For the past three-quarters of a century, superconductivity has done well on the science front. This is because superconductivity is interesting not only just in its own right but also in its ability to act as a probe to many exciting nonsuperconducting phenomena. For instance, it has continued to provide bases for vigorous activities in condensed matter science. Among the more recent examples are heavy-fermion systems[1] and organic superconductors.[2] During this same period of time, superconductivity has also performed admirably in the applied area.[3] Many ideas have been conceived and tested, making use of the unique characteristics of superconductivity—zero resistivity, quantum interference phenomena, and the Meissner effect. For example, powerful superconducting magnets and ultrasensitive devices have been constructed and used, sufficiently demonstrating the great technological potential of superconductivity.

However, in spite of this successful demonstration, the full impact of superconductivity on our technology has yet to be realized. This is mainly attributed to the so-called temperature barrier, referring to the unusually low temperature below which superconductivity occurs. In fact, it was not until late January 1987 that it became possible to achieve superconductivity with the mere use of liquid nitrogen—which is plentiful, cheap, efficient, and easy to handle—following the discovery of superconductivity[4] above 90 K in Y–Ba–Cu–O, the first genuine quaternary superconductor. Although the temperature barrier is now broken, the path to full understanding and full-scale application of superconductors at this unusually high temperature can remain long and tedious. Indeed, superconductivity above 90 K poses scientific and technological challenges not

CHING-WU CHU *has worked throughout his career in low-temperature physics, especially in superconductivity, where he and his group have made major contributions. He is Professor of Physics at the University of Houston, Texas, where he is affiliated with the Texas Center for Superconductivity and the Space Vacuum Epitaxy Center.*

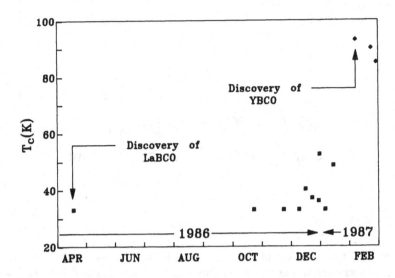

FIGURE 1   Number of papers on oxide superconductivity received by journals during the frantic months immediately after the discoveries of the 30 K and 90 K-superconductivity.

previously encountered: no existing theories can adequately describe superconductivity above 40 K and no known techniques can economically process the materials for full-scale applications. An avalanche of research activities on oxide superconductors during the ensuing frantic months is evident from the growth of the number of papers received by journals each month as displayed in Fig. 1. In this paper, therefore, we would like to recall a few events leading to the discovery of the new class of quaternary compounds with a superconducting transition temperature T_c in the 90 K range, to describe the current experimental status of high-temperature superconductivity and, finally, to discuss the prospect of very-high-temperature superconductivity, i.e., with a T_c substantially higher than 100 K. An excellent review on the discovery of the class of superconductors with a T_c in the 30 K range has previously been given by Müller and Bednorz[5] and a brief account of the discovery and physics of superconductivity above 90 K by us.[6]

II. A NEW CLASS OF SUPERCONDUCTORS WITH T_c IN THE 90 K RANGE

Although "a kick of luck," as pointed out by Alex Müller in his description of the discovery of the 30 K superconductors, is an important ingredient, a breakthrough usually does not occur in a vacuum. What makes the recent exciting rapid development in high-temperature superconductivity possible is the sound and broad knowledge of superconductivity and solids accumulated over the past 75 years by many in the fields of physics, chemistry, materials science, electrical engineering, etc. It would have taken months for scientists to solve the stoichiometry and structure of an unknown compound, instead of only a few days used for the complicated 90 K superconductors, had we not had the expertise and sophisticated analytic tools acquired in the course of studies of other solids over the past decades. In particular, I would like to recognize that the field of high-temperature superconductivity was formally inaugurated by Matthias, Geballe, and Hulm, due to their steady and persistent efforts on the so-called A15 intermetallic compounds.[7] A record T_c of 23.2 K was set[8] by Gavaller and Testardi, separately, in 1973; this record lasted until 1986. The unusual superconductivity in perovskite-based oxides can be traced far back to the pioneering work of Schooley, Hoster, and Cohen, on $SrTiO_{3-x}$ in 1964; Sweedler, Raub, and Matthias on $Na_x WO_3$ in 1965; Johnston, Prakash, Zachariasen, and Matthias on $Li_{1-x} Ti_{2-x} O_4$ in 1973; and Sleight, Gibson, and Bielstedt on $BaPb_{1-x} Bi_x O_3$ in 1975. $BaPb_{1-x} Bi_x O_3$ is particularly unusual because of its relatively high T_c of ~ 13 K, in spite of its extremely low density of states and the absence of any transition metal elements. In fact, more than ten years ago it was this compound that had initiated our interest[10] in superconducting oxides in our long search for high-temperature superconductivity. However, the T_c was not advanced until the exciting report of the observation of superconductivity in the 30 K range in the ternary $(La_{1-x} Ba_x)_2 CuO_4$ (LaBCO) or the so-called 2-1-4 compounds by Bednorz and Müller[11] in 1986. The discovery[4] in late January 1987 of superconductivity above 90 K in the quaternary $YBa_2 Cu_3 O_{7-\delta}$ (YBCO), or the so-called 1-2-3 compounds, drastically changed the scene of superconductivity both scientifically and technologically. We have, therefore, entered a new era of superconductivity, i.e., modern high-temperature superconductivity.

In their researches, Bednorz and Müller first observed the onset of superconductivity at 35 K, and zero resistance at 12 K, in LaBCO prepared by the co-precipitation method. The report came to our attention in the first week of November 1986. Within two weeks, we reproduced the Swiss results in the mixed phase LaBCO samples prepared by the solid-state reaction method. At the same time, we also observed[12] the suppression of the resistive transition by a magnetic field and an approximately 2% ac Meissner effect, characteristic of a superconducting transition. The preliminary resistance results were informally presented at the Materials Research Society Symposium on December 4, 1986. In the ensuing discussion period, Kitazawa of the Tokyo group showed results of their magnetic measurements at low temperature, clearly demonstrating the distinct 14–16% Meissner effect of a superconducting state. In addition, Kitazawa[13] announced that the superconducting component in LaBCO had been positively identified as the $K_2 NiF_4$ or the 2-1-4 phase.[14] On December 5, 1986, Kitazawa presented their results formally at the MRS Symposium. It should be noted that Bednorz and Müller had also pointed out the importance of the 2-1-4 phase in LaBCO in their first paper,[11] and reported about 2% Meissner effect in their mixed phase sample in their second paper,[15] but unfortunately the second report did not reach us until early 1987.

In an attempt to unravel the nature of this unusually high-temperature superconductivity, we examined the LaBCO under pressure. We found[12] that the T_c of LaBCO increases with pressure at an unprecedentedly high rate, almost $100 \times$ that for the previously known superconductors. The T_c in LaBCO was first raised to 40.2 K, then 52.5 K,[16] and finally, to 57 K[6] by the end of 1986, by both high pressure and improved stoichiometry. We learned[6] from these results that: (1) unusual superconducting mechanisms may exist in these oxides; (2) reduction in interatomic distance favors higher T_c; and (3) our faith in the possibility of superconductivity above 77 K was tremendously enhanced since the T_c finally exceeded the theoretically predicted 40 K limit. Both Müller and I are convinced[17] that a qualified faith plays an extremely important role in many discoveries. In the second week of December 1986, in collaboration with Wu at Alabama, we replaced Ba by the smaller Sr and observed superconductivity at 42.5 K without the application of pressure, which was reported in an added note in proof.[12] Unfortunately, further reduction of interatomic distance by the substitution of Ba by even smaller Ca results in a reduction of T_c to 20 K. During this period of time, AT&T Bell Laboratories[18] and the Tokyo[19] groups independently found La-Sr-Cu-O (LaSCO) 2–1–4 compounds superconducting at ~ 36 K with a very sharp transition at ambient pressure. The Beijing group[20] later reported a sharp resistive transition onset in LaSCO at 48.6 K at ambient pressure, which unfortunately dropped back to the 30 K range a few days later when detailed measurements were made. T_c in the high 40 K range in LaSCO at ambient pressure is yet to be confirmed.

By the end of 1986 we had collected a large volume of data. Careful examination of the data also revealed that (1) the onset superconducting transition temperature T_{co}, with a sharp transition at ~ 35 K, always shifted downward when the sample was made as the pure 2–1–4 phase and (2) a sharp resistance drop above 70 K, indicative of a superconducting transition, occurred[6] first on November 25, 1986 and many times later, but only in mixed-phase samples. Furthermore, as mentioned above, no efforts of ours, either physical or chemical, could raise T_c of the 2–1–4 compounds to above 60 K. Consequently, we decided to search for T_c above 70 K in compounds containing phases different from the 2–1–4 one. On January 12, 1987, a multi-phase Ba-rich LaBCO sample, purposely made in a reduced O_2 atmosphere, finally provided[6] sufficient evidence for superconductivity at temperatures up to ~ 100 K when the Meissner effect was detected below 100 K, increasing to 40% of the effect for a bulk superconductor at 4 K. Unfortunately, the signal disappeared completely the next day. By this time, we were convinced that superconductivity near 100 K must exist. The only remaining question to us was how to stabilize it. From the high-pressure study, we learned that smaller atoms tend to favor high T_c. Since La has a large atomic radius, we decided to replace La completely with the smaller but chemically similar Y. Samples of a wide range of compositions were made. Finally, on January 29, 1987, a sharp resistance drop starting at $T_{co} = 93$ K and a zero resistance state at $T_{c1} = 80$ K were unambiguously detected[4], as shown in Fig. 2, in a rather stable multi-phase YBCO sample consisting of two major phases, one black and one green. Next day, careful magnetic mea-

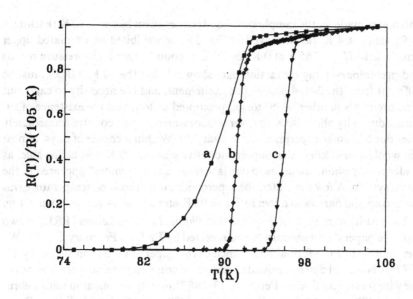

FIGURE 2 *R* vs. *T* for YBCO: (a) 29 January 1987; (b) 2 February 1987; (c) 5 February 1987.

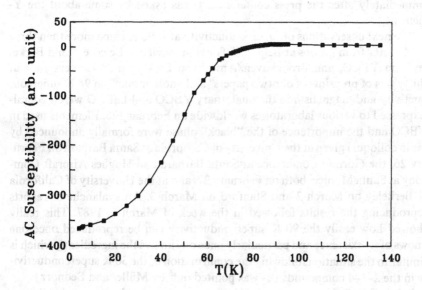

FIGURE 3 χ vs. *T* for YBCO, 30 January 1987.

surements made on the sample showed Meissner effect below ~ 90 K reaching a 24% value at 4 K (as displayed in Fig. 3) and exhibited an estimated upper critical field $H_{c2} \sim 165$ T at 0 K for the compounds. Our high-pressure results and preliminary x-ray diffraction data showed[4] that the 90 K YBCO must be different from the 2–1–4 phase. The excitement, and the necessity to carry out measurements in other laboratories, prompted us to reveal the existence of superconductivity above 90 K to several laboratories in this country immediately after our Meissner experiment on January 30. Within a couple of days, almost the whole world knew that superconductivity above 90 K was achievable, as evidenced by phone inquiries to our laboratory, and "rumors" appeared in the bitnet system. A few days later, the superconducting transition temperature was pushed up and narrowed down to $T_{c0} = 98$ K and $T_{c1} = 94$ K, as shown in Fig. 2. The results were Federal-Expressed to *Physical Review Letters* (PRL) in two separate papers[4] on February 5, and received by PRL on February 6, 1987. We were notified of the formal acceptance of the papers by PRL on February 11, 1987. A National Science Foundation news-release on the discovery was put in the wire service on Friday, February 13, 1987 but with an embargo until February 16, 1987, since Monday, February 16, 1987 was a Federal holiday—President's Day. The arrangement thus gave the time needed for the University of Houston and University of Alabama at Huntsville to arrange simultaneous press conferences on their respective campuses. It is interesting to note that the important use of element Y to stabilize the high-T_c phase was reported by the local newspaper, *The Houston Chronicle*, on both February 16 and 17, 1987. Immediately after the press conference, I was asked by some about the Y-element.

The next observations of superconductivity at ~ 90 K came more than three weeks later from groups at Beijing,[21] Tokyo–Komaba,[22] Belcore,[23] and Berkeley[24] on YBCO, and Brookhaven/Ames[25] on Lu-Ba-Cu-O. However, about thirty sets of preprints of our two papers[4] and another one[26] on 98 K superconductivity and magnetism in the quaternary YBCO and LaBCO were Federal-Expressed to various laboratories worldwide on February 25. Elements used in YBCO and the importance of the "black" phase were formally announced by me in colloquia given at the University of California at Santa Barbara on February 26, the Gordon Conference at Santa Barbara and Hughes Aircraft Company at Santa Monica, both on February 27, and at the University of California at Berkeley on March 2 and Stanford on March 3. An avalanche of reports reproducing the results followed in the week of March 3, 1987. This really showed how easily the 90 K superconductivity can be reproduced once one knows of its existence and particularly once one knows the ingredients, which is similar to the situation following the confirmation of the 30 K superconductivity in the 2–1–4 compounds (as was pointed out[5] by Müller and Bednorz).

In late February the two major phases were subsequently separated and identified:[27] $YBa_2Cu_3O_{7-\delta}$ (black) and Y_2BaCuO_5 (green). We also succeeded in making a single-phase compound and found that only the black phase is superconducting above 90 K. Similar but slightly different structure results were also obtained independently at about the same time by the Bell[28] and IBM[29] groups. Determination of the detailed structure was finally achieved by n-scattering by the Argonne group.[30] Looking back, the observation of 90 K superconductivity definitely involved a "kick of luck," because, as we found out later, the presence of the oxygen-deficient Y_2BaCuO_5 makes the formation of oxygen-rich $YBa_2Cu_3O_{7-\delta}$ much easier.

Immediately following the discovery of the 90 K-YBCO superconductor, we scanned through almost the whole periodic table to find new compounds and to identify the active elements in high-temperature superconductivity. During this period of time, we had learned that the synthesis conditions differ, although only slightly, for samples of different elements or different compositions, in contrast to the conventional A15-superconductors. By fine-tuning the synthesis conditions, and aided by the information on stoichiometry from EDAX, we successfully formed[31] the 1–2–3 Eu-Ba-Cu-O and Lu-Ba-Cu-O compounds and, on March 11, 1987, found them superconducting above 90 K. Within a few days, similar observations[31] were made by us on almost all other rare-earth replacements for Y. Some of these observations were also made independently during this period of time by other groups.[25,32] As a result, a new class of 90 K superconductors was discovered. This provides us with an excellent basis for the understanding of the occurrence of superconductivity in these unusual materials. It should be pointed out that the 1–2–3 LaBCO phase, which gave us the first sufficient evidence for superconductivity above 90 K in a multi-phase sample on January 12, was also stabilized in early March by us. However, the LaBCO 1–2–3 phase is more difficult to form, because of the large atomic size of La. Disorder due to mixing between La and Ba can also be a cause due to their similar atomic sizes. Although T_{c0} can be 95 K, T_{c1} usually is only ~ 75 K. However, we have occasionally observed[33] values of T_{c0} and T_{c1} in LaBCO 1–2–3 compounds as high as 110 and 94 K, respectively. To be able to achieve 90 K superconductivity by using La instead of Y or other rare-earth elements is significant both scientifically and technologically. This is because the contrast between the 2–1–4 and 1–2–3 LaBCO allows us to examine the role of structural symmetry in high-temperature superconductivity, and because the plentiful supply of La_2O_3 in the U.S. and other countries makes future full-scale application more realistic.

Some crucial publications on high temperature superconductivity in its early days are listed chronologically below and displayed in Fig. 4. Only the receiving dates of the articles (instead of the dates reported in press) are included, because of the difference in press-release practice of scientific results in different countries.

30 K Superconductors

35 K superconductivity in LaBCO: J. G. Bednorz and K. A. Müller, Z. Phys. B **64**, 189 (1986)—**Apr. 17, 1986.**

2% - Meissner effect in LaBCO at 4 K: J. G. Bednorz, M. Takashige, and K. A. Müller, Europhys. Lett. **3**, 379 (1987)—**Oct. 22, 1986.**

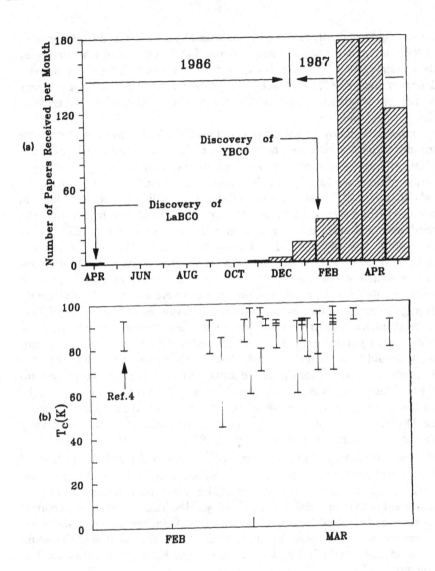

FIGURE 4   (a) T_c vs. time after the discovery of 30 K-superconductivity; (b) T_c vs. time after the discovery of 90 K-superconductivity (all data are from Section II).

10% - Meissner effect in LaBCO at 4 K: A. Uchida, H. Takagi, K. Kitazawa, and S. Tanaka, Jpn. J. Appl. Phys. **26**, L1 (1987)—**Nov. 22, 1986**.

2–1–4 structure of the 30 K LaBCO superconductor: H. Takagi, S. Uchida, K. Kitazawa, and S. Tanaka, Jpn. J. Appl. Phys. **26**, L123 (1987)—**Dec. 8, 1986**.

40.2 K superconductivity in LaBCO under pressure: C. W. Chu, P. H. Hor, R. L. Meng, L. Gao, Z. J. Huang, and Y. Q. Wang, Phys. Rev. Lett. **58**, 405 (1987)—**Dec. 15, 1986**.

37 K superconductivity in LaSrCuO and LaCaCuO: K. Kishio, K. Kitazawa, S. Kanabe, I. Yasuda, N. Sugii, H. Takagi, S. Uchida, K. Fueki, and S. Tanaka, Chem. Lett. **2**, 429 (1987)—**December 22, 1986**.

36 K bulk superconductivity in LaSrCuO: R. J. Cava, R. B. van Dover, B. Batlogg, and E. A. Rietman, Phys. Rev. Lett. **58**, 408 (1987)—**Dec. 29, 1986**.

52.5 K superconductivity in LaBCO under pressure: C. W. Chu, P. H. Hor, R. L. Meng, L. Gao, and Z. J. Huang, Science **235**, 567 (1987)—**Dec. 30, 1986**.

Glassy behavior of LaBCO: K. A. Müller, M. Takashiga, and J. G. Bednorz, Phys. Rev. Lett. **58**, 1143 (1987)—**Jan. 5, 1987**.

Superconductivity in LaSCO (48.6 K) and LaBCO (46.3 K): Z. H. Zhao, L. Q. Chen, C. G. Cui, Y. Z. Huang, J. X. Liu, G. H. Chen, S. L. Li, S. Q. Guo, and Y. Y. He, Kexue Tongbao (Journal of Physics, P.R.C.) **32**, 522 (1987)—**Jan. 14, 1987**.

90 K YBCO Superconductors

93 K superconductivity and high upper critical field $\sim$ 165 T in YBCO: M. K. Wu, J. R. Ashburn, C. J. Torng, P. H. Hor, R. L. Meng, L. Gao, Z. J. Huang, Y. Q. Wang, and C. W. Chu, Phys. Rev. Lett. **58**, 908 (1987)—**Feb. 6, 1987**.

93 K superconducting YBCO not the 2–1–4 phase: P. H. Hor, L. Gao, R. L. Meng, Z. J. Huang, Y. Q. Wang, K. Forster, J. Vassilious, C. W. Chu, M. K. Wu, J. R. Ashburn, and C. J. Torng, Phys. Rev. Lett. **58**, 911 (1987)—**Feb. 6, 1987**.

90 K superconductivity in YBCO: Z. X. Zhao, L. Q. Chen, Q. S. Yang, Y. Z. Huang, G. H. Chen, R. M. Tang, G. R. Liu, C. G. Cui, L. Chen, L. Z. Wang, S. Q. Guo, S. L. Li, and J. Q. Bi, Kexue Tongbao (Journal of Physics, P.R.C.) **32**, 661 (1987)—**Feb. 21, 1987**.

85 K superconductivity in YBCO: S. Hikami, T. Hirai, and S. Kagoshima, Jpn. J. Appl. Phys. **26**, L314 (1987)—**Feb. 23, 1987**.

90 K superconductivity in YBCO: J. M. Tarascon, L. H. Greene, W. R. McKinnon, and G. W. Hull, Phys. Rev. **35**, 7115 (1987)—**Feb. 27, 1987**.

100 K onset of superconductivity in YBCO: L. C. Bourne, M. L. Cohen, W. N. Creager, M. F. Crommie, A. M. Stacy, and A. Zettl, Phys. Lett. A **120**, 494 (1987)—**Feb. 28, 1987**.

98 K superconductivity, metal and magnetism in YBCO and LBCO: C. W. Chu, P. H. Hor, R. L. Meng, L. Gao, Z. J. Huang, Y. Q. Wang, J. Bechtold, D. Campbell, M. K. Wu, J. Ashburn, and C. Y. Huang, Phys. Rev. (1987)—**Mar. 2, 1987**.

90 K superconductivity in YBCO: S. J. Hwu, S. N. Song, J. Thiel, K. R. Poeppelmeier, J. B. Ketterson, and A. J. Freeman, Phys. Rev. B **35**, 7119 (1987)—**Mar. 3, 1987**.

90 K superconductivity in YBCO: J. Z. Sun, D. J. Webb, M. Naito, K. Chan, M. R. Hahn, J. W. P. Hsu, A. D. Kent, D. B. Mitzi, B. Oh, M. R. Beasley, T. H. Geballe, R. H. Hammond, and A. Kapitunik, Phys. Rev. Lett. **58**, 1574 (1987)—**Mar. 5, 1987.**

Glassy behavior in YBCO: P. H. Hor, R. L. Meng, C. W. Chu, M. K. Wu, E. Zirngiebl, J. D. Thompson, and E. Y. Huang, Nature **326**, 669 (1987)— **Mar. 18, 1987.**

1-2-3 Structure of YBCO

R. J. Cava, B. Batlogg, R. B. van Dover, D. W. Murphy, T. Siegrist, J. P. Remeika, E. A. Rietman, S. Zahurak, and G. P. Espinosas, Phys. Rev. Lett. **58**, 1676 (1987)—**Mar. 5, 1987.**

R. H. Hazen, L. W. Finger, R. J. Angel, C. T. Prewitt, N. L. Ross, H. K. Mao, C. G. Hadichiacos, P. H. Hor, R. L. Meng, and C. W. Chu, Phys. Rev. B **35**, 7238 (1987)—**Mar. 10, 1987.**

P. M. Grant, R. Beyers, E. M. Engler, G. Lim, S. S. P. Parkin, M. L. Ramirez, V. Y. Lee, H. Nazzal, J. E. Vasquez, and R. Savoy, Phys. Rev. B **35**, 7242 (1987)—**Mar. 10, 1987.**

T. Siegrist, S. Sunshine, D. W. Murphy, R. J. Cava, and S. M. Zahurak, Phys. Rev. B **35**, 7137 (1987)—**Mar. 16, 1987.**

M. A. Beno, L. Soderholm, D. W. Capone II, D. G. Hinks, J. D. Jorgensen, I. K. Schuller, C. U. Segre, K. Zhang, and J. D. Grace, J. Appl. Phys. Lett. **5**, 57 (1987)—**Mar. 27, 1987.**

90 K Superconductivity in $ABa_2Cu_3O_{7-\delta}$

A = Lu (multiphase): A. Moodenbaugh, M. Sueno, T. Asarro, R. N. Shelton, H. C. Ku, R. W. McCallum, and P. Klavins, Phys. Rev. Lett. **58**, 1885 (1987)—**Mar. 2, 1987.**

A = Yb and Lu (multiphase): S. Hosoya, S. Shamoto, M. Onoda, and M. Sato, Jpn. J. Appl. Phys. **26**, L325 (1987)—**Mar. 9, 1987.**

A = Yb (multiphase): K. Kitazawa, K. Koshio, H. Takagi, T. Hasegawa, S. Kanabe, S. Uchida, S. Tanaka, and K. Fueki, Jpn. J. Appl. Phys. **26**, L339 (1987)—**Mar. 10, 1987.**

A = Ho: S. Hikami, S. Kagoshima, S. Komiyama, T. Hirai, H. Minami, and T. Masumi, Jpn. J. Appl. Phys. **26**, L347 (1987)—**Mar. 11, 1987.**

A = Gd, Sm, Eu, Tb, Dy, Ho (multiphase): Z. Fisk, J. D. Thompson, E. Zirngiebl, J. L. Smith, and S. W. Cheong, S. S. Comm. **62**, 743 (1987)— **Mar. 13, 1987.**

A = Eu: D. W. Murphy, S. Sunshine, R. B. van Dover, R. J. Cava, B. Batlogg, S. M. Zahurak, and F. Schneemeyer, Phys. Rev. Lett. **58**, 1888 (1987)— **Mar. 13, 1987.**

A = La, Y, Nd, Sm, Eu, Gd, Ho, Er, Lu: P. H. Hor, R. L. Meng, Y. Q. Wang, L. Gao, Z. J. Huang, J. Bechtold, K. Forster, and C. W. Chu, Phys. Rev. Lett. **58**, 1891 (1987)—**Mar. 16, 1987.**

A = Er, Y: H. Takagi, S. Uchida, H. Sato, H. Ishi, K. Kishio, K. Kitazawa, K. Fueki, and S. Tanaka, Jpn. J. Appl. Phys. **26**, L601 (1987)—**Mar. 16, 1987.**

A = Tm: S. Kanabe, T. Hasegawa, M. Aoki, T. Nakamura, H. Koinuma, K. Kishio, K. Kitazawa, H. Takagi, S. Uchida, S. Tanaka, and K. Fueki, Jpn. J. Appl. Phys. **26**, L613 (1987)—**Mar. 20, 1987.**

Partial Substitution of Y by Rare-Earth Elements in $ABa_2Cu_3O_{7-\delta}$

A = Y-Eu, Y-La, Y-Sc, Eu-Sc: D. W. Murphy, S. Sunshine, R. B. van Dover, R. J. Cava, B. Batlogg, S. M. Zahurak, and F. Schneemeyer, Phys. Rev. Lett. **58**, 1888 (1987)—**Mar. 13, 1987.**

A = Y-Dy, Y-Lu, Lu-Dy, Lu-La, Lu-Y: Z. X. Zhao, L. Q. Chen, Q. S. Yang, Y. Z. Huang, G. H. Chen, R. M. Tang, G. R. Liu, Y. M. Ni, C. G. Cui, L. Chen, L. Z. Wang, S. W. Guo, S. L. Li, J. Q. Hua, and C. C. Wang, Kexue Tongbao (Journal of Science, P.R.C.) **11**, 817 (1987)—**Mar. 27, 1987.**

III. CURRENT EXPERIMENTAL STATUS OF HIGH-TEMPERATURE SUPERCONDUCTIVITY

Over the past seven months, immediately after the discovery[4] of superconductivity above 90 K in $ABa_2Cu_3O_{7-\delta}$ overcame the temperature barrier for practical applications of superconductivity, the intensity of research on the 1–2–3 oxide superconductors has been extremely high, almost unseen throughout the history of science. Extensive experimental studies have been carried out in various laboratories worldwide. They yielded results on this class of materials that were often uncharacteristic of the conventional inter-metallic superconductors with T_c below 23 K. Many theoretical models[34] have in consequence been proposed to account for the observations. They span a wide spectrum of ideas, ranging from the enhanced phonon-mediated electron-pairing to the delocalization of highly correlated magnetic singlet states. Various mechanisms have been proposed for enhanced superconducting pairing. However, due to the complications partially associated with the youth and the instabilities of the 1–2–3 high-T_c compounds, the experimental situation is far from being completely clear. Consequently, there are as yet no sufficient experimental data to distinguish unambiguously one model from another. For further experimental and theoretical development, it appears to be helpful to briefly summarize the "knowns" and "unknowns" on this wonderful class of oxide materials.

A. The "Knowns"

1. The Superconducting Transition Temperature T_c

As mentioned earlier, $ABa_2Cu_3O_{7-\delta}$ with A = Y, La, Nd, Sm, Eu, Gd, Dy, Ho, Er, Tm, Yb, and Lu are all superconducting. When prepared properly according to the published recipe, the compounds with an orthorhombic symmetry exhibit a stable and reproducible mid-point T_c in the 90 K to 100 K range, although deviation of resistance from linear temperature dependence has been observed at temperatures as high as ~ 200 K in some samples by us and others. There exists[6,31] no clear correlation between the T_c and the $4f$ electron configuration of the A-elements. However, we found[6,35] that high T_c and narrow transition seem to appear more easily for Eu and Gd. This implies[36,37] that there may exist an optimal size of the A-element for the high T_c 1–2–3 compound to form, as suggested from our high-pressure studies; i.e., neither oversized nor undersized A-atoms facilitate the formation of the 1–2–3 superconductor. This is consistent with the difficulty in forming the high T_c $LaBa_2Cu_3O_{7-\delta}$, due to the similar atomic size of La and Ba or the large size of La compared with that of Y. Usually, $T_{c0} \sim 95$ K and $T_{c1} \sim 75$ K, although occasionally a $T_{c0} \sim 110$ K and a $T_{c1} \sim 94$ K have been obtained by us.

Recently, superconductivity with a T_c between 80 K and 90 K has also been reported by one group[38] in the 1–2–3 compounds for A = Ce, Tb, and Pr after repeated pulverization and sintering; this completes the series of $ABa_2Cu_3O_{7-\delta}$ with A = Y, La and all rare earth elements. However, we ourselves have not been able to observe superconductivity in any of these three compounds, although the 1–2–3 phase for A = Pr is achieved.

Up to the time of this report, no effort of ours has been successful in making the correct phase of $ASr_2Cu_3O_{7-\delta}$.

2. The Upper Critical Field $H_{c2}(0)$

By measuring the magnetic field effect on the superconducting transition and using the WHH (Werthamer, Helfand, and Hohenberg) theory, the upper critical field at 0 K, $H_{c2}(0)$, for the 1–2–3 high T_c superconductors has been estimated[4,39] to be 150 to 200 T, the highest among all known superconductors. However, it should be pointed out that the WHH theory is based on a weak-coupling limit, which may no longer be valid for the present high T_c materials. A real experimental determination of $H_{c2}(0)$ at 0 K is presently out of the question, because it exceeds the highest field available in any laboratory.

3. Coherence Length $\xi(0)$

From H_{c2} one estimates[39,41] the coherence length at 0 K, $\xi(0)$, to be 10–30 Å, comparable to the unit cell dimension, although recent measurements[42] on thin films gave a value of 2 Å. The inherent weakness in extracting $\xi(0)$ is the same as that mentioned above for estimating $H_{c2}(0)$. However, if the estimated value of $\xi(0)$ is correct even qualitatively, this implies that there may exist intrinsic flux pinning due to the microscopically heterogeneous nature of the material, and that large critical current density (J_c) must be possible.

4. Critical Current Density, J_c

For bulk sintered samples, the maximum critical current density J_c reported[43] remains at $\sim 10^3$ A/cm^2 at 77 K in zero magnetic field. However, recently $J_c \sim 10^5$ to 10^6 A/cm^2 has been obtained[44] in epitaxially grown films on single crystalline SrTiO$_3$ substrates. This suggests that there exists no intrinsic limit to J_c of the 1–2–3 compounds for practical applications. Unfortunately,[45] at temperatures near T_c, J_c is drastically suppressed initially by a magnetic field.

5. Carrier Concentration n

The charge carriers in the 1–2–3 90 K compounds are of the p-type, similar to the 2–1–4 compounds. The carrier concentration n determined[46] from the Hall measurements is $\sim 10^{21}$/cm^3, again similar to the 2–1–4 30 K superconductors and the Ba-Pb-Bi-O 13 K superconductors. This low value of n is about one to two orders of magnitude smaller than that for the conventional high-temperature superconductors (except Nb$_3$Ge) with a $T_c < 23$ K. In addition, the Hall constant for the 1–2–3 superconductors displays a pronounced temperature dependence, attributable to the two-band effect, i.e., the Cu-O planes and the Cu-O chains.

6. Electron Densities of States N(0)

The electron density of states at the Fermi surface, $N(0)$, is proportional to the Sommerfeld coefficient γ of the electron specific heat of a solid in its normal state at low temperature. Direct measurement of γ for the 1–2–3 superconductors is not possible because the T_c is high and the superconductivity cannot be quenched at low temperature by any magnetic field available in the laboratory. Indirect estimation can be obtained[47] from the Pauli susceptibility χ or the temperature dependence, $- dH_{c2}/dT$, of $H_{c2}(T)$, without questioning the validity of their relationships to the γ of the high T_c material. Again, the relationship between γ and dH_{c2}/dT was derived on the basis of the weak coupling limit. The value of γ so obtained varies between 3 and 14 mJ(mole Cu)$^{-1}$K^{-2}, more than one order of magnitude smaller than that for the conventional superconductors. This is consistent with the low $N(0)$ inferred from photoemission experiments.[48]

7. The Universal Gap Parameter $2\Delta/kT_c$

The superconducting energy gap can be determined from measurements of the single-particle tunneling, the infra-red absorption, or the specific heat below T_c. Values of $2\Delta/kT_c$ so obtained[49,50] range from 1.3 to 20 in contrast to the 3.5 predicted by the BCS theory. In general, the infra-red absorption experiments give a lower value while the tunneling experiments give a higher one. Furthermore, the tunneling experiments sometimes exhibit multi-gaps. The unusual observations have been attributed partially to the anisotropic nature of the compounds. But the exact reason remains unknown. Possible resonance scattering of electrons between the Cu-O layers may lead to the appearance of multiple gaps.

The appearance of a linear T-term in the low-temperature specific heat below T_c of 1–2–3 compounds has raised questions concerning the very existence of Δ. However, we believe that the linear term may just be associated with an impurity phase and thus be an extrinsic character, or it may be associated with the microscopically heterogeneous nature of the 1–2–3 compounds.

8. Glassy Behavior

The magnetic properties of the 1–2–3 compounds were found[51] to depend on the history of the samples. Temperature-induced switch and time-dependence of the magnetization following warming have been observed, suggesting a glassy behavior of the type predicted for a granular superconductor. Similar behavior was reported[52] earlier for the 2–1–4 compounds.

9. Layered Structure

The x-ray diffraction experiments[27-29] on both the poly- and micro-single-crystals showed that the 1–2–3 compounds $ABa_2Cu_3O_{7-\delta}$ have a triple-layer perovskite-like structure, consisting of layers of Cu-O, Ba-O, and Y. After the determination of the O-locations by neutron diffraction,[30] the layer sequence is shown to be $CuO_{1-\delta}$: BaO: CuO_2: Y: CuO_2: BaO. The structure[53] has an orthorhombic symmetry when $\delta < 0.5$ and tetragonal symmetry when $\delta > 0.5$. In the orthorhombic structure, Cu and O form linear chains along the b-axes. Superconductivity has been suggested to occur only in the orthorhombic phase. Early reports of superconductivity in the tetragonal phase may be due to the poor resolution of the x-ray data. However, recent detailed data in our laboratory indicate that superconductivity indeed occurs only in the orthorhombic phase of undoped 1–2–3 compounds and that the orthorhombic phase does not have to be superconducting. On the other hand, in Fe-doped YBCO samples,[54] superconductivity has also been detected in the tetragonal phase although with a lower T_c. The T_c has been found to depend critically on the oxygen content. In addition, we observed[6,55] that there exists a discontinuous drop in T_c from ~93 K to ~60 K in $EuBa_2Cu_3O_{7-\delta}$ as the sample was heat-treated in vacuum for a fixed amount of time between 200 and 600 °C; this is in agreement with results obtained by the gas-getter technique.[56] Typical results are displayed in Fig. 5. Since it has been shown that evacuation removes oxygen preferentially from the Cu-O chains, the discontinuous change in T_c with evacuation demonstrates the importance of Cu-O chains to the 90 K superconductivity.

10. Location of Superconductivity

The complete replacement of Y in the $YBa_2Cu_3O_{7-\delta}$ compounds by the magnetic rare-earth elements without suppressing the T_c strongly demonstrates the negligibly small coupling between the A-layers and their surroundings. Recently, we have also measured the specific heat at low temperature calorimetrically[57] for both the superconducting and nonsuperconducting $GdBa_2Cu_3O_{7-\delta}$ samples. The latter was obtained by removing the oxygen from the former by evacuation at ~400 °C. We found that the low-temperature specific heats for both samples are identical, as shown in Fig. 6, where the anomaly is associated with the antiferromagnetic transition. The noninterference of the superconductivity with the antiferromagnetism again shows the almost complete isolation of the A-atoms from their surroundings. Therefore, superconductivity must be confined to the three Cu-O layer regions as shown in Fig. 7. The presence of the

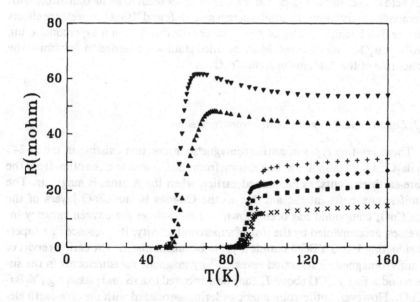

FIGURE 5 R vs. T for Eu-Ba-Cu-O vacuum-treated for 30 minutes at x—25, ■—100, ◆—200, + —250, ▲—300, and ▼—400 °C.

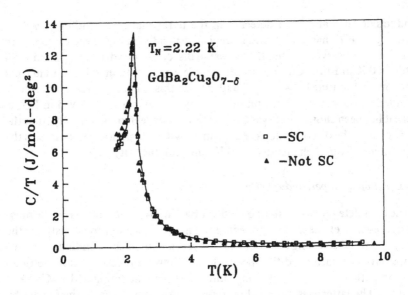

FIGURE 6 C_p vs. T for the superconducting and non-superconducting $GdBa_2Cu_3O_{7-\delta}$.

rare-earth atoms appears just to stabilize the 1–2–3 structure. The preliminary observations[58] of the g-shift and the different slopes[59] of the magnetization versus field curves in the Gd-doped YBCO samples seem to be incompatible with the above observations. In addition, recently we found[60] that the pressure effects on the Néel temperatures of the superconducting and nonsuperconducting $GdBa_2Cu_3O_{7-\delta}$ are different. More detailed studies are needed to determine the exact role of the A-atoms in $ABa_2Cu_3O_{7-\delta}$.

11. Superconductivity and Antiferromagnetism

There are two types of antiferromagnetic interaction existing in the 1–2–3 $ABa_2Cu_3O_{7-\delta}$ compounds: one arising from the Cu-ions and the other from the rare-earth elements, as mentioned earlier, when the A-atom is magnetic. The antiferromagnetic interaction due to the Cu-ions in the Cu-O layers of the La_2CuO_4 compound has been shown[61] to increase as the oxygen vacancy increases, accompanied by the loss of superconductivity. Its presence in a superconducting 1–2–3 YBCO sample is yet to be substantiated. The earlier report of antiferromagnetic interaction revealed[26] by magnetic measurements in the superconducting YBCO above T_c may be attributed to a second phase, e.g., Y_2Ba-CuO_5. However, antiferromagnetic ordering associated with the rare-earth elements in the 1–2–3 compounds has been shown[57] to coexist with superconductivity at ~ 2 K.

$$ABa_2Cu_3O_7$$

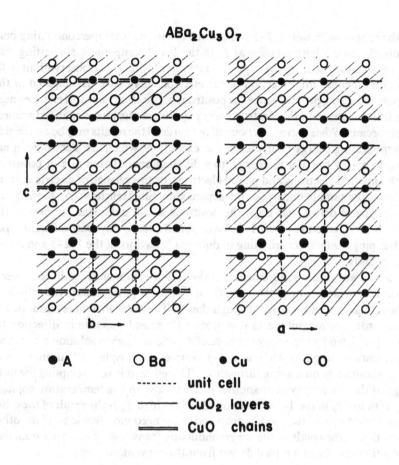

● A   ○ Ba   ● Cu   ○ O

------- unit cell

——— CuO_2 layers

═══ CuO chains

FIGURE 7   Three Cu-O layer regions of 1–2–3 superconductors (the shaded regions).

12. Electron-Pairing

Just as in conventional superconductors, the electrons in the super-conducting state of the 90 K 1–2–3 compounds form pairs, as is evident[62] from experiments on flux quantization and microwave-induced Shapiro steps in the *I–V* characteristics.

13. Substantial Electron and Associated Energy-Spectrum Change at T_c

Recently, the lifetime τ of positrons trapped by oxygen vacancies, and the associated Doppler broadening parameter S, have been measured[63] by us as a function of temperature in the superconducting and nonsuperconducting 1–2–3 YBCO, 2–1–4 LaSCO, and also the conventional 17 K superconductor V_3Si. In contrast to the case of V_3Si, τ exhibits a sharp drop and S a drastic increase at T_c

in the superconducting 1–2–3 sample but not in the nonsuperconducting one, although τ and S both increase at T_c in the 2–1–4 compounds, suggesting that the anomalies arise from the superconducting transition. The observations indicate that a substantial fraction of the electrons may have participated in the superconducting process, since the positrons probe electrons of all energies near the trap-centers, and the electron energy spectrum must have changed accordingly because S has varied. As pointed out earlier,[6] the results may be ascribed to a large scaling energy characteristic of excitations which are electronic in nature, to a very low Fermi energy, to a different annihilation rate of positrons with unpaired electrons and paired electrons, or to a change of electric current paths as the sample undergoes the superconducting transition. At present, the exact type of defect that traps the positrons is still unknown. In view of the complicated structure of the 1–2–3 compounds, there may exist different types of trapping centers, contributing to different behaviors in the 1–2–3 and 2–1–4 compounds.

Recent acoustic measurements on the 1–2–3 compounds exhibited several anomalies: a large drop in the attenuation[64] and a large increase in the velocity[65] when the sample was cooled through the T_c. The anomalies were about two to three orders of magnitude in size greater than, and opposite in direction to, those predicted by thermodynamic considerations. High-resolution x-ray measurements on single-crystalline 1–2–3 samples also displayed[66] an increase in orthorhombicity on cooling through T_c. Therefore, it is very tempting for us to suggest that there may exist another phase transition at a temperature coinciding with the T_c in the 1–2–3 compounds. If the high T_c is the result of the other transition, many of the models proposed so far need modifications. If the other transition is the result of the superconducting transition, the superconducting transition may be of a type different from the conventional one.

14. Isotope Effect

In some of the models proposed, the high T_c in both the 2–1–4 and 1–2–3 compounds has been attributed to the soft-modes associated with the O-bonds. An isotope effect on T_c varying as $M^{-\alpha}$ with $\alpha = 0.5$ is then expected, where M stands for the isotopic mass of oxygen. To determine the effect, samples of $La_{1.85}Sr_{0.15}CuO_4$ and $ABa_2Cu_3O_{7-\delta}$ with both ^{18}O and ^{16}O isotopes have been examined. α was found to be almost zero for the 1–2–3 90 K-superconductor[67] and to be ~ 0.2 for the 2–1–4 30 K-superconductor.[68] These almost zero isotope effects seem to suggest that the conventional electron–phonon interaction cannot play a dominant role in the occurrence of superconductivity, at least in the 90 K-superconductors.

15. Superconductor–Semiconductor–Magnet

It has been demonstrated[26] that both the 1–2–3 and 2–1–4 compounds can be made superconducting, semiconducting, or even magnetic just by varying the oxygen content without altering the chemical elements. Detailed information about these transitions should be invaluable in meeting both the scientific and technological challenges posed by this wonderful class of materials.

B. "Unknowns"

The unusually low $N(0)$ and n in contrast to the high T_c, the very low $\xi(0)$, and the multi-gap, all make the 1–2–3 compounds stand out from the conventional superconductors. The implications of all these unusual characteristics on superconductivity are still unknown. Now we have more unknowns than knowns about the 1–2–3 compounds. Some of these unknowns are listed below.

1. Superconducting Mechanism?

It is known that the electron–phonon interaction is sufficient to account for the superconducting electron pairing in the conventional superconductors. Since the electron–phonon interaction arises from the polarization of the lattice, it becomes stronger when the lattice is softer and thus more polarizable. It is therefore argued that strong electron–phonon interaction will eventually give rise to lattice instabilities, leading to the collapse of the lattice to a form no longer favorable for strong electron–phonon interaction, prior to the attainment of high-temperature superconductivity. In fact, in the conventional superconductors, the evidence for the simultaneous occurrence of high-temperature (> 15 K) superconductivity and lattice instabilities has long been well documented.[69] Recently, band structure calculations[70] on the oxide superconductors did show that a T_c exceeding 40 K became extremely difficult to account for solely on the basis of the electron–phonon interaction. Many nonconventional mechanisms have therefore been advanced to describe the high T_c, especially above 90 K, in the oxide superconductors. They include effects of bipolarons,[71] two-banded plasmons,[72] negative U-centers,[73] charge transfer fluctuations,[74] spin fluctuations,[75] resonating valence bonds,[76] or spin bag[77] for the pairing of electrons. The proposed models follow two general approaches to the superconducting pairing: one begins with free electrons plus pairing interaction between them; the others start from a highly correlated or even localized electron system plus proper interaction which is then delocalized into an itinerant superconducting state. In other words, electrons form superconducting pairs below T_c in one case, whereas they form nonsuperconducting pairs above T_c and then undergo a Bose–Einstein condensation at T_c to the superconducting state in the other case. Presently, there are no sufficient experimental data to differentiate one model from the other. However, based on the existing results, one may safely conjecture that electronic excitations instead of phonons will play a dominant role in the 90 K superconductivity, although the effect of phonons cannot be ignored. The small coherence length, the results of the positron experiment, and the appearance of antiferromagnetic correlation and of the metal–insulator transition in the oxide superconductors, all point to the importance of local micro-environment. This is consistent with the concepts in some of the models, such as charge transfer fluctuations, spin fluctuations, and resonating valence bonds. It should be noted that modification may be needed if later experiments really show another phase transition coinciding with the superconducting one, as suggested by us on the basis of the positron, acoustic, and x-ray experiments discussed earlier.

2. Dimensionality—1, 2, or 3?

The 1-2-3 oxide compounds are layer-like and highly anisotropic. The almost constant and A-independent T_c of $ABa_2Cu_3O_{7-\delta}$ (even if A is strongly magnetic) has led us to the suggestion[6] that the 90 K superconductivity is confined to the three Cu-O layer-assemblies, i.e. CuO_2-BaO-$CuO_{1-\delta}$-BaO-CuO_2, subtended by two A-layers. However, the exact location of superconductivity is still unknown. Several possibilities have thus been suggested concerning the dimensionality of the observed superconductivity: one-dimensional[78] along the $CuO_{1-\delta}$ chains in the b-direction; two-dimensional[79] in the CuO_2-planes perpendicular to the c-axis; or three-dimensional[6] in the combined layer-chain region perpendicular to the c-axis. Most of the models proposed so far are based on a low-dimensional system. In spite of the phase fluctuations inherently associated with a low-(< 3) dimensional system, which are known to be detrimental to the superconducting order, the high $N(0)$ and the soft phonon modes usually accompanying the instabilities of such a system have long been proposed[79] to be potential sources for high-temperature superconductivity. However, the low $N(0)$ in $ABa_2Cu_3O_{7-\delta}$ that is evident[47] from the calorimetric and magnetic measurements, and the lack of drastic mode softening[80] as indicated by the optical and acoustic experiments, appear to be inconsistent with the above proposition, although one may argue that the "real" high $N(0)$ of the $CuO_{1-\delta}$-chains or the CuO_2-planes may have been loaded down because of the inactive parts in the compounds, such as BaO- and Y-layers, owing to the bulk nature of the specific heat measurements to determine $N(0)$. On the other hand, the sensitive dependence of the oxygen content in the $CuO_{1-\delta}$-chains observed[78] in samples rapidly quenched from high temperature seems to support the conjecture that the 90 K-superconductivity may indeed occur in these chains. Unfortunately, later experiments have shown[55,56] that T_c of $ABa_2Cu_3O_{7-\delta}$ is independent of δ for $0 < \delta \leqslant 0.2$, provided that the oxygen removal is carried out at low temperatures by evacuation or the gas-getter technique. In addition, tetragonal $ABa_2Cu_3O_{7-\delta}$ slightly doped with Fe or Zn, where $CuO_{1-\delta}$ chains no longer exist, remains superconducting with a $T_c \sim 80$ K.[54] The relatively large $\xi(0)$ (i.e., > 7 Å, the dimension of the three Cu-O layer assemblies) and the ease in reaching a zero-resistivity state in a polycrystalline sample, are not compatible with a one-dimensional superconducting system. Recent studies on the paraconductivity in the 1-2-3 compounds showed three-dimension fluctuations above T_c, suggesting[81] the three-dimensional nature of the 90 K-superconductivity. The large electrical anisotropy[82] between directions parallel and perpendicular to the c-axis does not rule out the possibility of three-dimensional superconductivity, since electrical conduction can still take place mainly in the three-dimensional chain layer assemblies, but not across them. As indicated earlier, the importance to superconductivity of the oxygen in the $CuO_{1-\delta}$ chains is rather obvious. However, the exact role of the chains remains unknown. Preliminary data of ours indicate that there may exist two T_c's or gaps in the 1–2–3 compounds.

The insensitivity of superconductivity in the 1–2–3 compounds to the A-atoms implies the complete isolation of the A-layers from the superconducting environment.[6,31,32] This is consistent with the same low-temperature specific heat[57] (including the Néel transition at 2.2 K) for both the superconducting and nonsuperconducting $GdBa_2Cu_3O_{7-\delta}$ samples. However, the pressure effects[60] on the Néel transition temperatures for these two types of samples are different. Furthermore, recent neutron scattering data showed[83] that there exists inter-A-layer coupling across the three Cu-O layer assemblies to maintain the stability of the antiferromagnetic structure of the A-atoms. The implication of these observations for the dimensionality of superconductivity is not clear. The short coherent length of the Cooper pairs in comparison with the long correlation length of the antiferromagnetic state further complicates our understanding of the coexistence of superconductivity and antiferromagnetism in the 1–2–3 compounds at low temperature.

3. Valence State of Cu—1+, 2+, 3+ or mixed 2+/3+?

Strictly speaking, the valence state of a certain element in a metallic compound such as the 1–2–3 superconductors loses its ordinary meaning when hybridization of various atomic orbitals occurs, leading to the formation of conduction band or bands. However, for the convenience of discussion and the possible covalency nature of the high-T_c compounds, the concept of valence of the Cu-atoms in the 90 K superconductors is still widely used, especially in the consideration of various electron-pairing mechanisms. One of the most common ways of determining the valence of Cu-atoms in $ABa_2Cu_3O_{7-\delta}$ is by charge-neutrality, since the respective valence states of A, Ba, and O, and the value of δ, are known. Mixed Cu^{2+}/Cu^{3+} valence states of Cu were therefore deduced and suggested[18,28] to be responsible for the observed high T_c. However, $ABa_2Cu_3O_{7-\delta}$ is prone to the formation of defects of different types,[84] e.g., vacancies, antisites, interstices, etc. In the presence of defects of complicated nature, the general count of charge neutrality can no longer provide a reliable value for the valence state of Cu.

In principle, the valence states of Cu-ions in the 2–1–4 and 1–2–3 compounds can be determined by experiments on the x-ray absorption near edge structure (XANES), x-ray photoelectron spectroscopy (XPS), or electron spin resonance (ESR). Conflicting claims concerning the valence states of Cu have been made, sometimes even based on the identical spectra. For example, XANES experiments[85] have resulted in reports of Cu^{2+} and Cu^{3+}, or only Cu^{2+} in the 2–1–4 superconductors; and Cu^{1+}, Cu^{2+}, and Cu^{3+} or mixed Cu^{2+}/Cu^{3+} in the 1–2–3 superconductors. XPS results[86] have led to reports of Cu^{1+}, Cu^{2+} and Cu^{3+}, or mixed Cu^{2+}/Cu^{3+} in the 2–1–4 superconductors and only Cu^{2+}, mixed Cu^{2+}/Cu^{3+}, or Cu^{2+} and $Cu^{2+}O^-$ in the 1–2–3 superconductors. The $Cu^{2+}O^-$ species was supposed to arise from the holes in the conduction band derived from oxygen. The ground state of Cu^{2+} should be a doublet and therefore is expected to display a distinct ESR signal with a g-value around 2.1–2.4. Although the ESR-line corresponding to Cu^{2+} has been observed[6,87] in the 2–1–4 LaBCO and 1–2–3 YBCO compounds, its strength decreases with increase of the size of the superconducting signal. In addition, no ESR-line has been detected[6] in the 1–2–3 Eu–Ba–Cu–O superconductor, for which a perfect or close-to-perfect stoichiometry, i.e., $EuBa_2Cu_3O_7$, is easier to achieve. Consequently, it has been suggested[87] that the well-defined Cu^{2+} valence state with a local moment does not exist in the stoichiometric 1–2–3 90 K superconductors, and that the ESR-signal observed is associated with the Cu^{2+}-ions in the impure phase, e.g., Y_2BaCuO_5, $BaCuO_3$ or the oxygen-deficient 2–1–4 or 1–2–3 compounds. The absence of Cu^{2+}-ions in the stoichiometric 1–2–3 superconductors is also in agreement with the absence[88] of the Curie-behavior in their magnetic susceptibility below 300 K. A mixed (or fluctuating) Cu^{2+}/Cu^{3+}-state does predict a non-Curie behavior with only a Pauli susceptibility at low temperatures, as in the case of the 90 K-superconductors, but a Curie (or even Curie–Weiss) behavior at high temperature. The loss of oxygen of the 1–2–3 compounds at high temperatures makes the interpretation of the high-temperature magnetic susceptibility data difficult. To a certain extent, we[6] try to understand the different results of ESR vs. XANES or XPS by taking into consideration the possible short lifetime of the Cu^{2+}-state due to its fluctuation to the Cu^{3+}-state and the difference in measuring times (10^{-10} sec for ESR vs. 10^{-19} sec for XANES). The absence of the ESR-line has also been attributed[87] to the itinerant nature of the Cu^{2+}- and/or Cu^{3+}-ions, in direct conflict with their localized character inferred from the XPS experiments.

It should be noted that, for XANES and XPS measurements, the oxide superconductors are exposed to high vacuum and high-energy photon conditions and only a very thin surface (sometimes ~ 5 Å) of the sample is probed. Given the known high mobility of oxygen of the compound (particularly on the surface) and the high sensitivity of the superconducting characteristics to oxygen content, it is reasonable to ask if results obtained by such high-energy microscopic probes are really representative of the properties of the bulk materials. It has therefore been suggested[6] that extreme attention should be paid to the possible strong interaction between samples and characterizing probes and, thus, to the interpretation of results so obtained.

In summary, the valence-state of Cu in the 1–2–3 90 K-superconductors is still an open question, arising from difficulties in material, experimental techniques, and/or interpretation of data.

4. Implications of Partial or Nonexistent Isotope Effect?

The presence or absence of isotope effect on $T_c \propto M^{-\alpha}$ with $\alpha = 0.5$ has been considered to be a rather clean-cut indication of the presence or absence of the conventional electron–phonon interaction for electron pairing. However, it was found that, with few exceptions, most of the conventional low-T_c superconductors display only a partial ($\alpha < 0.5$) isotope effect. For some cases, α is zero[89] (Ru and Zr) or even negative[90] (-2 for α-U under pressure). Without abandoning the electron–phonon interaction, the results have been explained[91] in terms of a large retardation effect and/or a large Coulomb repulsion. It is not clear at the present time if the partial isotope effect in the 30 K-superconductors and the almost complete absence of isotope effect in the 90 K-ones can be understood in a similar way to that for the low-T_c compounds. It appears that a full-fledged electron–phonon interaction is not sufficient to account for superconductivity above 40 K. The introduction of a large Coulomb repulsive term to explain the less-than-normal isotope effect will suppress T_c to an even lower value. However, judging from the perovskite-like crystal structure of the 2–1–4 and 1–2–3 compounds, which is known to harbor complex phonon behavior, it is difficult to ignore a phonon effect completely. We believe that phonons, associated with oxygen–copper bonds not along the $CuO_{1-\delta}$ chains or on the CuO_2 planes, still play an important role in affecting excitations of an electronic nature, responsible for superconductivity at 90 K and possibly above. However, only more experimental and theoretical work can find a more satisfactory answer to the question concerning the roles of phonons in this new class of materials.

5. Glassy Behavior—Intrinsic or Extrinsic?

Glassy superconducting behavior, associated with the percolative character of the material, has been suggested[92] to exist in granular superconductors. Consequently, the question is raised if the glassy behavior detected[51,52] in the 2–1–4 and 1–2–3 superconductors is associated with the percolative nature of an imperfectly processed sample consisting of a nonsuperconducting component (in the grain and/or twin boundaries, for instance) or with the microscopically nonhomogeneous nature of the compounds as discussed earlier. An unambiguous answer will be important to both the understanding of the occurrence of superconductivity and the application of this material. Scientifically, it helps us to pin-point the active constituents and thus the mechanism of superconductivity: technologically it provides information leading to the enhancement of J_c and the reduction of ac loss. Magnetic study on stoichiometric and off-stoichiometric single-crystalline samples will be crucial in providing an answer. Preliminary data of a single-crystalline sample appeared still to exhibit the glassy behavior. However, it is as yet unclear whether or not the sample is perfectly stoichiometric without twinning.

6. Superconductivity and Magnetism?

Both superconductivity and antiferromagnetic correlation originating from Cu-ions have been observed[61] in the 2–1–4 and 1–2–3 compounds with different oxygen content, although not simultaneously. Strong superconducting interaction arising from antiferromagnetic correlation has therefore been suggested in various models such as spin-fluctuations,[75] resonance-valence-bond,[76] and spin bag.[77] However, it has also been shown[61] that the antiferromagnetic interaction diminishes with increase in the oxygen-content of the compound and disappears eventually, followed by the onset of superconductivity. Similar observation in the organic superconductors[2] is taken as evidence for the mutual exclusion of superconductivity and antiferromagnetism (or spin-density-waves), i.e., the former occurs at the expense of the latter under pressure. Therefore, the interesting question arises whether the high-temperature superconductivity is the consequence of antiferromagnetism, or whether both phenomena just occur accidentally in the same material system under different conditions, e.g., different oxygen contents for the oxides and different pressures for the organic superconductors. No definitive answer is yet available.

It should be pointed out that antiferromagnetism due to the ordering of the rare-earth elements in the 1–2–3 compounds is known to coexist with superconductivity at low temperatures but not to interact strongly with superconductivity. Antiferromagnetism of this type is hence not of concern to us in this section.

7. Superconducting and Metal–Insulator Transitions?

Both the superconducting 2–1–4 and 1–2–3 compounds are shown[93] to be very close to a metal–insulator phase boundary. Superconductivity appears when the sample becomes metallic with high oxygen-content but disappears when the sample becomes semiconducting with low oxygen-content. The close proximity of the superconducting compound to the metal–insulator phase boundary, observed previously in other oxide superconductors,[10] led people to the suggestion that mode-softening associated with such a transition may play a significant role in enhanced superconductivity. Unfortunately, determining the exact role of metal–insulator phase-transition in the high temperature superconductivity of oxides is hampered by material problems. For example, superconductivity has been reported in 2–1–4 and 1–2–3 samples which show a semiconducting behavior, but the homogeneity of the samples is still in doubt. The question whether high-temperature superconductivity is the consequence of the close proximity of the 1–2–3 superconductor to the metal–insulator phase-boundary, or whether the two just occur accidentally in the same compound system under different conditions, remains unanswered.

8. Exact Roles of Oxygen and Defects?

It is generally agreed[93] that oxygen content is crucial to the superconducting properties of both the 2–1–4 and 1–2–3 compounds. Although we now know that the oxygen in the $CuO_{1-\delta}$ is least strongly bound, the exact role of oxygen in the enhancement or degradation of superconductivity is unknown. For instance, it is unclear if oxygen affects superconductivity through change in stoichiometry chemically, or through change in crystal symmetry physically. Since the charge carriers in the 2–1–4 as well as the 1–2–3 superconductors are p-types, holes associated with a $Cu^{2+}O^-$ hole-complex have also been proposed[86] for pairing. It is an often observed but neglected fact that oxygen-deficiency reduces the Meissner signal of the samples while it alters other superconducting properties. This implies that an oxygen-deficient sample may be a nonbulk or a two-T_c superconductor even if its oxygen-rich, perfectly stoichiometric, counterpart is a bulk one. Can the 1–2–3 superconductor be a two-superconducting system whose parts are coupled to each other and are strongly influenced by oxygen content? All these add complexity to our understanding of the exact role of oxygen in high-temperature superconductivity.

There are defects of different kinds in the 1–2–3 compounds, e.g., vacancies, antisites, and interstices. Disordered oxygen-vacancies closely associated with the oxygen content and thus with the orthorhombicity are clearly detrimental to superconductivity. Partial substitution of the A-atoms[94] in $ABa_2Cu_3O_{7-\delta}$ suppresses T_c but not significantly, suggesting that interstitial defects may not be so critical to T_c. This may be due to the small ξ in comparison with the electron mean free path. However, antisite defects appear to have a large negative effect on T_c, as evident from the difficulty in making 90 K-LaBCO where both the La- and Ba-atoms have similar atomic radii. Partial substitution of Cu in the 1–2–3 compounds by magnetic and nonmagnetic elements such as Ni, Fe, Zn, etc. has been found to suppress[95] the T_c but sometimes to retain superconductivity even in the tetragonal phase,[95] in contrast to the undoped 1–2–3 compounds whose tetragonal form[78] is not superconducting. More studies are needed to elucidate the defect-role in superconductivity: stoichiometry, symmetry, or linear chain?

9. Chemical Stabilities?

We have found that thermal cycling, gas environment, humidity, temperature, electric field, stress, and strain, often can generate an irreversible change in the normal and superconducting state properties of the nominal 1–2–3 compound samples. We believe that chemical reactions induced by the variation of oxygen content due to the above factors may be responsible for the observed changes. The high mobility of oxygen ions together with the susceptibility to defect formation in these compounds may have been the culprits. The reduction of defects and oxygen diffusion rate by synthesizing the compounds in higher quality and in better forms, i.e., dense, single-crystalline or epitaxial films, was found to enhance the stability and thus to reduce the aging process. By comparing the conditions of synthesis and the properties of materials in the various forms, we concluded that the presence of water vapor and carbon-containing complexes during compound synthesis are detrimental to the stability and to the performance (such as J_c and T_c) of the superconducting products. Controlling the growth of grain boundaries is also important. Detailed studies of the stability problem of the materials are being carried out in our laboratory and are expected to provide crucial insights to the understanding of the chemistry and physics of the materials. As we have pointed out previously, the interest in and the application of this class of materials should not be restricted to their super-conductivity. The high oxygen mobility, high sensitivity of electrical properties to gas environment, susceptibility to defect formation, possible mixed-valence character, and instabilities, make this class and related materials promising candidates for gas sensors, humidity detectors, battery cells, catalysts, etc.

IV. PROSPECTS OF VERY-HIGH-TEMPERATURE SUPERCONDUCTIVITY

Only a few days after we stabilized the 90 K-superconductivity in YBCO, a sharp resistance (R) drop to zero (within our experimental resolution) was detected at 180 K in a multi-phase YBCO sample specifically prepared on a Cu-substrate (which was removed for R-measurements) and rapidly quenched to room temperature in air on February 2, 1987. Unfortunately, the size of the R-drop decreased continuously upon thermal cycling and was accompanied by a growing out-of-phase signal of R, meaning the appearance of a capacitive com-ponent and the deterioration of the sample. The sample became completely semiconducting the next day, without any trace of the R-drop, before any mag-netic measurements could be attempted. No further efforts of ours to date have succeeded in producing a sample with the same zero R-drop. However, we did succeed in obtaining R-drops at temperatures as high as 240 K, but often not to zero (and even less stable in many samples) with stoichiometric and off-stoi-chiometric 1–2–3 compounds of YBCO, Sc-doped YBCO, Gd-Ba-Cu-O, Eu-Ba-Cu-O, and La-Ba-Cu-O prepared not at optimal conditions prescribed for the 90 K-superconductors. These R-drops usually could hardly survive repeti-tive thermal cyclings. A few of these samples also exhibited a small diamagnetic component ($<0.1\%$ for a bulk superconductor). Preliminary results were re-ported first in the February 16 press release and then at the March 16 APS Meeting. Later, laboratories in different parts of the world reported similar observations. A comprehensive review and assessment of the prospect of very-high-temperature superconductivity has already been presented[35] at the XVIII International Conference on Low-Temperature Physics at Kyoto, Japan, in August 1987. The situation in very-high-temperature superconductivity between August and October this year has changed little except that a few more reports and rumors have since surfaced. These have included reports from Aus-tralia[96] of superconductivity at ~140 K in YBCO after being stored in a N_2-atmosphere and, from Philips–North America[97], of superconductivity at tem-peratures up to ~159 K in YBCO and fluorinated-YBCO after being "trained" in a cryostat up to 239.5 K. Therefore, in this section we shall only summarize the main points dealing with this exciting and yet unsettled problem.

There are several common features of the reported superconductivity at very high temperatures: R-drops sometimes to zero but more often not, Meissner effect (always $<0.1\%$ of a bulk superconductor), reverse Josephson effect, poor stability, and poor reproducibility. As a result, it is impossible, to date, to carry out definitive checks on these reports concerning the very existence of superconductivity at very high temperatures. All observations of this kind to date can only be considered as unstable superconducting anomalies (USAs) at best.

In view of the unusual characteristics of the 1–2–3 and related oxide compounds, we have proposed four criteria[6,35] for the establishment of superconductivity, namely (1) zero-resistivity, (2) Meissner effect, (3) high stability, and (4) high reproducibility. By high stability, we mean that the anomaly (whether associated with an intrinsically stable or metastable phase) is stable enough to survive the thermal cycling for definitive diagnoses for superconductivity; and by high reproducibility, we mean that the anomaly is reproducible enough to be observed not just on one sample but in many, and not just by one laboratory but by many. To satisfy these four criteria, the highest T_c at present remains in the 90 to 100 K range. Based on preliminary data in our laboratory, 100 to 120 K appears to be an achievable range, but only by proper and as yet incompletely identified treatments of the 1–2–3 compounds and others. A 10–20 K enhancement in T_c will have a significant impact on applications, even if based on the liquid-nitrogen technology. A word of warning: we should keep in mind the question, "Can these USAs be associated with a new phenomenon, perhaps related to but not identical to the commonly known superconducting phenomenon?"

In spite of the current uncertainty around the existence of very-high-temperature superconductivity, there exists no experimental or theoretical evidence to exclude superconductivity at temperatures substantially higher than 100 K. Based on the existing information, the USAs at very high temperatures occur only in off-stoichiometric 1–2–3 compounds or in the stoichiometric 1–2–3 compounds processed under nonoptimal conditions; but never in the optimally prepared 90 K 1–2–3 compounds. They apparently are independent of the A elements in $ABa_2Cu_3O_{7-\delta}$. The results seem to suggest that there may exist a competition between the 90 K-superconducting phase and the cause for USA's at very high temperatures. The anomalies must be associated with filaments consisting of interfaces, twin boundaries, or a metastable phase at the grain-boundaries. Thermal cycling or aging can change the physical connectivity of the filaments due to the differential thermal expansion coefficients of the filaments and the matrix material, or to the chemical change arising from the introduced stress and strain. We have found[35] that thermal cycling can be either detrimental or advantageous to the observations of USAs, depending on some not-yet-identified subtle parameters.

Currently, we are performing systematic studies on the stability of the existing 1–2–3 compounds by examining the effects of gas environment, the heat treatment, the electric field, vacuum, etc. on the materials. We are also investigating the nature of the superconducting transition and the occurrence of 90 K-superconductivity, searching for dissimilarities and similarities between the 2–

1–4 and 1–2–3 superconducting systems, and attempting to synthesize new high-T_c compounds. Some of the results will have been presented at the Michelson–Morley Centennial Symposium at Cleveland in late October 1987. We strongly believe that information obtained from the above and related studies will shed light on the existence of very-high-temperature superconductivity and may lead to the identification and eventual stabilization of the phase, if it exists.

V. ACKNOWLEDGMENT

I am extremely grateful for the continuous inspiring discussions with and generous preprints from many colleagues in the field. In particular, I would like to thank M. K. Wu, P. H. Hor, R. L. Meng, Y. Q. Wang, C. Y. Huang, Y. C. Jean, L. Gao, Z. J. Huang, J. Bechtold, D. Campbell, T. Lambert, and A. Testa for their collaboration and dedication, without which our discovery and understanding of the 90 K-superconductivity, no matter how preliminary, would not have been possible.

The work is supported in part by the Low Temperature Physics Program of NSF, Grant No. 86-16539, NASA Grant No. NAGW-977, the Texas Center for Superconductivity at the University of Houston, and the T. L. L. Temple Foundation.

REFERENCES

1. For a review, see G. R. Stewart, Rev. Mod. Phys. **56**, 755 (1984); Z. Fisk, H. R. Ott, T. M. Rice, and J. L. Smith, Nature **320**, 124 (1986); P. A. Lee, T. M. Rice, J. W. Serene, L. J. Sham, and J. W. Wilkins, Comm. Cond. Matter Phys. **12**, 99 (1986); P. W. Anderson, Lecture series at Varenna Summer School (July 1987).
2. For a review, see D. Jerome and H. J. Schulz, Adv. in Phys. **31**, 299 (1982); J. M. William and K. Carneiro, Adv. Inorg. Chem. Radiochem. **29**, 249 (1985); D. Jerome and L. G. Caron, Proc. NATO Adv. Study Inst. on Low-Dimen. Cond. Supercond. (Plenum, New York, 1986).
3. For applications, see A. P. Malzemoff, W. J. Gallagher, and R. E. Schwall, ACS Symposium Series (1987).
4. M. K. Wu, J. R. Ashburn, C. J. Torng, P. H. Hor, R. L. Meng, L. Gao, Z. J. Huang, Y. Q. Wang, and C. W. Chu, Phys. Rev. Lett. **58**, 908 (1987); P. H. Hor, L. Gao, R. L. Meng, Z. J. Huang, Y. Q. Wang, K. Forster, J. Vassilious, C. W. Chu, M. K. Wu, and C. J. Torng, Phys. Rev. Lett. **58**, 911 (1987).
5. K. A. Müller and J. G. Bednorz, Science **237**, 1134 (1987).
6. C. W. Chu, Proc. Natl. Acad. Sci. (USA) **84**, 4681 (1987); C. W. Chu, P. H. Hor, R. L. Meng, L. Gao, Z. J. Huang, Y. Q. Wang, and J. Bechtold, Proc. Intl. Workshop on Novel Superconductivity (Berkeley, California, June 22–26, 1987); C. W. Chu, Fed. Conf. Commercial Appl. Superconductivity (Washington, DC, July 28–29, 1987).
7. See, for example, J. K. Hulm, J. E. Kunzler, and B. T. Matthias, Phys. Today **34**(1), 34 (1981).
8. J. R. Gavaler, Appl. Phys. Lett. **23**, 480 (1973); L. R. Testardi, J. H. Wernick, and W. A. Roy, Solid State Commun. **15**, 1 (1974).

9. SrTiO$_{3-x}$—J. F. Schooley, W. R. Hosler, and M. L. Cohen, Phys. Rev. Lett. **12**, 474 (1964); Na$_x$WO$_3$—A. R. Swedler, C. J. Raub, and B. T. Matthias, Phys. Lett. **15**, 108 (1965); Li$_{1-x}$Ti$_{2-x}$O$_4$—D. C. Johnston, H. Perakash, W. H. Zachariasen, and R. Viswanathan, Mater. Res. Bull. **8**, 777 (1973); BaPb$_{1-x}$Bi$_x$O$_3$—A. W. Sleight, J. L. Gibson, and F. E. Bielstedt, Solid State Commun. **17**, 27 (1975).

10. C. W. Chu, S. Huang, and A. W. Sleight, Solid State Commun. **18**, 977 (1976); and C. W. Chu and M. K. Wu, *High Pressure Physics and Technology*, edited by C. Homan, R. K. MacCrone, and E. Whalley (North–Holland, Amsterdam, 1983), p. 3, Part 1.

11. J. G. Bednorz and K. A. Müller, Z. Phys. B **64**, 186 (1986).

12. C. W. Chu, P. H. Hor, R. L. Meng, L. Gao, Z. J. Huang, and Y. Q. Wang, Phys. Rev. Lett. **58**, 405 (1987).

13. S. Uchida, H. Takagi, K. Kitazawa, and S. Tanaka, Jpn. J. Appl. Phys. **26**, L1 (1987); H. Takagi, S. Uchida, K. Kitazawa, and S. Tanaka, Jpn. J. Appl. Phys. **26**, L123 (1987).

14. I. S. Shaplingin, B. G. Kahan, and V. B. Lazarev, Zh. Neorg. Khimii **24**, 1478 (1979); C. Michel and B. Raveau, Rev. Chim. Miner. **21**, 407 (1984).

15. J. G. Bednorz, M. Takashige, and K. A. Müller, Europhys. Lett. **3**, 379 (1987).

16. C. W. Chu, P. H. Hor, R. L. Meng, L. Gao, and Z. J. Huang, Science **235**, 567 (1987).

17. Private Conversations, Intl. Workshop on Novel Mechanisms of Superconductivity (Berkeley, California, June 22–26, 1987).

18. R. J. Cava, R. B. van Dover, B. Batlogg, and E. A. Rietman, Phys. Rev. Lett. **58**, 408 (1987).

19. K. Kishio, K. Kitazawa, S. Kanabe, I. Yasuda, N. Sugii, H. Takagi, S. Ushida, K. Fueki, and S. Tanaka, Chem. Lett. **2**, 429 (1987).

20. Z. X. Zhao, L. Q. Chen, C. G. Cui, Y. Z. Huang, J. X. Liu, G. H. Chen, S. L. Li, S. Q. Guo, and Y. Y. He, Kexue Tongbao **32**, 522 (1987).

21. Z. X. Zhao, L. Q. Chen, Q. S. Yang, Y. Z. Huang, G. H. Chen, R. M. Tang, G. R. Liu, C. G. Cui, L. Chen, L. Z. Wang, S. Q. Guo, S. L. Li, and J. Q. Bi, Kexue Tongbao **32**, 661 (1987).

22. S. Hikami, T. Hirai, and S. Kagoshima, Jpn. J. Appl. Phys. **26**, L314 (1987).

23. J. M. Tarascon, L. H. Greene, W. R. McKinnon, and G. W. Hull, Phys. Rev. **35**, 7115 (1987).

24. L. C. Bourne, M. L. Cohen, W. N. Greager, M. F. Crommie, A. M. Stacy, and Z. Zettl, Phys. Lett. A **120**, 494 (1987).

25. A. Moodenbaugh, M. Sueno, T. Asaro, R. N. Shelton, H. C. Ku, R. W. McCallum, and P. Klavins, Phys. Rev. Lett. **58**, 1885 (1987).

26. C. W. Chu, P. H. Hor, R. L. Meng, L. Gao, Z. J. Huang, Y. Q. Wang, J. Bechtold, D. Campbell, M. K. Wu, J. Ashburn, and C. Y. Huang, Phys. Rev. (1987); results were also included in the Feb. 16 news release, the March 18 APS meeting, and Ref. 6.

27. R. H. Hazen, L. W. Finger, R. J. Angel, C. T. Prewitt, N. L. Ross, H. K. Mao, C. G. Hadichiacos, P. H. Hor, R. L. Meng, and C. W. Chu, Phys. Rev. B **35**, 7238 (1987).

28. R. J. Cava, B. Batlogg, R. B. van Dover, D. W. Murphy, T. Siegrist, J. P. Remeika, E. A. Rietman, S. Zahurak, and G. P. Espinsosa, Phys. Rev. Lett. **58**, 1676 (1987).

29. P. M. Grant, R. Beyers, E. M. Engler, G. Lims, S. S. P. Parking, M. L. Ramirez, V. Y. Lee, H. Nazzal, J. E. Vasquez, and R. Savoy, Phys. Rev. B **35**, 7242 (1987).

30. M. A. Beno, L. Soderholm, D. W. Capone II, D. G. Hinks, J. D. Jorgenson, I. K. Schuller, C. U. Segre, K. Zhang, and J. D. Grace, Appl. Phys. Lett. **51**, 57 (1987).

31. P. H. Hor, R. L. Meng, Y. Q. Wang, L. Gao, Z. J. Huang, J. Bechtold, K. Forster, and C. W. Chu, Phys. Rev. Lett. **58**, 1891 (1987).

32. D. W. Murphy, S. Sunshine, R. B. van Dover, R. J. Cava, B. Batlogg, S. M. Zahurak, and F. Schneemeyer, Phys. Rev. Lett. **58**, 1888 (1987); S. Hosoya, S. Shamoto, M. Onoda, and M. Sato, Jpn. J. Appl. Phys. **26**, L325 (1987); K. Kitazawa, K. Koshio, H. Takagi, T. Hasegawa, S. Kanabe, S. Uchida, S. Tanaka, and K. Fueki, Jpn. J. Appl. Phys. **26**, L339 (1987); S. Hikami, S. Kagoshima, S. Komiyama, T. Hirai, H. Minami, and T. Masumi, Jpn. J. Appl. Phys. **26**, L347 (1987); Z. Fisk, J. D. Thompson, E. Zirngiebl, J. L. Smith, and S. W. Cheong, Solid State Commun. **62**, 743 (1987); H. Takagi, S. Uchida, H. Sato, H. Ishi, K. Kishio, K. Kitazawa, K. Fueki, and S. Tanaka, Jpn. J. Appl. Phys. **26**, L601 (1987); S. Kanabe, T. Hasegawa, M. Aoki, T. Nakamura, H. Koinuma, K. Kishio, K. Kitazawa, H. Takagi, S. Uchida, S. Tanaka, and K. Fueki, Jpn J. Appl. Phys. **26**, L613 (1987).

33. Y. Q. Wang, P. H. Hor, R. L. Meng, L. Gao, Z. J. Huang, Y. Y. Sun, and C. W. Chu (to be published).
34. See, for example, Proc. Intl. Workshop on Novel Superconductivity (Berkeley, California, June 22–26, 1987).
35. C. W. Chu, J. Bechtold, L. Gao, P. H. Hor, Z. J. Huang, R. L. Meng, and Y. Q. Wang, Proc. LT-18, Kyoto, Japan, August 20–26, 1987.
36. R. L. Meng, P. H. Hor, and C. W. Chu (to be published).
37. C. W. Chu, Proc. Yamada Conference on Superconductivity in Highly Correlated Fermion Systems (Sendai, Japan, 31 August–3 September, 1987).
38. N. E. Alekseevskii, Private Communications (July 1987).
39. P. H. Hor, R. L. Meng, Z. J. Huang, C. W. Chu, and C. Y. Huang, Appl. Phys. Comm. 7, 129 (1987).
40. T. P. Orlando, K. A. Delin, S. Foner, E. J. McNiff, J. M. Tarascon, L. H. Greene, W. R. McKinnon, and G. W. Hull, preprint; A. P. Ramirez, B. Batlogg, R. J. Cava, L. Schneemeyer, R. B. van Dover, E. A. Rietman, and J. V. Wanzizak, Proc. Intl. Workshop on Novel Superconductivity (Berkeley, California, June 22–26, 1987).
41. See Ref. 40.
42. T. H. Geballe *et al.*, Proc. LT-18 (Kyoto, Japan, August 20–26, 1987).
43. B. Batlogg, Am. Phys. Soc. Meeting (New York, 18 March 1987).
44. P. Chaudhari, R. H. Koch, R. B. Leibowitz, T. R. McGuire, and R. J. Gambino, Phys. Rev. Lett. 58, 2684 (1987).
45. B. Oh, M. Naito, S. Arnason, P. Rosenthal, R. Barton, M. R. Beaseley, T. H. Geballe, R. H. Hammond, and A. Kapituknik (preprint).
46. Z. Z. Wang, J. Clayhold, N. P. Ong, J. M. Tarascon, L. H. Greene, W. R. McKinnon, and G. W. Hull (preprint).
47. For a summary, see M. V. Nevitt, G. W. Crabtree, and T. E. Klippert, to appear in Phys. Rev.; N. E. Phillips, R. A. Fisher, S. E. Lacy, C. Marcenat, J. A. Olsen, W. K. Hair, A. M. Stacy, J. E. Gordon, and M. L. Tan, Proc. Yamada Conference (Sendai, Japan, August 1987).
48. F. C. Brown, T. C. Chiang, T. A. Friedmann, D. M. Ginsberg, G. N. Kwawer, T. Miller, and M. G. Mason (preprints).
49. For tunneling results on YBCO, see K. E. Gray, M. E. Hawley, and R. E. Moog, Proc. Intl. Workshop on Novel Superconductivity (Berkeley, California, June 22–26, 1987); M. D. Kirk, B. P. E. Smith, D. B. Mitzi, D. J. Webb, K. Char, M. R. Hahn, M. Naito, B. Oh, M. R. Beaseley, T. H. Geballe, R. H. Hammond, A. Kapituknik, and C. F. Quate (preprint); I. Iguchi, H. Watanabe, Y. Kisai, T. Mochiku, A. Sugishita, and E. Yamaka, Jpn. J. Appl. Phys. 26, L645 (1987).
50. See, for example, Z. Schlesinger, R. T. Collins, D. L. Kaiser, and F. Holtzberg (preprint).
51. P. H. Hor, R. L. Meng, C. W. Chu, M. K. Wu, E. Zirngiebl, J. D. Thompson, and C. Y. Huang, Appl. Phys. Comm. 7, 123 (1987); *ibid.* Nature 326, 669 (1987).
52. K. A. Müller, M. Takashige, and J. G. Bednorz, Phys. Rev. Lett. 58, 1143 (1987).
53. For a review, see J. D. Jorgensen, Proc. LT-18 (Kyoto, Japan, August 20–26, 1987); E. Salomons, N. Koeman, R. Brouwer, D. G. de Groot, and R. Griesen (preprint).
54. G. Xiao, F. H. Streitz, A. Gavin, Y. W. Wu, and C. L. Chien, Phys. Rev. B 35, 8782 (1987); Y. Oda, H. Fujita, H. Toyoda, T. Kanchko, T. Kohara, I. Nokada, and K. Asayama, Jpn. J. Appl. Phys. (1987); Y. Maeno, M. Kato, Y. Aoki, and T. Fujita (preprint).
55. C. W. Chu, Z. J. Huang, R. L. Meng, L. Gao, and P. H. Hor (preprint).
56. B. Batlogg, R. J. Cava, C. H. Chen, G. Kourouklis, W. Weber, A. Jayaraman, A. E. White, K. T. Short, E. A. Rietman, L. W. Rupp, D. Werder, and S. M. Zaburak, Proc. Intl. Workshop on Novel Superconductivity (Berkeley, California, June 22–26, 1987).
57. J. C. Ho, P. H. Hor, R. L. Meng, L. Gao, Z. J. Huang, J. Bechtold, and M. K. Wu, Proc. Spring Meet. Mater. Res. Soc. (Anaheim, California, April 21–24, 1987); to appear in Solid State Commun.
58. C. W. Chu, P. H. Hor, R. L. Meng, L. Gao, Z. J. Huang, J. Bechtold, and M. K. Wu, Proc. Spring Meet. Mater. Res. Soc. (Anaheim, California, April 21–24, 1987) and C. Y. Huang, Z. J. Huang, P. H. Hor, R. L. Meng, and C. W. Chu (to be published).

59. C. Y. Huang, L. J. Dries, F. A. Junga, P. H. Hor, R. L. Meng, and C. W. Chu, Proc. Intl. Workshop on Novel Superconductivity (Berkeley, California, June 22–26, 1987).

60. Z. J. Huang, P. H. Hor, R. L. Meng, C. W. Chu, and J. M. Ho (to be published).

61. See, for example, S. Mitsuda, G. Shirane, S. K. Sinha, D. C. Johnston, M. S. Alvarez, D. Vaknin, and D. C. Moncton (preprint).

62. C. E. Gough, M. S. Colclough, E. M. Gorgan, R. G. Jordan, M. Keene, C. M. Muirhead, A. I. M. Rae, N. Thomas, J. S. Abell, and S. Sutton, Nature 326, 855 (1987); W. R. McGrath, H. K. Olsson, T. Claeson, S. Eriksson, and L. G. Johaneson (to appear in Europhys. Lett.).

63. Y. C. Jean, S. J. Wang, H. Nakanishi, W. N. Hardy, M. E. Hayden, R. F. Kiefl, R. L. Meng, P. H. Hor, Z. J. Huang, and C. W. Chu, Phys. Rev. 36, 3994 (1987); and Y. C. Jean *et al.* (to be published).

64. K. J. Sun, M. Levy, B. K. Sarma, P. H. Hor, R. L. Meng, Y. Q. Wang, and C. W. Chu (preprint).

65. D. J. Bishop, A. P. Ramirez, P. L. Gammel, B. Batlogg, E. A. Rietman, R. J. Cava, and A. J. Mills; A. Migliori, T. Chen, B. Alvi, and G. Grüner (preprint).

66. P. M. Horn, D. T. Keane, G. A. Held, J. L. Jordan–Sweet, D. L. Kaiser, and F. Holtzberg (preprint).

67. B. Batlogg, R. J. Cava, A. Jayaraman, R. B. van Dover, G. A. Kourouklis, S. Sunshine, D. W. Murphy, L. W. Rupp, H. S. Chen, A. White, K. T. Short, A. M. Mujsce, and E. A. Rietman, Phys. Rev. Lett. 58, 2333 (1987); L. C. Bourne, M. F. Crommie, A. Zettl, H. C. Zurloye, S. W. Keller, K. L. Leary, A. M. Stacy, K. L. Chang, M. L. Cohen, and D. E. Morris, Phys. Rev. Lett. 58, 2337 (1987); A. Stacy, Fed. Conf. on Commercial Applications of Superconductivity (Washington, DC, July 28–29, 1987).

68. M. L. Cohen, D. E. Morris, A. Stacy, and A. Zettl, Proc. Int. Workshop on Novel Superconductivity (Berkeley, California, June 22–26, 1987); Ref. 56.

69. C. W. Chu, *High Pressure and Low Temperature Physics*, edited by C. W. Chu and J. A. Wollam (Plenum, New York, 1978), p. 359.

70. W. Weber and L. F. Mattheiss, Yamada Conference on Superconductivity (Sendai, Japan, 31 August–3 September, 1987).

71. C. S. Ting and D. Y. Xing, Proc. Intl. Workshop on Novel Mechanisms of Superconductivity (Berkeley, California, June 22–26, 1987), and references therein.

72. J. Ruvalds, Phys. Rev. B 35, 8869 (1987); V. Z. Kresin, Phys. Rev. B 35, 8716 (1987).

73. P. W. Anderson, Phys. Rev. Lett. 34, 953 (1975); C. S. Ting, D. N. Talwar, and K. L. Ngai, Phys. Rev. Lett. 45, 1213 (1980).

74. C. M. Varma, S. Schmitt-Rink, and E. Abrahams, Phys. Rev. Lett. 58, 2691 (1987); J. Yu, A. J. Freeman, and S. Massida, Proc. Intl. Workshop on Novel Superconductivity (Berkeley, California, June 22–26, 1987).

75. P. A. Lee and N. Reed, Phys. Rev. Lett. 58, 2691 (1987); V. J. Emory, Phys. Rev. Lett. 58, 2794 (1987).

76. P. W. Anderson, Science 235, 1196 (1987).

77. J. R. Schrieffer, Fed. Conf. on Commercial Appl. of Superconductivity (Washington, DC, July 28–29, 1987).

78. J. D. Jorgensen, M. A. Beno, D. G. Hinks, L. Soderholm, K. J. Volin, R. L. Hitterman, J. D. Grace, I. K. Schuller, C. U. Segre, K. Zhang, and M. S. Kleefisch, Phys. Rev. B 36, 3608 (1987); R. J. Cava, B. Batlogg, C. H. Chen, E. A. Rietman, S. M. Zahurak, and D. Werder, Nature 329, 423 (1987).

79. L. F. Mattheiss, Phys. Rev. Lett. 58, 1028 (1987); J. J. Yu, A. J. Freeman, and J. H. Zu, Phys. Rev. Lett. 58, 1035 (1987).

80. M. Levy, B. K. Sarma, P. H. Hor, R. L. Meng, Y. Q. Wang, and C. W. Chu (preprint).

81. P. P. Freitas, C. C. Tsuei, and T. S. Plaskett (preprint); C. Laurent, M. Laguesse, S. R. Patapis, H. W. Vanderschueren, and G. Lecomte (preprint); M. Ausloos and C. Laurent, preprint; N. Goldenfeld, P. D. Olmsted, T. A. Friedmann, and D. M. Ginzburg (preprint).

82. T. R. Dinger, T. K. Worthington, W. J. Gallagher, and R. L. Sandstrom, Phys. Rev. Lett. 58, 2687 (1987); S. W. Tozer, A. W. Kleinsasser, T. Penney, D. Kaiser, and F. Holtzberg (pre-

print); G. W. Crabtree, J. Z. Liu, A. Umezawa, W. K. Kwok, C. H. Sowers, S. K. Malik, B. W. Veal, D. J. Lam, M. B. Broadsky, and J. W. Downey (preprint).

83. F. W. Young, LBL Workshop (Berkeley, California, December 13–14, 1987).

84. B. Raveau and C. Michel, Proc. Intl. Workshop on Novel Superconductivity (Berkeley, California, June 22–26, 1987).

85. J. M. Tranquada, S. M. Heald, A. Moodenbaugh, and M. Suenaga, Phys. Rev. B **35**, 7187 (1987); E. E. Alp, G. K. Shenoy, D. G. Hinks, D. W. Capone II, L. Soderholm, H. B. Schattler, J. Guo, D. E. Ellis, P. A. Montano, and M. Ramanathan, Phys. Rev. B **35**, 7199 (1987); H. Oyanagi, H. Ihara, T. Matsubara, M. Tokumoto, T. Matsushita, M. Hirobayashi, K. Murata, N. Terada, T. Yao, H. Iwasaki, and Y. Kimura, Jpn. J. Appl. Phys. **26**, L1561 (1987); A. Bianconi, A. Castellano, M. D. Santis, C. Politis, A. Marcelli, S. Mobilio, and A. Savoia, Z. Phys. B (1987).

86. H. Ihara, M. Hirabayashi, N. Terada, Y. Kimura, K. Senzaki, M. Akimoto, K. Bushida, F. Kawashima, and R. Uzuka, Jpn. J. Appl. Phys. **26**, L460 (1987); H. Ihara, M. Hirabayashi, N. Terada, Y. Kimura, K. Senzaki, and M. Takumoto, *ibid.*, **26**, L463 (1987); P. Steiner, J. Albers, V. Kinsinger, I. Sander, B. Siegwart, S. Hufner, and C. Politis, Z. Phys. B **66**, 275 (1987); *ibid.*, **67**, 19 (1987); A. Bianconi, A. Congiu Castellano, M. de Santis, P. Delogu, A. Gargano, and R. Giorgi, S. S. Comm. **63**, 1135 (1987); D. E. Ramaker, N. H. Turner, J. S. Murday, L. E. Toth, M. Osofsky, and F. L. Hutson (preprint); T. Takahashi, F. Maeda, A. Arai, H. Katayama–Yoshida, Y. Okabe, T. Suzuki, S. Hosoya, A. Fujimori, T. Shidara, T. Koide, T. Miyahara, Mo Onoda, S. Shamoto, and M. Sato (preprint).

87. C. Y. Huang, Z. J. Huang, P. H. Hor, R. L. Meng, C. W. Chu, and L. Kevin (to be published); S. Sugawara, K. Kita, K. Akagi, Y. Taniguchi, I. Nakajima, Y. Kataoka, S., LT-18 (Kyoto, Japan, August 22–26, 1987); T. Kohara, H. Kohara, H. Yamagata, M. Matsumura, Y. Yamada, I. Nakada, E. Sakagami, Y. Oda, and K. Asayama (preprint); H. Thomann, D. C. Johnston, P. H. Tindall, D. P. Goshorn, and R. A. Klemm (preprint); D. Shaltiel, J. Genossar, A. Grayevsky, Z. H. Kalman, B. Fisher, and K. Kaplan (preprint).

88. S. Uchida, H. Tagaki, T. Hasegawa, K. Kishio, S. Tajima, K. Kitazawa, K. Feuki, and S. Tanaka (preprint).

89. See, for example, B. T. Matthias, NRP Report 6962(1969), p. 32.

90. R. D. Fowler, J. D. G. Lindsay, R. W. White, H. H. Hill, and B. T. Matthias, Phys. Rev. Lett. **19**, 892 (1967).

91. J. W. Garland, NRL Report 6962 (1968) p. 1.

92. C. Ebna and D. Stroud, Phys. Rev. B **31**, 165 (1985).

93. See, for example, L. H. Greene *et al.*, LT-18 (Kyoto, Japan, August 1987).

94. J. M. Tarascon, L. H. Greene, B. G. Bagley, W. R. McKinnon, P. Barboux, and G. W. Hill, Proc. Intl. Workshop on Novel Superconductivity (Berkeley, California, June 22–26, 1987).

95. Y. Maeno, T. Tomita, M. Kyogoku, S. Awaji, Y. Aoki, K. Hoshino, A. Minami, and T. Fujita, Jpn. J. Appl. Phys. **328**, 512 (1987).

96. D. N. Matthews, A. Bailey, R. A. Vaile, G. J. Russell, and K. N. R. Taylor, Nature **328**, 786 (1987).

97. R. N. Bhargava, S. P. Herko, and N. N. Osborne, Phys. Rev. Lett. **59**, 1468 (1987).

Superconductivity and its Applications (Modern and Traditional Approaches) / YU. A. OSSIPYAN

INTRODUCTION

The year 1987 has been marked by rapid-fire discoveries in high-temperature superconductivity around the world. They were initiated by the now world-renowned work of K. A. Müller and G. Bednorz,[1] who discovered the superconducting properties in a mixed oxide of lanthanum, barium, and copper, whose transition temperature T_c exceeded by 10 K the highest of any known at that time. Their work stimulated an intensive search for new superconductors with high T_c among complex oxides of rare-earth metals with the layered-perovskite structure, and it was crowned with success. By the efforts of several groups of American[2] and Japanese[3] physicists, the superconducting transition temperature was raised within a very short time to 90–95 K (Y–Ba–Cu–O). Thus 75 days of intensive work in the beginning of 1987 did what many thousands of researchers had failed to do during the past 75 years (1911–1986). I shall not dwell at length on the chronology of the recent events associated with the discovery of high-temperature superconductors (HTSC). They have been reviewed in this conference by Prof. C. W. Chu, who was one of the most active participants in these events. I shall only show a graph which, it seems to me, strikingly reflects both the dynamics and the tenseness of the recent events in this field (Fig. 1). Today the problem of high-temperature superconductivity and its applications is, undoubtedly, the most popular in current physics and engineering. It has become the focus of attention of many thousands of researchers in hundreds of laboratories in universities, state research centers, and firms and corporations in all the world's leading industrialized countries.

The number of scientific publications dealing with the HTSC is growing at a record speed. The world's most prestigious scientific journals alone published about a thousand scientific papers on this topic during six months of 1987. Individual groups of researchers and scores of laboratories in universities and

YU. A. OSSIPYAN *is a former Vice-President of IUPAP and, as of the XIX General Assembly, its President-Designate (1987–1990). He is a member of the USSR Academy of Sciences and is Director of its Solid-State Physics Institute.*

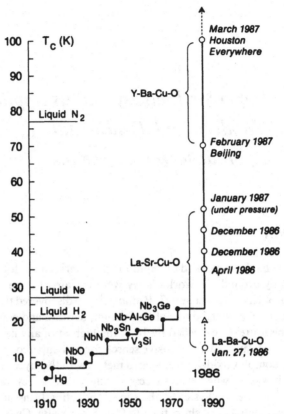

FIGURE 1   Chronology of the discoveries of superconductors with different T_c.

research centers are in an international scientific and technological competition of historic proportions. In the intensity of the scientific and engineering work, in the enthusiasm of the scientists, in the grandiosity of the practical achievements expected in this field, the situation with the HTSC strongly reminds one of the situation in the 1940s, when the world's scientists were striving to penetrate into the secrets of nuclear physics and nuclear energy. Not only the scientific journals, but also the press (with its vast circulation), carry the information on sensational scientific and practical results. Scientists and engineers, businessmen, politicians, and government leaders are all engaged in discussions about possible applications of the HTSC discoveries. The leading industrial countries are working out state programs for research and development, with the largest firms and corporations joining in. Billions are expected to be invested. So, the problem of a practical use of the HTSC materials becomes one of the most important technical problems of today. President Reagan, at his meeting with American physicists and engineers on 29 July 1987, called this problem a "technological breakthrough into the 21st century."

To conclude the introduction to my report, I should like to say that the problem of the development and application of the HTSC materials fits exceedingly well into the theme of our conference: *Physics in a Technological World*. It is actually a brillant example, demonstrating the outstanding role of physics in modern engineering.

APPLICATIONS OF SUPERCONDUCTIVITY

Now, coming to the specifics of the matter, I should first like to note that superconductivity has long found its application in scientific devices and electrical engineering. Well-known examples are Josephson junction devices, highly sensitive magnetometers based on SQUIDs, and superconducting solenoids for generating strong magnetic fields.

A large-scale "Tokamak-15" with superconducting magnets is being fabricated in the Soviet Union for investigations in thermonuclear energetics. The largest superconductor accelerator in the world—the "Supercollider"—is being designed in the United States. There are other examples of the use of superconductivity in science and engineering. However, a wide practical application of superconductors has hitherto faced serious obstacles. One of the most serious has been the need to achieve deep cooling by bathing the materials in liquid helium, which is costly and technically difficult, and commercially limits the practical use of superconductors.

The discovery of the new superconductors has drastically changed this situation because these materials can be cooled by liquid nitrogen, which is much cheaper than liquid helium (from various estimates it is about 200 times cheaper and its sources on the Earth are inexhaustible, whereas the amount of helium is rather limited); and also because nitrogen exhibits a much greater heat of vaporization and heat capacity, which means that it is a much better coolant than helium.

Comparative characteristics of nitrogen and helium are presented in Table 1. Using this table one can construct a simple parameter $Q = HET/P$. If one adopts this as a "figure of merit," then nitrogen as a coolant is more advantageous than helium by the factor $Q_N/Q_{He} = 10^6$. Thus if these new superconductors can be turned into usable forms (wire, ribbon, etc.), then the practical application of superconductors becomes technically attainable and commercially advantageous. This promises a real revolution in many fields of engineering.

TABLE 1   Comparative characteristics of nitrogen and helium.

	Characteristic	Nitrogen	Helium
1	Boiling temperature (T)	77.4	4.2
2	Heat capacity (H)	1.6	0.17
3	Evaporation heat (E)	160	3
4	Arbitrary cost per liter (P)	0.05	11

$Q = EHT/P$ $Q_N/Q_{He} = 10^6$

Researchers are now facing several serious scientific and technological problems. First of all, there is the basic problem of understanding the physical mechanism of high-temperature superconductivity. Here is a vast field for both theorists and experimenters. The theorists are now considering several other fundamental models besides the usual electron-phonon mechanism: excitonic, bipolaronic, plasmonic, etc. The experiments are conceived and performed to verify the correlation between different manifestations of these models and the real facts and relationships, observed experimentally. The searches for a connection of the specific features of the atomic crystalline structure and its defects to the parameters of the electric and magnetic properties of the HTSC materials are very important. It is very important to understand what the composition of the materials must be and how to treat them in order to raise the critical temperature T_c up to room temperature and higher. The majority of theorists consider that this is, in principle, attainable, and the world's scientific community is now ready to accept this possibility as quite natural.

But the new materials are rather impractical technologically on a macroscopic scale. They are very brittle, easily disintegrate when deformed, are rather sensitive to heat treatment and to the ambient environment (especially to oxygen), and not sufficiently stable. Fabricating wire of the HTSC ceramics would be like trying to get flexible wires from a piece of solidified cement or concrete.

Thus the problems facing engineers are as serious as those facing physicists or chemists. A very important practical achievement of recent work has been the fabrication of thin films of HTSC materials capable of passing high critical currents of 10^7 A/cm^2 or more at liquid helium temperature, and of 10^6 A/cm^2 at liquid nitrogen temperature, while remaining superconducting. The first to succeed with this was the IBM team led by Dr. P. Chaudhari.

These achievements are paving the road to many uses of the new materials in microelectronics—in particular, to creating the component base for the new generation of computers.

Today, one can consider the problem of application of the HTSC materials in electronics quite concretely. I shall try to do it later.

SUPERCONDUCTIVITY IN ENGINEERING

I shall now simply enumerate some possibilities of a large-scale use of the new materials in electrical engineering and energetics if, for example, we assume that the materials with T_c above room temperature have become available and the associated technological problems have been solved.

1. *Power energetics (superpowerful generators and transformers, electrical transmission lines, and electric energy storage)*

Presently, generators, transformers, and electric transmission lines waste 10–20% of the electric energy in the form of heat. The application of superconductors will make it possible to lower these losses by hundreds of times, as a minimum, and to remarkably miniaturize the generators and transformers. Solenoidal energy storages of vast capacity can be created. One very attractive possibility would be ecologically clean systems combining solar cells with a superconducting energy storage system—much more progressive, in my view, than thermonuclear energy. These miniaturized storage systems will be able to store up to 10^5 kWh of energy in a volume of 1 m^3.

2. Creation of superpowerful engines

The main problems stemming from the creation of superpowerful engines are now due to a disproportionate growth of their size and weight because of increasing the engine power above 100 kW (Fig. 2). Superconducting engines could increase the power and reduce the size and weight.

3. Mechanical engineering

Here there are many examples of the possible application of superconductors. One of these is a form of levitation based on a frictionless bearing (Fig. 3). Such a bearing may be of any capacity, from large heavy shafts to the smallest gyroscopic systems. In all of these bearings the energy dissipation is minimal.

Another example is a new type of device for transforming electrical into mechanical energy. It is based on the principle of a cyclic transition of superconductors to the normal state (Fig. 4).

These engines can be small, very powerful, and easily operated. They can be used to produce both linear and rotational motion. This principle can be used to

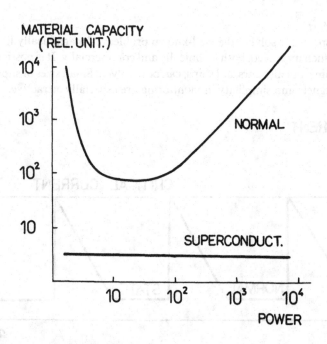

FIGURE 2   Specific consumption of materials in electric generators versus the power for traditional generators and superconductor generators.

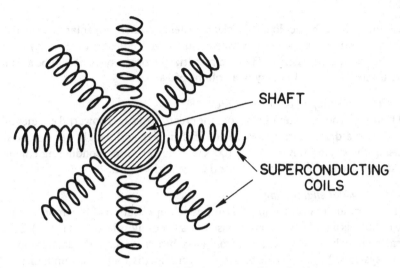

FIGURE 3   Schematic representation of levitation superconductor bearing.

create different kinds of generators, electric gates, presses, propulsion facilities, etc.

Because of lack of time, I have given only these two examples. However, any experienced engineer can propose a great many mechanical arrangements using the superconductivity principles.

4. *Transport*

The prospects for solving the well-known problem of magnetically levitated trains are much improved, both technically and commercially. It is possible that not only trains, but also cars and ships, can be involved. Small sizes of superconducting magnets and simplicity in monitoring are especially attractive.

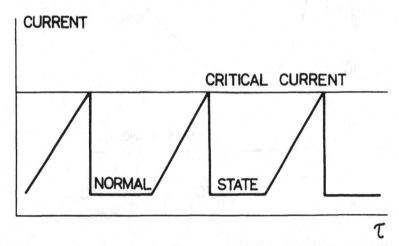

FIGURE 4   Schematic representation of a superconductor mechanical engine.

5. *Instrument making; medical and domestic equipment*
The input circuits of amplifiers can be made highly sensitive and almost noise-free by using superconducting elements. In medicine, compact magnetic-resonance tomographs based on superconducting magnets may be available to every doctor.

In conclusion, it may be noted that superconductors can be widely used for strategic purposes, including space and military applications.

SUPERCONDUCTIVITY IN ELECTRONICS

Now I shall analyze at greater length some possible applications of the HTSC materials in electronics. It should be remembered here that, on the one hand, the possibility of working at liquid nitrogen temperature (and subsequently, at room temperature) eliminates rather important technical and economic problems. On the other hand, some other problems of superconductor electronics remain and may even become aggravated. Therefore, the prospects for the use of the HTSC materials in electronics should be analyzed separately for each class of devices. In what follows, I shall try to give such a preliminary analysis.

A. *Passive Components of Electronic Devices*

1. *Electromagnetic screens*
If some material is confined within a superconducting screen of even a small thickness (1μm or greater), it becomes almost completely shielded from external electromagnetic radiation of any frequency, provided that the radiation density P is not too large ($P < P_c$). Estimates show that for the yttrium oxides the value of P_c is rather large—at least 10^4 W/cm^2.

Practically, it would be appropriate to achieve screening by placing a thin superconducting coating on the surfaces of nonsuperconductors. An especially promising possibility would be screening from relatively low-frequency radiations ($f \lesssim 10^2$–10^3 Hz) for which the skin depth in normal metals exceeds 0.1–1 cm.

2. *Signal transmission lines and resonators*
Some people suggest that high-temperature superconductivity will make it possible to create signal transmission lines (including combinations of elements in microelectronics) and low-loss resonators, operating at liquid nitrogen temperatures. However, if one allows for the fact that so far these temperatures T are close to T_c, and that the resistance of the known ceramics is high (even for crystals of Y–Ba–Cu-O, ρ_N is of the order of 100 $\mu\Omega$ cm), one cannot expect that the losses will be very low. If we cannot obtain materials with T_c of the order of 300–400 K, then the way out will be helium cooling with its high cost.

However, such cooling may be justifed for the recently developed superconductor stripline analog information processing devices and resonators. The use of high-temperature superconductors in these devices will reduce their losses even for operations at not too low temperatures (to ~ 30 K). The main problem here is the development of thin-film ($t \sim 0.5$–3μm) or thick-film ($t \cong 3$–50μm) coverages with small excess losses (due to pinning of the residual magnetic flux, etc.).

B. Functional Electronics

1. SQUIDs

Possibly, the main practical advantage of cryoelectronics has been the development and industrial production (since 1970) of SQUIDs (superconducting quantum interference devices) to measure very weak magnetic fields. SQUIDs can be used as pickups in a number of low-frequency measuring devices—namely, magnetometers, gradiometers, galvanometers, etc. Since their introduction the sensitivity of SQUIDs, usually measured in units of energy, has been improved from $\sim 10^{-27}$ J/Hz to $\sim 10^{-30}$ J/Hz for quantity production, and to $\sim 10^{-32}$ J/Hz for test samples; this has enabled them to retain their position as the most sensitive of the existing electrical measuring devices. SQUIDs are used in biomedicine, geophysics, physical experimental techniques, and in several other fields, but a wide application of these simple tools is inhibited by the necessity of cooling them down to low temperatures.

The application of high-temperature superconductors in SQUIDs can have a great practical effect. Their use in a number of new fields, such as medicine, controlling the quality of products, reconnaissance of natural resources, etc., will be economically advantageous.

The main problem here is the production of thin-filmed SQUID pickups containing one or two special components—Josephson junctions. Unfortunately, the production of a bulk version of SQUID for liquid nitrogen temperatures is scarcely feasible, because of an increase in thermal fluctuations that limits the maximum value of such SQUID interferometer inductance to about 10^{-10} H. This corresponds to an outer diameter for the coupling coil of less than 100μm. Though the presence of macroscopic interference in high-temperature superconductors, including liquid nitrogen temperatures, is an experimental fact, the production of reproducible and safe Josephson junctions, wherein such interference occurs, is a basic technological problem.

If this problem is solved so successfully that the main parameter of the junctions (the so-called characteristic voltage V_c) becomes close to its theoretical maximum $V_c = 3 \, kT_c / e$, then an increase of T_c will compensate the increase of the working temperature, and the sensitivity of "nitrogen" SQUIDs will be approximately at the same level as of "helium" ones. However, even an insignificant drop (by approximately one order) of V_c and, as a consequence, an increase of the intrinsic noises of SQUIDs, is not very important for the many prospective practical applications.

2. Stroboscopic transducers

The same significant broadening of the scope of utilization can be expected for stroboscopic transducers—devices for measuring faint picosecond electric pulses. The existing superconducting strobe-transducers involve relatively simple integrated circuits having of the order of ten Josephson junctions. Their time resolution is several picoseconds at a sensitivity of tens of microvolts. This is much better than in any other measuring devices of this type.

The application of high-temperature superconductors and, hence, liquid nitrogen temperatures, will make it possible to employ these measuring devices in pulse techniques, and especially in microelectronics, in the development of new generations of high-speed digital devices.

3. *Voltage standards*

One more field in which cryoelectronics already gives its practical yield is quantum metrology. Back in the beginning of the 1970s, constant voltage natural standards were fabricated on the basis of quantum macroscopic effects in superconducting Josephson junctions. These standards exhibited a record relative accuracy of the output voltage (about 10^{-8}), but the smallness of the voltage (several millivolts) required the utilization of bulky voltage multipliers. This limited the scope of application of superconducting voltage standards to primary (national and international) voltage standards.

Nowadays the output voltage of these standards has been raised to approximately one volt. This makes voltage multipliers unnecessary and increases the potential accuracy of the standards to at least 10^{-13}. This development in itself will undoubtedly broaden the scope of application of superconductor voltage standards in measuring devices. Their utilization will be noticeably facilitated by employing high-temperature superconductors and liquid nitrogen temperatures. To this end, however, one has to solve a number of technological problems typical of Josephson-junction integrated circuits (see below).

4. *Microwave receivers*

Many different superconductor microwave receivers were proposed in the 1970s. These devices have record sensitivity in millimetric wave bands and in a short-wave part of the centimetric band. They are already employed to solve some problems of radio astronomy and experimental physics. However, a small saturation power (typically less than 10^{-9} W) prevents them from having a wide application for radiolocation and radio communication.

This disadvantage is unlikely to be removed by using HTSC. Moreover, the application of nitrogen temperatures will reduce the sensitivity of these devices—a consequence that, seemingly, can only be partly compensated for by increasing the parameter V_c (see above). Such sensitivity deterioration will lower the possibility of competition with semiconducting cooling arrangements of the same range (mainly with Schottky-barrier-diode mixers).

Therefore in this field the efforts will most likely be concentrated on the application of HTSC for improving the parameters of helium receivers and, in particular, on broadening their operating range to the far infrared range. At the initial stage, the most promising types of receivers are, seemingly, quadratic wide-band detectors and, possibly, quasi-partial mixers. The second possibility may be realized only in the case of a successful development of high-quality tunneling junctions with small leakage currents. However, even if these problems can be successfully solved, the scope of application of these superconductor microwave receivers will probably be confined to scientific and special devices.

5. Signal commutators

Microwave and pulsed-signal commutators based on a phase transition in thin superconducting films will seemingly find a wider application. With an optimal geometry of the thin-film component this transition, which may easily be initiated by a relatively small (of the order of 10 mA) control current, leads to a significant (by several orders) alteration of the component impedance and, therefore, to the redistribution of the signal intensity between the channels of the device. Besides a great modulation range of the parameters, these devices are characterized by small losses in the superconducting state and rapid response (of the order of units and fractions of a nanosecond).

The application of HTSC will markedly reduce the cost of these simple devices, and at the same time preserve their working characteristics. This will make them useful in radio electronics.

C. Digital Microelectronics

An important practical application of HTSC would be the creation of new basic components for microelectronic high-speed computers and information machines. A preliminary study shows that there are at least three different promising approaches.

1. Semiconductor transistors with superconductor interconnections

A new family of silicon and gallium arsenide semiconductor transistors have been developed now. Their response times are exceedingly short—tens and even units of picoseconds—which is an improvement by three orders of magnitude as compared with those used in currently practical devices. However, this colossal speed of response cannot be exploited practically because normal conductor interconnections, with their high resistance, do not allow picosecond pulses to be transmitted even between transistors of the same integrated circuit. The application of the superconductor microstrip line, exhibiting very small attenuation and dispersion, may eliminate this problem.

An important advantage of this combination is its relative simplicity. However, there are disadvantages: namely, the difference of technologies for active (semiconductor) and passive (superconductor) components, and considerable losses by microstrip lines due to the nearness of the working temperatures ($T \sim 77$ K) to a critical temperature ($T_c \sim 100$ K) for the known high-temperature superconductors. (The latter factor will not allow pulses of shorter than 100 psec to be transmitted in a standard digital integrated circuit.) Besides, it is very important that the impedance of the microstrip lines (about 10 Ω for typical dimensions) should be compatible with the effective impedance of semiconductor transistors (from hundreds of ohms to tens of kilohms). Finally, the relatively great power dissipated by these transistors (the parameter P_d is of the order of 10^{-15} W or greater) will probably not allow the creation of high-speed, high-integration logic microcircuits (greater than 10^5–10^6 transistors per crystal).

2. Superconductor integrated circuits based on quantum macroscopic effects

Many of the above disadvantages may be eliminated by using superconductor microstrip lines together with superconductor Josephson-junction active components. Recently, a number of circuit-engineering solutions (so-called resistor and high-speed, single-quantum logics) have been proposed, enabling one to increase the speed of these active components. Thus the operation of a universal logic and memory cell at clock frequencies up to 30 GHz has been demonstrated experimentally. The analysis shows that even with low-temperature superconductors this frequency may be increased by one order of magnitude. The application of high-temperature superconductors may give a further increase of this frequency (provided that sufficiently high-frequency Josephson junctions can be based on them). Moreover, the power dissipated by these active components ($P_d \cong 10^{-18}$ W) is much lower than that for their semiconducting rivals, which will allow attainment of the integration of about 10^7 active components on one crystal.

However, we encounter some problems here. First, going to active working temperatures and the associated increase of thermal fluctuations will necessitate a greater than one-order increase of the critical currents I_c of Josephson junctions in order to preserve their stability. At the same time, the values of the inductances L should be decreased by the same factor (in order to preserve the relationship $I_c L \cong \phi_0$, where ϕ_0 is a flux quantum value). Preliminary estimates show that such a decrease imposes severe restrictions on microcircuit designers. Secondly, the aforementioned attenuation of microstrip lines with $T \cong T_c$ will not allow a full utilization of the potential speed of Josephson junctions.

Therefore it might be more advantageous to use large high-temperature superconductor integrated circuits, not at liquid nitrogen temperature, but at still lower temperatures—possibly, at liquid helium temperatures. This solution may appear paradoxical, but it may be real, because in the near future the cost of such a complicated electronic system as a high-speed computer will considerably exceed the cost of cryogenic provision for its work at helium temperatures.

3. Single Electronics

The last consideration also applies to one more promising version of high-speed digital devices—the combination of the same microstrip lines with active components on the basis of the recently discovered effect of discrete single-electron tunneling.[4] Preliminary estimates show that these active elements may exhibit approximately the same small delay times and dissipated power as Josephson-junction components. However, the sizes of "single-electron" active components may be much smaller than those of any other semiconductor or superconductor components (a few nanometers instead of hundreds of nanometers). This paves the way to microcircuits of exceptionally high integration—of the order of 10^8–10^9.

However, these high parameters can be attained only at sufficiently low (helium) temperatures, because of the growth of dissipated power as T^2. The second problem of "single electronics" is the need for reproducible production of a great number of minute tunneling junctions (of the order of 10×10 nm^2 with $T \approx 3$ K). The development of modern nanolithography gives hope that these operations will be possible in the very near future (in 5–10 years). In this case single electronics may become a most promising direction of digital microelectronics.

In conclusion, I have to note that even if one of the directions described above has a considerable success, it will require a rapid development of new ideas in computer architecture. In fact, base components with delay times τ of the order of 10^{-10} sec cannot be used in the context of traditional architecture or its slight modification. It will require an extensive application of production-line (conveyer) principles of digital information processing. These ideas should be explored immediately.

CONCLUDING REMARKS

The examples presented in this report have demonstrated how the HTSC can improve existing devices or arrangements or reduce their cost. This approach is a conservative one, based on currently available information.

Possibly the discovery of the HTSC will lead to changes in engineering not yet imagined. To illustrate that this is possible, I shall quote the words of Robert Schrieffer, our well-known colleague: "When transistors were discovered, everybody was thinking that they would be a successful substitute for a tube. No one could imagine at the time that the invention of a transistor would lead to the age of integrated circuits with its far-reaching consequences".

The author is very grateful to Professor K. K. Likharev (Moscow State University) for helpful discussions on many problems touched on in this paper.

REFERENCES

1. J. G. Bednorz and K. A. Müller, Z. Phys. **B64**, 189 (1986).
2. M. K. Wu *et al.*, Phys. Rev. Lett. **58**, 98 (1987); P. H. Hor *et al.*, *ibid.* 911 (1987).
3. K. Kishio *et al.*, Chem. Lett. (Japan) No. 2, 429 (1987); H. Takagi *et al.*, Jpn. J. Appl. Phys. **26**, L123 (1987).
4. L. S. Kusmin and K. K. Likharev, Pisma JETP (USSR) **45**, 389 (1987); T. A. Fulton and D. J. Dolan, Phys. Rev. Lett. **59**, 109 (1987); K. K. Likharev, "Correlated Discrete Transfer of Single Electrons in Ultrasmall Tunnel Junctions," IBM J. Res. Devel. **32**, 144 (1988).

Physics and Biology/ HANS FRAUENFELDER

Physics and Biology has been scheduled under the heading "Frontier Applications of Physics". It could, however, just as well be placed in the section on "Frontiers of Modern Physics". The coupling between physics and biology is becoming closer as time goes on. I will try to show that physical studies on biological systems not only yield insight into biology but also provide results of interest to physics.

1. INTRODUCTION

Ideally we would like to understand the behavior of a complex biological system, such as a "simple" organism like a microbe, in terms of the behavior of the constituent atoms. Such organisms consist of the order of 10^{20} atoms. These are not simple many-body systems, but are built from subunits that are arranged in a hierarchy as shown in Fig. 1. Where does life begin in this chain? There is no unambiguous answer; the complexity of the systems increases as we move up from atoms towards organisms. Unique characteristics of life begin with the *biomolecules*. Some of these perform essentially the same function when isolated as they do when incorporated in the living system. I will therefore mainly discuss biomolecules.[1,2]

Two types of biomolecules dominate: *nucleic acids* and *proteins*. The first store and transmit information and direct the construction of proteins; the second are the machines that perform the functions necessary for life. The essential features are shown in Fig. 2. (In reality the functions are not so neatly separated between nucleic acids and proteins, but the complications are unimportant here.)

The information is stored in the form of "three-letter words" on a very long linear unbranched deoxyribonucleic acid (DNA) molecule. The information on the DNA is read and transcribed onto a ribonucleic acid (RNA) molecule and transported to a ribosome, where the protein assembly takes place. The protein is also built as a linear chain, but the building blocks of nucleic acids and

HANS FRAUENFELDER, *chairman of the Governing Board of the American Institute of Physics, has worked in nuclear physics (Mössbauer Effect) and, more recently, in biophysics. He began his career in Switzerland, but since 1952 has been at the University of Illinois, Urbana, Illinois.*

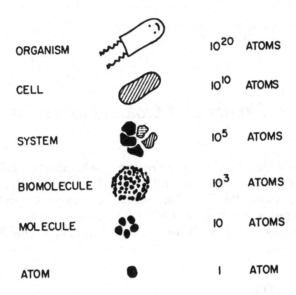

FIGURE 1 From atoms to organisms.

proteins differ. Nucleic acids are built from four different nucleotides; proteins from twenty different amino acids. The RNA instructs the ribosome in which order to connect the nucleic acids to form the primary sequence of the protein. The instruction involves the translation from the DNA and RNA language (three-letter words formed from an alphabet of four letters) to the protein

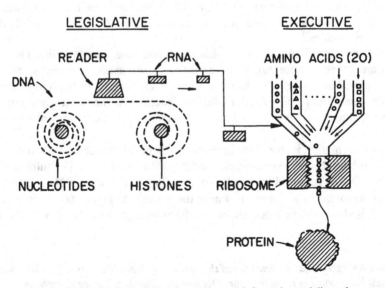

FIGURE 2 *Biomolecules.* Nucleic acids store and transmit information and direct the construction of proteins. The information is stored on DNA and transported to the factory (ribosome) by RNA.

language (consisting of an alphabet of twenty amino acids). When the primary chain emerges from the ribosome it folds into the functionally active three-dimensional structure.

A special class, globular proteins, fold into nearly close-packed structures. They consist typically of about 200 amino acids and have a diameter of a few nanometers (Fig. 3). Many of their physical properties are determined by their unique construction: Along the backbone, the bonds are covalent and hence so strong that they cannot be broken by thermal fluctuations. The cross connections that hold the protein together are hydrogen bonds, disulfide bridges, and Van der Waals forces; these are "weak" and can be broken by thermal fluctuations. This asymmetry in cohesion leads to unique characteristics that are

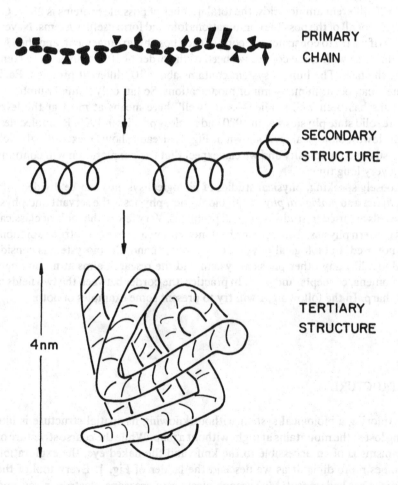

PRIMARY CHAIN

SECONDARY STRUCTURE

TERTIARY STRUCTURE

4nm

FIGURE 3 The covalently linked primary chain of amino acids (backbone) folds into the working tertiary protein structure. The globular protein is surrounded by a hydration shell and functions only in its presence.

important for the function of proteins: In contrast to solids, proteins are flexible and coupled to the solvent. Their size is such that they are at the border between classical and quantum systems. The proteins that exist now are not randomly constructed, they are the result of more than 3 Gy of R&D (research and development). Their ground state is structurally not unique: Most of the side chains of the amino acids can assume a number (n) of positions without changing the total energy of the protein appreciably. For a protein with N amino acids we consequently can expect a total number of approximately isoenergetic states of the order of n^N. With $n = 2$, $N = 200$, the number of possible ground states is of the order of 10^{60}. Proteins are highly degenerate systems.

The construction of proteins leads to the realization that *biological numbers* are far larger than astronomical numbers. Consider a protein with $N = 200$. With 20 different amino acids, the total number of possible proteins is 20^{200}. Of course, not all of the possible combinations fold and form useful systems. Nevertheless if we try to construct one copy of each possible arrangement, and then fill the universe with these copies, we need of the order of 10^{100} universes to complete the task. The human system contains about 10^5 different proteins. Each protein can occur in many minor modifications. So far, only a small number of proteins has been well studied—and "well" here means at most at the level where solid state physics was in 1920 and nuclear physics in 1930. Biomolecules, at the lowest biological level shown in Fig. 1, already show an exceedingly rich diversity and complexity and we can expect that their exploration will continue for a very long time.

Loosely speaking, physical studies of biological systems can be divided into *biophysics* and *biological physics*. In biophysics, physics is the servant and physical tools are used to study biological problems. Very few of the tools of classical and modern physics, from specific heat measurements to synchrotron radiation, are not used. In biological physics, on the other hand, the biosystem is considered just like any other physical system and the research aims at finding new phenomena, concepts, and laws. In practice, the border between the two fields is not sharp. In the following, I will try to present some examples of both.

2. STRUCTURE

Exploring a biological system without knowing its spatial structure is like being lost in the mountains at night without a map. While the coarse structure of organisms is often accessible to the knife and the naked eye, the exploration becomes more difficult as we descend the ladder of Fig. 1. Every tool of the physicist has had impact: Microscope, electron microscope, electron, x-ray, and neutron diffraction, EXAFS, and NMR all have added greatly to the knowledge of biological structures. Here only one approach, x-ray diffraction, will be sketched because it yields a rich variety of information.

A. X-Ray Diffraction Studies of Proteins

In principle the electron distribution in a protein can be determined without a protein single crystal. Assume that monochromatic x rays are scattered by a single oriented protein molecule. The scattering intensity $I(\theta)$ at the scattering angle θ is then proportional to $|f|^2$, the absolute square of the scattering amplitude f, f in turn is proportional to the Fourier transform of the charge density. If $I(\theta)$ is measured for a number of orientations of the protein molecule, the charge distribution can be found by Fourier transformation. The principle is the same as in the structure determination of particles and nuclei. One hurdle in all these structure determinations is the phase problem: In order to invert directly, the scattering amplitude f is required, but the experiment yields only the absolute square. The scattering phase must consequently be determined in a separate experiment.[3,4]

In practice, the single protein molecule is replaced by a protein single crystal which can be oriented and which yields a larger scattering intensity and suffers less radiation damage. The scattering intensity then is no longer a smooth function of the scattering angle; the interference from the many essentially identical protein molecules in the crystal leads to the discrete Laue–Bragg pattern. The positions of the spots in the diffraction pattern depend on the wavelength of the monochromatic x rays and on the lattice parameters of the single crystal; the intensities of the spots are determined by the electron distribution within each unit cell. The protein structure, i.e., the charge distribution or the arrangement of the nonhydrogen atoms in the protein, is consequently found from the spot intensities.

Proteins are rather complicated systems and it is difficult or even impossible to visualize their structure from the data in numerical form. For many years elaborate models built with great care from sticks and balls served as guides. They had many disadvantages; if the model was space-filling, the interior could not easily be examined; if the atoms were represented by dots, the impression was misleading. Computer graphics now gives beautiful representations which permit enlargement of important details, "walks" inside proteins, and examination of the effect of small changes in composition.[5]

B. Information from X-Ray Diffraction

X-ray diffraction experiments are capable of yielding far more than just the average or static structure of proteins. We sketch here some of the additional information that can be extracted from scattering experiments.

(i) *Dynamic structure.* In the simplest treatment of x-ray diffraction from a single crystal it is assumed that all equivalent atoms are at the correct periodic position. In this ideal situation, constructive interference is maximized and the Laue–Bragg spots possess maximum intensities. If the atoms are spread out over a linear dimension $\langle x^2 \rangle^{1/2}$, the constructive interference is reduced and the contribution of a particular atom to the total intensity of a Laue–Bragg spot is multiplied by the Debye–Waller factor[6]

$$T = \exp\{ - 8\pi^2 \langle x_i^2 \rangle \sin^2\theta / \lambda^2 \}$$

where $\langle x_i^2 \rangle$ is the mean-square displacement (msd) of the atom i and λ the wavelength of the x rays. If the x-ray diffraction data are taken with sufficient accuracy, they yield both the average position and the msd of all nonhydrogen atoms in a protein.[7]

(ii) *Thermal Expansion Coefficient.* If the average positions of all nonhydrogen atoms in a protein have been determined at a number of temperatures, a linear expansion coefficient can be determined between any two atoms.[8] In a liquid or an isotropic solid, the coefficient is a scalar; in a protein, it depends on the positions of the atoms and becomes a tensor field. Because the expansion coefficient is related to the fluctuations in volume and energy, such a measurement can yield information about dynamic aspects of proteins.

(iii) *Time evolution.* In some cases, a protein reaction can be started by a rapid process, for instance initiated by a laser flash. In general, the structure of the protein will be changed by the reaction. In principle, the evolving structure can be studied by fast x-ray diffraction.

C. Structure: The Future

Even though structure studies have already yielded rich information about proteins and nucleic acids, it is likely that the future results will far surpass what is now known. This prognosis is based on a number of current developments: Synchrotron radiation sources are beginning to be widely available; their high intensities, widely variable wavelengths, and pulse structure make them ideal sources for biomolecular studies. Area detectors are speeding up data collection and make digital data acquisition feasible. Adequate computer power allows a rapid evaluation of the data. Data gathering is further speeded up by the use of the original Laue technique: Most of the work performed so far has been done with monochromatic x rays, yielding well-defined Laue–Bragg spots. It turns out that the Laue technique, where the incident x-ray beam is *not* monochromatic, still produces resolvable spots, but increases the speed enormously. These various improvements together make structure studies of biomolecules far more powerful than before and may well lead to another revolution in structural biology.

Additional insight into biomolecular phenomena is coming from the combination of x-ray diffraction with cryogenics and with high pressure. Since pressure and temperature are the two most important thermodynamic variables, knowledge of the protein structure and of the msd of the atoms participating in the dynamics as functions of the two variables adds greatly to the understanding of biological processes. Such studies, also, are only at a beginning.

3. COMPLEXITY

Traditionally, physicists study simple systems and attempt to extract generally valid laws from the simplest system that exhibits certain characteristic properties. The two prototypes of simple systems are the perfect gas (complete disorder) and the ideal solid (perfect order). We can assign complexity 0 to both of these systems, because neither can be used to store or transmit information. Studies of such simple systems have contributed greatly to physics. Many interesting systems, from proteins to brains, are neither fully ordered nor completely disordered; they are truly complex.[9] While some physicists have always been attracted to complex systems (turbulence!), most have stayed away from them. Within the past few decades, however, studies of disordered systems have increased markedly and complexity has become an acceptable concept even to physicists. The following examples show that complexity in proteins may connect these systems to glasses and spin glasses, the disordered systems that are better known to many physicists.

A. Nonexponential Time Dependence

We usually assume that the time dependence of a physical process is, to a first approximation, either sinusoidal or exponential. A large class of phenomena, however, deviate profoundly from either of these simple forms. These processes, sometimes called "endless", were first observed and described by W. Weber in 1835.[10,11] They typically extend over many orders of magnitude in time and can often be described by a power law or a "stretched exponential". Such processes have been measured in mechanical creep, discharge of capacitors, dielectric relaxation, and many other processes, and they always imply complexity.[12,13]

In proteins, nonexponential time dependence characterizes a number of processes. The best studied example is the rebinding of a small molecule such as dioxygen (O_2) to myoglobin.[14] Myoglobin (Mb) contains 153 amino acids, has a molecular weight of about 18 000 dalton, and has dimensions of about $3 \times 3 \times 4$ nm^3.[1] It contains a small organic molecule, protoheme, with an iron atom at its center. Mb acts as oxygen (O_2) carrier; O_2 binds reversibly at the iron atom: Mb $+ O_2 \leftrightarrow MbO_2$. The binding process can be studied by photodissociation. Starting with MbO_2, the bond between the heme iron and the dioxygen molecule is broken by a laser flash and the subsequent rebinding is monitored in the visible[14] or the infrared.[15] It turns out that rebinding is not exponential in time, but approximately follows a power law,

$$N(t) = (1 + t/t_0)^{-n}$$

where $N(t)$ is the survival probability, i.e., the fraction of Mb molecules that have not rebound an O_2 at time t after photodissociation, and where t_0 and n are two temperature-dependent parameters. Similar nonexponential time developments are found in all heme proteins that have been studied.[16]

A simple explanation for the observed time dependence is based on the assumption that binding is governed by a barrier of height H at the heme iron and that each protein molecule possesses a different H. Denote with $g(H)dH$ the probability of having a protein with barrier between H and $H + dH$ and assume further that the rate coefficient $k(H)$ for binding is connected to H by an Arrhenius relation, $k(H) = A \exp\{-H/RT\}$. The survival probability then is given by

$$N(t) = \int dH \, g(H) \, \exp[\, -k(H)t \,] \,.$$

With a suitably determined probability distribution $g(H)$, the experimental data can be reproduced well between about 40 and 160 K. Below about 40 K, quantum-mechanical tunneling becomes important and the Arrhenius relation is no longer valid. Above about 160 K, competing processes obscure the direct rebinding.

B. Conformational Substates

Why do different myoglobin molecules possess different barrier heights at low temperatures? In the introduction we noted that a given primary sequence may not always fold into the same final globular structure. We now use this idea to introduce the concept of conformational substates (CS): We assume that a given protein in a given state (for instance MbCO) can assume a large number of structurally slightly different substates as sketched in Fig. 4. The energy surface of a protein consequently is postulated to be not smooth, but to contain a very large number of valleys, separated by energy mountains. The different CS or valleys then can possess different barrier heights H. At low temperatures, say below about 180 K, a protein remains frozen into a particular CS with corresponding barrier height H, and rebinding after photodissociation is nonexponential in time. At high temperatures, say 300 K, a given protein fluctuates rapidly from CS to CS and rebinding can be exponential in time.[14]

Corroborating evidence for CS comes from the Debye–Waller factor of myoglobin.[17,18] If all proteins were identical, and if no lattice imperfections existed in the protein single crystal, the msds of all atoms would be given by the vibrational contribution. If CS exist and have structural meaning, then the same atom should occupy slightly different positions in different protein molecules, leading to msd larger than the vibrational ones. Figure 5 shows the msd for Mb at two different temperatures. As expected for CS, the msd depends strongly on the positions of the atoms and is much larger than predicted by a purely vibrational contribution.

C. Proteins, Glasses, and All That

The existence of a highly structured energy landscape with many deep and approximately isoenergetic valleys may well be one significant characteristic of many (all?) complex systems. Highly degenerate ground states have been pos-

CONFORMATIONAL SUBSTATES

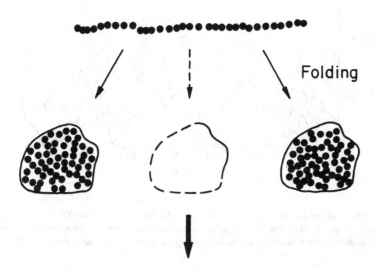

CONFORMATIONAL SUBSTATES

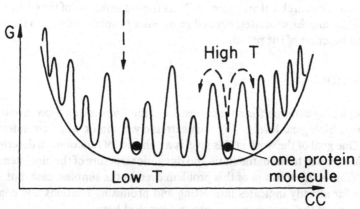

FIGURE 4 *Conformational substates.* A given primary sequence does not lead to a unique tertiary structure, but to a very large number of related, but in detail different, protein structures. These can be represented as points in a conformational space where each CS corresponds to a valley in the energy landscape.

tulated to explain the characteristics of spin glasses,[19,20] glasses,[21] and neural networks,[22–24] evolutionary biology,[25,26] optimization problems,[27] and learning.[28] All of these fields are at a beginning and many problems are unsolved or not even clearly recognized yet. In the case of the conformational substates of proteins, some questions can be posed: Is there just one tier of substates or do the

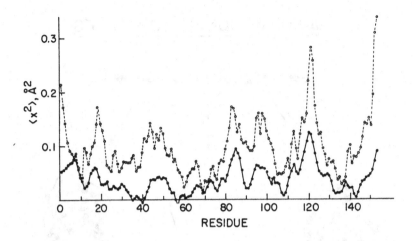

FIGURE 5   Mean-square displacements $\langle x^2 \rangle$ of the backbone atoms of metMb as a function of the amino acid number. The values are averages over the backbone atoms of each amino acid; the amino acid (residue) number labels the amino acids along the primary sequence. (After Ref. 18.)

CS fall into a hierarchical arrangement?[29] Is the organization of the CS ultrametric?[20] How are the substates, defined in the energy space, related to the structure and function of the protein?

4. FUNCTION

Biological systems perform functions. Even the simplest function executed at the lowest biological level of Fig. 1 is extremely complex when considered in detail. One goal of the work at this level is a "physics of function," a description of the function in terms of the static and dynamic structure of the biosystem. We are far from a full solution of this problem even in the simplest case, but work done so far already indicates interesting and promising relations and connections. Some of the emerging concepts are sketched here.

A. Equilibrium Fluctuations and "Fims"

Consider again myoglobin as example. Its function is dioxygen storage and it therefore must have at least two *states*, MbO_2 and Mb. In each of these states it possesses a very large number of conformational substates. The existence of states and substates leads to two types of motions, equilibrium fluctuations (EF) and functionally important motions, or fims. In a given state at high temperature, a protein fluctuates from CS to CS. These EF are governed by the general principles of equilibrium thermodynamics: A small system such as a protein does not have sharp values of internal energy, entropy, and volume.

These quantities fluctuate about their mean values.[30] The protein function, the transition from Mb to MbO_2 or its inverse, is performed through fims. Fims are nonequilibrium processes.

B. The Fluctuation–Dissipation Relation

EF and fims are often related. Studies of EF can consequently help elucidate the function. The first connection between an equilibrium and a nonequilibrium property was found by Einstein in his celebrated relation of the diffusion coefficient D to the friction coefficient f.[31] Later Nyquist related the voltage fluctuations across a resistor to the resistance.[32] General relations between equilibrium fluctuations and dissipative processes were formulated by Onsager,[33] Callen and Welton,[34] and many others.[35-37] It is likely that fluctuation–dissipation relations will be central for studying and understanding biological phenomena.

C. Proteinquakes

In any biomolecular reaction, the structures of the initial and final states are usually different. During and after the reaction, the biomolecule must rearrange its structure. Consider myoglobin again as an example. When it binds O_2, a stress is established at the active center, the heme group. Upon photodissociation, the stress is relieved and the strain energy is dissipated in the form of waves and through the propagation of a deformation. The process is similar to an earthquake and we call it a "proteinquake".[29] A proteinquake can happen because of the special structure of proteins as shown in Fig. 3: While the released energy is not sufficient to break the covalent bonds that link the backbone, the cross connections can be broken and rearranged, thus leading to a slightly different structure. Proteinquakes can be monitored by following changes in spectroscopic markers and, just as with earthquakes, we expect that they can give information about dynamic features.

D. Theories

So far, the discussion has been based predominantly on experimental observations. Ultimately there may be a fundamental theory that describes much of biology and biological processes, based on physical principles. We are far from such a theory. Nevertheless, progress has been rapid in the description and understanding of the phenomena on the lowest biological level: nucleic acids and proteins. The most successful direction so far has been the application of molecular dynamics, pioneered by Karplus and his collaborators.[38,39] One main limitation with this technique has been the restricted time range that could be covered. Most existing computations extend only to about 100 psec. Typical biologically important processes are much slower. New techniques promise to overcome the time limitations.[39,40]

5. OUTLOOK

Despite the vast amount of observational material already in existence, biophysics and biological physics are only at a beginning. We can expect that physics will continue to interact strongly with biology. Actually, the connection also includes chemistry and mathematics. New tools that become available in physics will continue to be applied to biological problems. We can expect that the flow of information will not be one way; biological systems will provide new information on many old and new parts of physics, from reaction theory and transport phenomena to complexity, cooperativity, and nonlinear processes.

ACKNOWLEDGMENTS

This contribution was written while the author was a Senior U.S. Scientist Awardee of the Alexander von Humboldt Foundation. The work was supported in part by U.S. National Institutes of Health grant GM 18051, National Science Foundation grant DMB 82-09616, and Office of Naval Research grant N00014-86-K-00270.

REFERENCES

1. L. Stryer, *Biochemistry* (Freeman, San Francisco, 1981).
2. H. Frauenfelder, Helv. Phys. Acta **57**, 165 (1984).
3. T. L. Blundell and L. N. Johnson, *Protein Crystallography* (Academic, New York, 1976).
4. J. D. Dunitz, *X-Ray Analysis and the Structure of Organic Molecules* (Cornell University Press, Ithaca, NY, 1979).
5. R. Langridge, T. E. Ferrin, I. D. Kuntz, and M. L. Connolly, Science **211**, 661 (1981).
6. B. T. M. Willis and A. W. Pryor, *Thermal Vibrations in Crystallography* (Cambridge University Press, Cambridge, 1975).
7. G. A. Petsko and D. Ringe, Ann. Rev. Biophys. Bioeng. **13**, 331 (1984).
8. H. Frauenfelder, H. Hartmann, M. Karplus, I. D. Kuntz, Jr., J. Kuriyan, F. Parak, G. A. Petsko, D. Ringe, R. F. Tilton, M. L. Connolly, and N. Max, Biochemistry **26**, 254 (1987).
9. B. A. Huberman and T. Hogg, Physics **22D**, 376 (1986).
10. W. Weber, Ann. Phys. Chem. (Poggendorf) **34**, 247 (1835).
11. J. T. Bendler, J. Stat. Phys. **36**, 625 (1984).
12. E. W. Montroll and J. T. Bendler, J. Stat. Phys. **34**, 129 (1984).
13. J. Klafter and M. E. Shlesinger, Proc. Nat. Acad. Sci. USA **83**, 848 (1986).
14. R. H. Austin, K. W. Beeson, L. Eisenstein, H. Frauenfelder, and I. C. Gunsalus, Biochemistry **14**, 5355 (1975).
15. A. Ansari, J. Berendzen, D. Braunstein, B. R. Cowen, H. Frauenfelder, M. K. Hong, I. E. T. Iben, J. B. Johnson, P. Ormos, T. B. Sauke, R. Scholl, A. Schulte, P. J. Steinbach, J. Vittitow, and D. Young, Biophys. Chem. **26**, 337 (1987).
16. H. Frauenfelder, F. Parak, and R. D. Young, Ann. Rev. Biophys. Biophys. Chem. (1988).
17. H. Frauenfelder, G. A. Petsko, and D. Tsernoglou, Nature **280**, 558 (1979).
18. H. Hartmann, F. Parak, W. Steigemann, G. A. Petsko, D. Ringe Ponzi, and H. Frauenfelder, Proc. Natl. Acad. Sci. USA **79**, 4967 (1982).

19. M. Mezard, G. Parisi, N. Sourlas, G. Toulouse, and M. Virasoro, Phys. Rev. Lett. **52**, 1156 (1984).
20. R. Rammal, G. Toulouse, and M. A. Virasoro, Rev. Mod. Phys. **58**, 765 (1986).
21. J. Jäckle, Rep. Prog. Phys. **49**, 171 (1986).
22. J. J. Hopfield, Proc. Natl. Acad. Sci. USA **79**, 2554 (1982).
23. J. J. Hopfield and D. W. Tank, Science **233**, 625 (1986).
24. *Neural Networks for Computing*, edited by J. S. Denker, Am. Inst. Phys. Conf. Proc. **151** (1986).
25. M. Eigen, in *Emerging Syntheses in Science*, edited by D. Pines (Santa Fe Institute, Santa Fe, NM, 1985).
26. P. W. Anderson, Proc. Natl. Acad. Sci. USA **80**, 3386 (1983).
27. G. Baskaran, Y. Fu, and P. W. Anderson, J. Stat. Phys. **45**, 1 (1986).
28. G. Toulouse, S. Dehaene, and J. P. Changeux, Proc. Natl. Acad. Sci. USA **83**, 1695 (1986).
29. A. Ansari, J. Berendzen, S. F. Bowne, H. Frauenfelder, I. E. T. Iben, T. B. Sauke, E. Shyamsunder, and R. D. Young, Proc. Natl. Acad. Sci. USA **82**, 5000 (1985).
30. A. Cooper, Proc. Natl. Acad. Sci. USA **73**, 2740 (1976).
31. A. Einstein, Ann. Phys. **17**, 549 (1905).
32. H. Nyquist, Phys. Rev. **32**, 110 (1928).
33. L. Onsager, Phys. Rev. **37**, 405 (1931).
34. H. B. Callen and T. B. Welton, Phys. Rev. **83**, 34 (1951).
35. L. Onsager and S. Machlup, Phys. Rev. **91**, 1505 (1953).
36. R. Kubo, Rep. Progr. Phys. **29**, 255 (1966).
37. P. Hanggi, Helv. Phys. Acta **51**, 202 (1979).
38. M. Karplus and J. A. McCammon, CRC Crit. Rev. Biochem. **9**, 293 (1981).
39. J. A. McCammon and S. C. Harvey, *Dynamics of Proteins and Nucleic Acids* (Cambridge University Press, Cambridge, 1987).
40. R. Elber and M. Karplus, Science **235**, 318 (1987).

Physics and the
Information Age/ PAUL A. FLEURY

INTRODUCTION—PHYSICS ROOTS OF
INFORMATION-AGE TECHNOLOGY

The changing nature of our society is illustrated by the changing composition of the U.S. work force over the past century. Agricultural, manufacturing, and informational occupations accounted for 40%, 30%, and 5%, respectively, in 1870. By 1970, the same fields commanded 4%, 30%, and 45%, respectively.[1] This dramatic shift has accelerated following the Second World War and has transformed virtually every aspect of our lives. The ease and efficiency with which every person in this country can communicate with other people is an important ingredient in our continually increasing productivity and national wealth. Our ability to generate, manipulate, transmit, store, and retrieve astounding amounts of information has been growing exponentially for nearly 30 years. This convolution of computing and communications is at the heart of the "Information Age." The rapid advance and international impact of information technology make it important that planners, politicians, and scientists themselves recognize the basis in past scientific research for present information-age technologies and acquire some appreciation of future trends which today's scientific research makes possible.

The expansion of information-age technology can be characterized by the sustained exponential growth in (a) telecommunications network capacity (Fig. 1) and (b) computing power per unit cost. Although largely invisible to the user, this growth has required a succession of new systems based on new materials, new devices, and sometimes even new physical phenomena emerging from physics research laboratories. Within the last decade and a half, since John Bardeen's IUPAP discussion[2] of the accomplishments and prospects of solid state physics, the pace of scientific discovery has, if anything, quickened. Our ability to fabricate, interrogate, and control materials with atomic scale precision has begun to move from the laboratory into the factory. Photonics is beginning to vie with electronics for center stage in information-age technology, and

PAUL FLEURY works in the area of solid-state physics and spectroscopy. His whole professional career has been with AT&T Bell Laboratories, where he has been Director of Physics Research since 1984.

CAPACITY OF MAJOR TELECOMMUNICATIONS HIGHWAYS

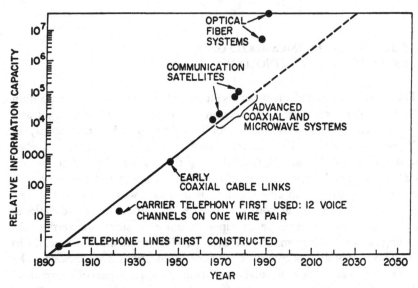

FIGURE 1 Historical growth in relative capacity of U.S. telecommunication networks.

quite recently the discovery of high-temperature superconducting materials has rekindled interest in the possibilities of superconducting logic devices, transmission lines, etc.

In this paper I shall attempt to trace the origins and impact of recent technological advances upon the information age, first from the viewpoint of devices and systems already in use, and second from the perspective which recent scientific discoveries promise for the future.

MICROELECTRONICS

Ever since the invention of the transistor, electronics technology has been based in silicon. The exponential growth in computing power was made possible by integrated circuits and relentless improvements in their size and speed. Limits to device performance have been raised dramatically, largely by understanding how to grow ever more perfect silicon crystals and then how to dope and/or damage them controllably. The tools of nuclear physics have been married first with solid-state physics and then with device processing. Ion implantation is now routinely used to put just the right amount of dopant in just the right place. Literally dozens of chemical, physical, and thermal processing steps are required to transform a crystalline silicon wafer into the thousands of integrated circuits which define today's microchips. Defects and damage can creep in at every step. Understanding their effects on device performance has benefited greatly from physics research. Epitaxial thin-film growth, the goal of silicon-on-insulator fabrication, metallization, oxidation, etc. on length scales rapidly falling below 1 micron have provided incentives to, and have benefited from, research in surface physics. Indeed the surface of silicon is perhaps the most thoroughly studied two-dimensional entity of all.

Physics tools such as Rutherford backscattering, x-ray standing wave analysis, electron microscopy, and vacuum tunnelling microscopy are providing structural information at the atomic scale of detail. The first stages of epitaxial growth are being modeled by molecular dynamics (surface roughening phase transition) and monitored during molecular beam epitaxy fabrication by LEED and Auger spectroscopy. Synchrotron radiation and computer-controlled ion and electron beams are driving design rule dimensions well below a micron. Simple silicon devices have been fabricated with thousand-angstrom features—which would correspond to over a hundred million gates per chip when fully developed. Yet even this density does not approach the fundamental limit. For example, a scanning-tunnelling microscope (Fig. 2) has been used in the laboratory to store and retrieve a bit of information within a single unit cell on a

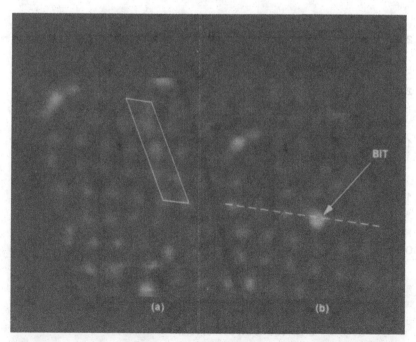

FIGURE 2   Scanning tunnelling microscope images of the same area of a germanium crystal surface (a) before voltage elevation above the marked area and (b) after voltage elevation; illustrating the storage of a bit of information in the displacement of a single atom. (After Ref. 3.)

germanium crystal surface.[3] Layered structures of entirely new materials which either enhance known properties or give rise to entirely new ones have been produced by MBE (Molecular Beam Epitaxy). Recent research has given even more emphasis to the compound semiconductors (particularly the III–V's) than to Si and Ge. Although more complex than their elemental cousins, the III–V's offer attractive possibilities for high speed electronics, for entirely new kinds of electronic devices, and for optoelectronics and photonics.

QUANTUM WELLS AND SUPERLATTICES

Not so long ago, electronic band structures concerned only the research physicist. Today advances in both theory and materials fabrication have given us the ability to practice "band structure engineering." [4] The variety of energy gaps and lattice constants exhibited by the III–V family (Fig. 3) makes possible astonishing flexibility in device design and performance. Layered heterostructures of GaAs and GaAlAs exemplify the advantages of "modulation doping" whereby the dopant impurity atoms which contribute the carriers are selectively introduced into one species (where they remain) while the carriers themselves

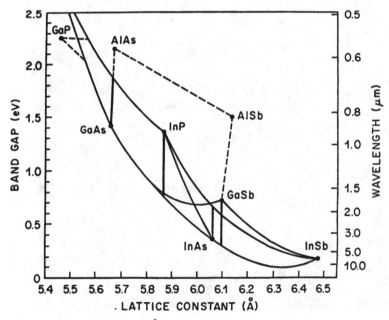

FIGURE 3   Lattice constants (Å) and energy gaps (eV) in several III–V semiconductor compounds. Right-hand scale gives corresponding optical wavelengths and illustrates, for example, why compounds of InP:GaAs are suited for device operation in the 1.3–1.5 μm regime.

MODULATION - DOPING

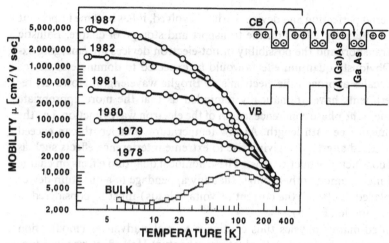

FIGURE 4   Temperature-dependent electron mobilities in bulk GaAs and in successively more perfect modulation-doped material. The record values of five million cm²/V sec. represent a 5000-fold improvement over room-temperature silicon. Inset illustrates the modulation-doping idea.

reside in the other, dopant-free material. This technique, together with continued improvement in the structural perfection of the layers themselves, has permitted electron mobilities in excess of 5 million cm²/V sec (Fig. 4). Picosecond speeds have been achieved in superlattice transistors.

With this level of materials perfection and control a new range of quantum phenomena becomes important. Real quantum wells show the kinds of resonant tunnelling behavior formerly seen only in textbook exercises. They form the basis for new families of devices including multivalued-logic electronics. The perfection of such planar materials has led to new discoveries in physics as well, most spectacularly the fractional quantum Hall effect—a manifestation of a new many-body state of matter in two dimensions.[5]

Heterogeneous epitaxy involving silicon and metal silicides may permit realization of the long-sought ballistic metal-base transistor promising order-of-magnitude speed increases. Our new understanding and experience with strained-layer superlattices offers promise for a range of infrared detectors, and possibly even for "optical" silicon. Progress toward integrating silicon and III–V epitaxially is particularly promising as we move from electronics to optoelectronics as a prelude to a possible fully photonic technology.

Magnetic materials and phenomena have a long tradition in information technologies—particularly on the memory side. MBE has now begun to influence magnetics as well—with the first epitaxial magnetic superlattices[6] being fabricated only within the last two years.

NONLOCAL ELECTRON TRANSPORT

As material perfection and device size have evolved, it has become important to consider quantum effects on the transport and storage of charge. Ballistic electron transport and the possibility of hot-electron devices are emerging examples. Obviously quantum effects would be expected to dominate as device structure scales approach the electron's de Broglie wavelength. However, recent experiments have dramatically reminded us that the more appropriate length scale is the phase coherence length of the electron wave function, i.e., the inelastic mean free path length. At low temperatures this length can exceed several hundred angstroms, giving rise to extreme interference effects such as universal conductance fluctuations.[7] These nonlocal quantum effects on charge transport mean, among other things, that any appendage to a quantum device (even if placed "outside" the current or voltage leads) must be considered as part of the sample (Fig. 5).

Condensed-matter physics thus continues to make advances (modulation doping, universal conductance fluctuations, quantum Hall effect, vacuum tunnelling microscope. . .) which both determine the ultimate limits to the size and speed of electronic devices and suggest alternatives that may obviate these limits. Such effects must be understood and accounted for as the size of electronic circuits shrinks further.

LIGHTWAVE COMMUNICATIONS

An important new dimension to the information age is represented by photonics. Electrons interact strongly with electric and magnetic fields and of course with each other. Photons do not. Of the major functions required for information technology, at first glance electrons appear better suited to some (logic, storage, switching) and photons better suited to others (generation, transmission, detection).

Optical (lightwave) communication technology[8] is today a multibillion dollar reality, despite the fact that little more than a decade ago all of its major components (semiconductor light sources and detectors, optical fibers) were firmly confined to the research laboratory. Reliable room-temperature operation of the semiconductor laser[9] was first demonstrated in 1970. Since then both laser physics and materials science have made great strides. Lasers engineered to operate at 1.3 or 1.5 μ wavelengths (the regime of lowest optical loss in silica fibers) are in mass production. They combine miniature size, low threshold powers, precise wavelength control, long life, and high modulation rates (Fig. 6). Lightwave communication systems operating at super gigabit rates are being installed. Glass fibers with attenuations below 1 dB/km are being manu-

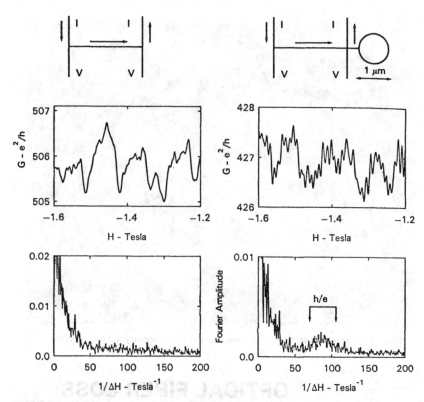

FIGURE 5 Nonlocal quantum effects on electron transport in small structures can be dramatic. On the top left is a schematic structure (polycrystalline gold line 2 μm long; 0.1 μm wide; 0.04 μm thick held at 40 mK); directly below are plots of its magnetoconductance (exhibiting the so-called universal conductance fluctuations) and corresponding Fourier transform. On the right-hand side is a nearly identical structure with a small ring appended *outside* the leads that define the classical current path. The added structure in the *G* vs. *H* curve is due to the quantum interference of the electronic wave function with period h/e due to the Aharonov–Bohm effect in the ring.

factured in the millions of kilometers each year (Fig. 7). An appropriate figure of merit for lightwave systems, analogous to the device density for electronics, is the "bit-rate-distance product" (Fig. 8). This has increased one-thousand fold in just over a decade. The same astounding growth rate applies both to commercial and to laboratory experimental systems. And the interval between invention and commercial implementation is remarkably short.

THE PHOTONIC FUTURE

Physics discoveries of the past few years (femtosecond pulses, solitons, squeezed radiation) are likely to sustain the impressive improvement in infor-

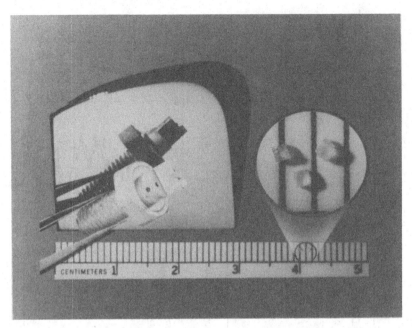

FIGURE 6   Photograph of semiconductor laser used in contemporary lightwave system.

OPTICAL FIBER LOSS

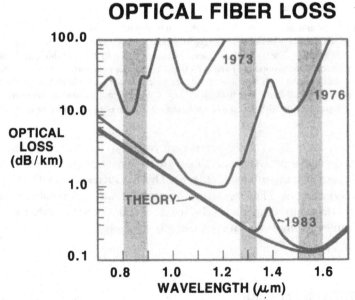

FIGURE 7   Optical loss vs. wavelength for several generations of fused-silica-based optical fiber. The dates correspond to achievements in the research laboratory. By 1983 the losses due to impurities had been virtually eliminated (~ 1 ppb OH^- causes the < 1 dB/km peak at 1.4 μm). Intrinsic Rayleigh scattering and infrared absorption combine to place the theoretical loss minimum of < 0.2 dB/km at 1.55 μm.

THE BIT-RATE-DISTANCE DERBY

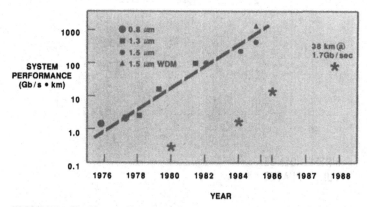

FIGURE 8 The bit-rate-distance product measures the figure of merit for an optical transmission system. Dashed line shows improvement in research experiments by 1000-fold in ten years. Asterisks show similar improvement in commercially installed systems with only about a five-year time lag.

mation transmission systems for the forseeable future. Femtosecond pulses were first produced at AT&T Bell Laboratories in 1981.[10] Using colliding-pulse mode locking, physicists were able to produce transform-limited pulses of about 90 fsec in duration. Combining the phenomenon of self-phase modulation with fiber dispersion and grating compensation has permitted dramatic reductions in pulse width since then. The current record is an optical pulse whose full width at half maximum is only 6 fsec—just three cycles of the optical field. The femtosecond laser is certain to influence future information technologies—if not directly, then surely through the discoveries which this remarkable new source will make possible. It is worth noting that such pulses can in principle carry 10^{14} bits of information per second on a single fiber channel, about equal to the information content of the entire Library of Congress.

Materials dispersion will prevent direct use of femtosecond pulses for long-distance transmission, at least so long as one operates in the linear regime. Physicists may have found a way around this limitation by turning dispersion into an advantage. A soliton is a stable finite-amplitude solution to a *nonlinear* dispersive wave equation. The physics of guided waves, nonlinear optics and short pulses has been combined recently not only to demonstrate the stable formation and transmission of optical solitons in glass fibers, but also to operate a new kind of laser[12]—the soliton laser (Fig. 9). Indeed a full soliton-based long-distance transmission system may be possible which sends picosecond pulses over thousands of kilometers without electronic regeneration by utilizing a distributed optical amplification scheme involving stimulated Raman scattering from the core glass of the optical fiber.

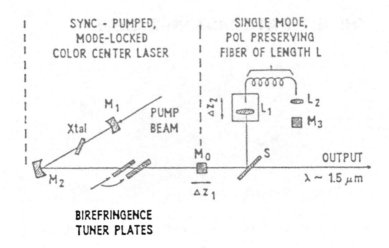

SYNC - PUMPED,
MODE-LOCKED
COLOR CENTER LASER

SINGLE MODE,
POL PRESERVING
FIBER OF LENGTH L

M_1

Xtal

PUMP
BEAM

L_2

L_1

M_3

M_0

S

OUTPUT

$\lambda \sim 1.5 \, \mu m$

M_2

Δz_1

Δz_2

BIREFRINGENCE
TUNER PLATES

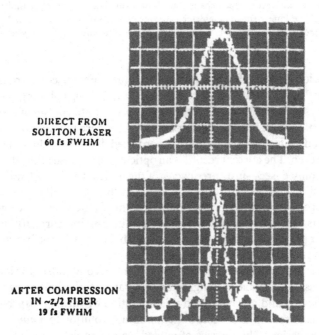

DIRECT FROM
SOLITON LASER
60 fs FWHM

AFTER COMPRESSION
IN ~z/2 FIBER
19 fs FWHM

FIGURE 9   The soliton laser is a promising new source of short pulses ideally matched to the nonlinear propagation characteristics of single-mode optical fibers. The optical soliton generated by launching a picosecond pulse of the correct wavelength into a fiber light guide of the correct length and dispersion is fed back by the mirror M_3 to produce a stable soliton oscillation. The resulting pulse is 60 fsec in duration but is compressible to less than 20 fsec.

SQUEEZED LIGHT

Exploring the ultimate limitations on the sensitivity of optical systems as well as the precision of fundamental measurements has recently led physicists to examine and modify the nature of light itself. Even with perfectly noiseless amplifiers and balanced homodyne detectors, an optical signal will carry noise due to the vacuum fluctuations. The uncertainty principle dictates that the product of variances in the quadrature components $(\Delta X_1, \Delta X_2)$ of the electric field must equal or exceed 1/4. Coherent light (from a laser) achieves the equality with noise equally distributed between the two quadrature components $(\Delta X_1 = \Delta X_2 = 1/2)$. It had been predicted theoretically for more than 20 years that one should be able to redistribute or "squeeze" the noise from one quadrature to the other. The result would be a reduction of quantum noise at some parts of the cycle and an increase in others, both of which could be used in phased signaling or measurements to increase precision or reduce error. Experimentally squeezed light was first produced[13] in 1985 in sodium vapor. By the end of 1986, squeezing had also been demonstrated in transparent crystals of $LiNbO_3$[14] and in ordinary fused-silica glass fibers.[15] Squeezing of picosecond pulses[16] (Fig. 10) and operation of the first squeezed-light interferometer[17] are

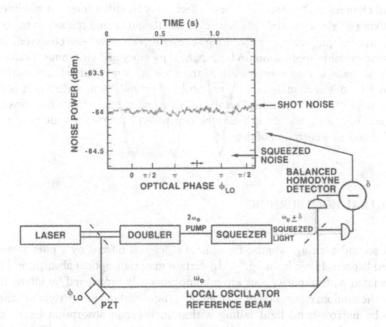

FIGURE 10 Squeezed light is produced by noiseless phase-sensitive parametric nonlinear optical processes. Shown here is an apparatus which utilizes parametric down-conversion with a pump from a frequency-doubled, mode-locked YAG laser and with $KTiOPO_4$ as the nonlinear medium. The noise output of a balanced homodyne detector is shown in the accompanying graph to rise above and fall below the vacuum noise level as the phase of the local oscillator is varied. (After Ref. 16.)

even more recent accomplishments. Systematic assessments of the potential of squeezed light in the transmission of optical information are already under way. Although the increase in channel capacity which squeezed light offers over simple direct detection schemes is only a surprisingly modest factor of two, it can in principle reduce bit error rates by an order of magnitude. Again, as with femtosecond lasers, squeezed light may have greater impact on information technology for the discoveries it produces than in its direct application.

PHOTOREFRACTIVE EFFECTS

Other novel optical phenomena discovered by physicists are somewhat more likely to find near-term use. The photorefractive effect, optical bistability, and spectral hole burning may help photons compete with electrons for switching, memory, and logic applications. In the photorefractive effect,[18] exposure causes a change in the local refractive index of the medium. Direct photorefractive effects change the electronic state of a material by direct optical transition. Indirect photorefractive effects involve local heating and consequent configurational changes in the material. These effects may be either linear or nonlinear, involving single or multiple photon exposure. Response and relaxation times as short as a few picoseconds and as long as several years have been observed, thus opening possible applications in both optical memory and switching. Tradeoffs must be made among response time, archivability, and read–write sensitivity offered by different materials. A particularly promising new direction is the photorefractive effect in doped semiconductors where the decay time, sensitivity, and operating wavelength may be controlled over rather wide ranges by choice and concentration of dopants.

SPECTRAL HOLE BURNING

A second exciting promise for optical storage is offered by a phenomenon called "spectral hole burning." [19] In certain materials optical absorption lines associated with impurity ions are inhomogeneously broadened by interaction with the random environment surrounding the absorbing ion. Strong illumination by narrow-band light falling within such broad absorption bands can bleach out the absorption over the same narrow frequency band (Fig. 11). By bleaching different spatial configurations with several different narrow-band illuminations, storage of many holograms within the *same* spatial volume may be achieved. These may be separately read without interference, by light of the appropriate wavelengths. Thus it may be possible to increase substantially the

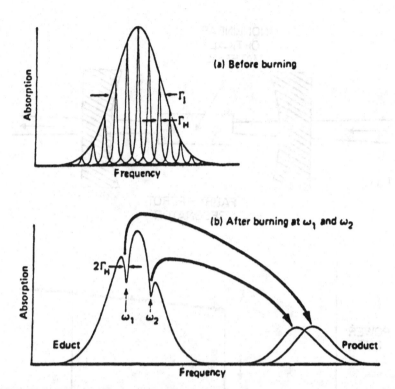

FIGURE 11 Spectral hole-burning is illustrated by the effects of laser irradiation at frequencies ω_1 and ω_2 which lie within the inhomogeneously broadened linewidth shown before burning in (a). The final spectrum (b) shows reduced absorption at ω_1 and ω_2 and a corresponding appearance of product absorption in another region of frequency space.

$10^{12}/cm^3$ bit density possible in a volume hologram which is sensitive at only one wavelength. Several hundred trillion bits (corresponding to more than ten million encyclopedia volumes) might be stored in a sugar-cube-sized material. Optimizing the relaxation times and the durability of spectrally hole-burned optical memories will require much more research and development, but the amounts of information that such memories may handle are truly prodigious.

OPTICAL BISTABILITY

Optical bistability[20] was first observed in 1976, and has already been exploited to fashion miniature optical analogs of the electronic transistor. It is perhaps most clearly illustrated by considering a resonant optical structure such as a Fabry–Perot cavity in which a nonlinear optical material is placed (Fig. 12). The resonance properties of the cavity are (by virtue of the nonlinear materials) strongly dependent upon the frequency and intensity of the light incident upon

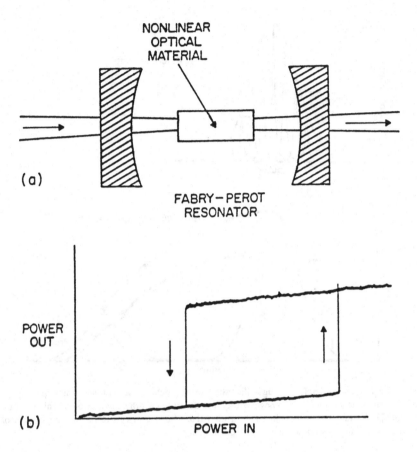

FIGURE 12 Optical bistability can be produced by irradiating a nonlinear optical material contained within a resonant cavity (a). As the incident power is increased the overall transmission (power out) suddenly increases as the cavity is driven onto resonance. Operation is bistable as shown in (b) because, once driven onto resonance, the optical intensity within the cavity is sufficient to maintain resonance until the pump intensity is reduced substantially below the onset threshold.

the cavity. The cavity may be considered a periodic pass-band filter whose transmission at a given wavelength depends upon the optical status of the material within the resonator. If this status is sensitively dependent upon the intensity of the light present, the possibility of bistable operation exists. If the total intensity of the light in the resonator comes from two sources, a "signal" beam and a "control" beam, varying the intensity in the control beam can control the transmission or reflection of the signal beam quite accurately. The device may be regarded as an optically activated optical switch. Variations on this theme permit implementation of any basic logic operation. Other related optoelectronic schemes like SEEDs[21] and QWESTs[22] are now also under investigation.

The sensitivities and response times of optical bistable devices have improved rapidly. For example, by working at wavelengths close to the band gap of a semiconductor such as gallium arsenide, the optical nonlinearities are greatly enhanced, so the optical power required for switching is correspondingly decreased. Advances in materials physics have permitted the fabrication of "multiquantum well" structures in the III–V semiconductor family, wherein the sharp and extremely strong exciton adsorption resonances in the vicinity of the semiconductor band gap may be precisely tuned and controlled. Operating a bistable device based upon a multiquantum well exciton resonance has further reduced power requirements for optical switching. Arrays of micron-sized optical logic etalons have been fabricated[23] which approach picosecond switching speeds at sub-picojoule switching energies (Fig. 13).

NEURAL NETWORKS

Not only has physics research of the past decade explored the fundamental issues of size and speed connected with information technology; it has also begun to provide new viewpoints on the behavior of complex collections of such

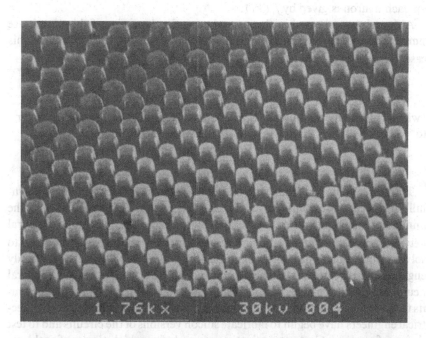

FIGURE 13 Large arrays of miniature optical etalons have been fabricated using molecular beam epitaxy to fashion multiquantum-well nonlinear materials which are then photolithographically patterned into micron diameter cylinders. Shown here are several hundred such "optical logic etalons" composed of GaAs-GaAlAs multilayers designed for maximum nonlinear response at 1.5 μ. Switching energies of less than 1 pJ and switching times shorter than 70 psec have been demonstrated for these individually addressable elements.

devices. The result may be new architectures for tomorrow's computers. A particularly exciting approach has arisen from the disparate centers of neurobiology and the physics of spin glasses.

Characteristics of mammalian brain operation include: shallow level of computation, massive parallelism, rapid convergence to "almost optimum" answers, ability to reconstruct memories from incomplete input data, content addressability, fault tolerance, etc. Many of these attributes are reproduced to some degree in highly interconnected analog circuits called "neural networks." [24] In these, each neuron is modeled by a device whose output can be either "positive" or "negative." Call it V_i. The output V_i may be fed into any or all of the other neurons in the circuit. V_i is determined in turn by the sum of all the inputs which neuron "i" receives from its neighbors, and the gain function $f(V_i)$. The state of the network of N neurons is then specified by a vector in N space, each element of which is $+1$ or -1. There are of course 2^N distinct state vectors possible in this system.

A "memory" corresponds, for example, to a stable state-of-the-network—one in which the output vector matches as closely as possible the input vector. A simplified electronic manifestation is shown in Fig. 14, where the T_{ij}'s describe the interconnection between the ith and jth neurons and the input-output curve for each neuron is given by $f_i(V_i)$.

Although complicated in the general case, the dynamics of this circuit are simple and monotonic relaxation in the special case where $T_{ij} = T_{ji}$. In this case, the quantity

$$E \equiv \sum_{ij} T_{ij} V_i V_j$$

(which is formally reminiscent of a spin-glass Hamiltonian) evolves according to

$$\frac{dE}{dt} \leqslant 0.$$

Thus after being launched with an initial input the system will "flow downhill" to the nearest minimum in the energy function E. The specifics of the minima or stable memories are dictated by the array T_{ij}. This type of neural network acts as a content-addressable memory and can be easily modified to solve optimization or pattern recognition problems as well. Physicists are busily engaged in exploring the statistical mechanics and the dynamics of more general neural nets, with some fascinating results. Neurobiologists are probing the limits to the analogies between neural nets and real biological neural systems. Electrical engineers have begun to fabricate silicon versions of the circuits and to test their performance. Computer scientists are exploring which classes of problems may be more or less well suited to this architecture.

Even as many of the key questions are still under exploration, it is not too early to examine possible optical implementations. The advantage of optics, of course, is that light beams may be sent from one point in space to any other while

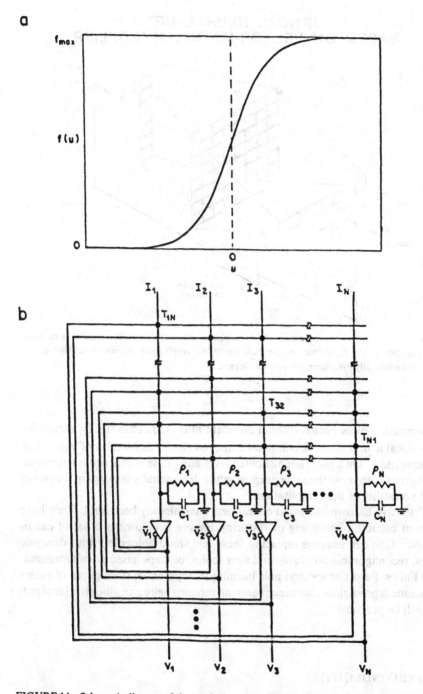

FIGURE 14   Schematic diagram of electronic "neural network" circuit. Each neuron is modeled by a small resistor-capacitor-operational amplifier combination. The normal (V_i) or inverted ($\overline{V}_i$) output of each neuron is fed back to the inputs of other network members through the resistive interconnection matrix T_{ij}.

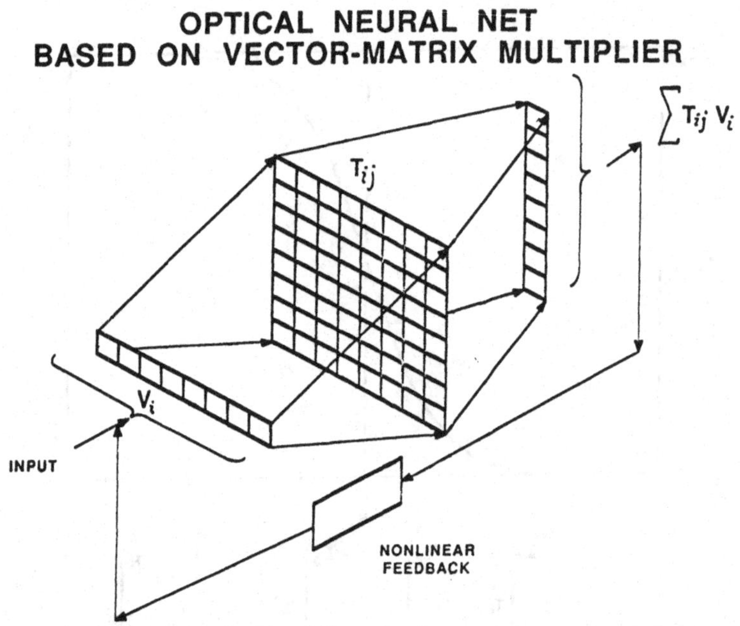

OPTICAL NEURAL NET
BASED ON VECTOR-MATRIX MULTIPLIER

FIGURE 15   Neural networks may also be implemented optically as illustrated schematically here. In this case the T_{ij} interconnection matrix may be a simple mask, a phase object (hologram) or even an optically programmable array with gain.

crisscrossing other beams without crosstalk (Fig. 15). Thus the T_{ij} matrix for an optical neural network computer can be as simple as an array of holes in an opaque mask. On a more sophisticated level any phase object (even a dynamic hologram) can serve the same purpose while dramatically leveraging the power and versatility of the computation.

Primitive hardwired optical neural nets have already been built. They have proven capable of selecting or restoring images from among a set of candidates.[25] One can imagine replacing these few static holograms with dynamic ones, reconfigurable on picosecond time scales, perhaps based on microresonator Fabry–Perots or spectral hole burning. The potential throughput of such a machine is prodigious, but much research remains before an all-optical computer will be practical.

SUPERCONDUCTIVITY

While semiconductor-based electronics has dominated information technologies for decades and continues to advance rapidly, it is being complemented if

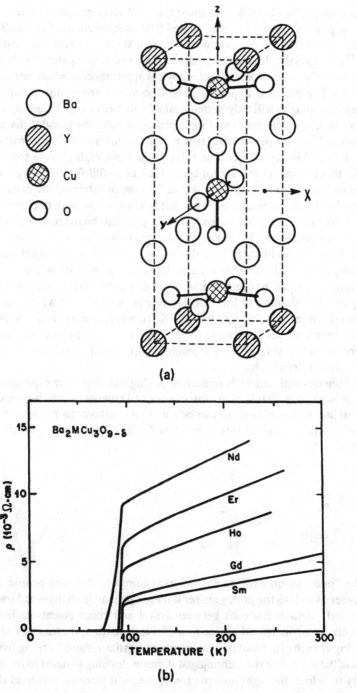

(a)

(b)

FIGURE 16   (a) Unit cell of $Ba_2YCu_3O_7$ with superconducting $T_c = 93$ K; (b) graphs of resistivity vs. temperature for several closely related materials, illustrating insensitivity of T_c toward substitution for yttrium.

not challenged by photonics—a trend that will accelerate for the next decade. How large a role the newly rejuvenated field of superconductivity will play in the future information age is uncertain at this stage. The remarkable discoveries[26,27] of materials which remain superconducting to temperatures above 90 K (Fig. 16) offer the prospect of technological application at liquid-nitrogen temperature. The practical impact of these discoveries upon information (or any other) technologies will only be determined with better understanding and control of the known materials and their processing and the possible discovery of new ones. The hope persists that new materials with still higher transition temperatures will be found. Possible roles range from high-bit-rate transmission media, through energy storage for central offices, to full-fledged Josephson logic systems. Early assessments suggest that for long or intermediate distances, superconducting transmission lines will not be competitive with optical fibers. On very short runs such as intercircuit, interchip or interboard connectors in high-speed semiconductor electronic systems, superconductors may offer advantages. This is more likely in high-density applications where speed and power consumption are at such a premium that it becomes worthwhile to cool the electronics to liquid nitrogen temperature (e.g., the next generation of supercomputers). Thin films of $Ba_2YCu_3O_7$ have been fabricated which can sustain current densities of $\sim 10^6$ A/cm^2 at 77 K. However, compatibility with silicon and electronic circuit processing conditions has yet to be demonstrated. Other papers[28,29] in this volume treat superconductivity and its technological potential in greater detail.

The physics and materials science of the high-temperature superconductors are just beginning. Surely many discoveries will emerge before they reach commercial use. But as always we can be sure that whatever technology will drive the information age in the future, it will be firmly based on research advances in physics.

CONCLUSION

The "information age" means different things to different people. For the customer as well as the policy maker it means primarily changes in how people work and interact: tradeoffs between travel and communication; leveraging productivity by increased access to useful information; increasing our understanding of each other and our environment by sharing and learning from large bases of data; etc. For the technologist it means devising systems to increase our ability to deliver the right information to the right place or person at the right time.

In this paper we have seen that the role of physics underlying present information technologies has been pervasive and essential. A cursory view of some current physics research areas has shown clear possibilities for extending these technologies in several directions in the foreseeable future by orders of magnitude. The interdependent relationships among materials, devices, and systems are becoming even more intimate as their operation approaches more closely the fundamental limits set by physical laws. At the same time, the delay between research invention and technological implementation is shrinking. Both of these trends ensure that physics research will be even more crucial for the future of the information age than it has been for its history to date. Leadership in these endeavors will continue to require a growing knowledge base generated by new generations of highly skilled and well-trained physicists.

REFERENCES

1. *The Microelectronics Revolution*, edited by T. Forester (MIT Press, Cambridge, MA, 1981), p. 381.
2. John Bardeen, in *Physics 50 Years Later*, edited by S. C. Brown (National Academy Press, Washington, DC, 1973), pp. 165–193.
3. R. S. Becker, J. A. Golevchenko, and B. S. Swartzentruber, Phys. Rev. Lett. **54**, 2678 (1985).
4. A. Cho, these Proceedings, pp. 163–175.
5. See H. L. Stormer, Surf. Sci. **132**, 519 (1983); see also these Proceedings, pp. 185–209.
6. J. R. Kwo, E. M. Gyorgy, D. B. McWhan, M. Hong, F. J. Di Salvo, C. Vettier, and J. E. Bower, Phys. Rev. Lett. **55**, 1402 (1985).
7. A. D. Benoit, C. P. Umbach, R. B. Laibowitz, and R. A. Webb, Phys. Rev. Lett. **58**, 2343 (1987); W. J. Skocpol, P. M. Mankiewich, R. E. Howard, L. D. Jackel, D. M. Tennant, and A. D. Stone, Phys. Rev. Lett. **58**, 2347 (1987).
8. P. A. Fleury, "Optical Information Technologies", in *Scientific Interfaces and Technological Applications of Physics* (National Academy Press, Washington, DC, 1986), pp. 161–185.
9. I. Hayashi, M. B. Panish, P. W. Foy, and S. Sumski, Appl. Phys. Lett. **17**, 109 (1970).
10. R. L. Fork, B. I. Greene, and C. V. Shank, Appl. Phys. Lett. **38**, 671 (1981).
11. R. L. Fork, C. H. Brito–Cruz, P. C. Becker, and C. V. Shank, Opt. Lett. **12**, 483 (1987).
12. L. F. Mollenauer and R. L. Fork, Opt. Lett. **9**, 13 (1985).
13. R. E. Slusher, L. W. Hollberg, B. Yurke, J. C. Mertz, and J. F. Valley, Phys. Rev. Lett. **55**, 2409 (1985).
14. L. Wu, H. J. Kimble, and J. L. Hall, Phys. Rev. Lett. **57**, 2520 (1986).
15. R. M. Shelby, M. D. Levenson, S. H. Perlmutter, R. G. DeVoe, and D. F. Walls, Phys. Rev. **57**, 691 (1986).
16. R. E. Slusher, P. Grangier, A. LaPorta, B. Yurke, and M. J. Potasek, Phys. Rev. Lett. **59**, 2566 (1987).
17. P. Grangier, R. E. Slusher, B. Yurke, and A. LaPorta, Phys. Rev. Lett. **59**, 2153 (1987).
18. A. M. Glass, Science **226**, 657 (1985).
19. W. E. Moerner, J. Molec. Elec. **1**, 55–71 (1985).
20. H. M. Gibbs, S. L. McCall, and T. N. C. Venkatesan, Phys. Rev. Lett. **36**, 1135 (1976).
21. D. A. B. Miller, J. E. Henry, A. C. Gossard, and J. H. English, Appl. Phys. Lett. **49**, 821 (1986).
22. L. C. West and S. J. Eglash, Appl. Phys. Lett., **46**, 1156 (1985).
23. J. L. Jewell, *et al.*, Appl. Phys. Lett. **51**, 94 (1987).
24. J. J. Hopfield and D. W. Tank, Science **233**, 625–632 (1986).
25. Y. S. Abu-Mostofu and D. Psaltis, Sci. Am. **257**, 88–95 (March 1987).

26. M. K. Wu, J. R. Ashburn, C. J. Torng, P. H. Hor, R. L. Meng, L. Gao, Z. J. Huang, Y. Q. Wang, C. W. Chu, Phys. Rev. Lett. **58**, 908 (1987); R. J. Cava, B. Batlogg, R. B. van Dover, D. W. Murphy, S. Sunshine, T. Siegrist, J. P. Remeika, E. A. Rietman, S. Zahurak, and G. P. Espinosa, Phys. Rev. Lett. **58**, 1676 (1987).
27. D. W. Murphy, S. Sunshine, R. B. van Dover, R. J. Cava, B. Batlogg, S. M. Zahurak, and L. F. Schneemeyer, Phys. Rev. Lett. **58**, 1888 (1987).
28. C. W. Chu, these Proceedings, pp. 211–242.
29. Yu. Ossipiyan, these Proceedings, pp. 243–254.

Towards the Limits of
Precision and Accuracy in
Measurement/ BRIAN W. PETLEY

INTRODUCTION

As we have made scientific and technological progress over the centuries, so too our ability to make measurements to ever higher accuracy has kept pace with it. If we look at the progress in the measurement of a typical physical quantity, the acceleration due to gravity (Fig. 1), we can see several features which are also common to other measurements. First, the precision of the relative measurements is greater than the accuracy of the absolute methods; second, each method has a limited accuracy; and third, the method replacing it leads to more rapid progress. We may generalize this as in Fig. 2. We see that (i) there is a region that is inaccessible to measurement; (ii) at any time we view a section through this; (iii) there is a gap between the frontier region, where the standards laboratories and the universities operate close to the measurement frontier, and the upper region in which measurements are increasingly common and accessible to all. Few scientists pursue accuracy for its own sake, and in most cases there are still many intermediate obstacles to overcome before we reach the ultimate limits. Accurate measurement permeates the whole of physics and, while discussing the ultimate limits, we will also highlight some of the areas of present activity, beginning with our system of measurement.

SOME ASPECTS OF OUR MEASUREMENT SYSTEM

Our System of Measurement

The subdivisions of physics are characterized by their length scale (Fig. 3). To a certain extent these subdivisions may remain separate, but there are many occasions when the connections between them become important—at which

BRIAN W. PETLEY *has devoted his career to precision measurements at the National Physical Laboratory of the UK at Teddington, Middlesex. He has worked on the exploitation of the Josephson effect for fundamental metrology, and is head of the Division of Quantum Metrology at the NPL.*

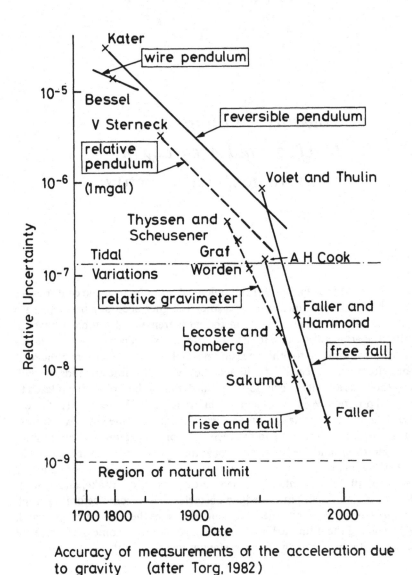

Accuracy of measurements of the acceleration due
to gravity   (after Torg, 1982)

FIGURE 1 Measurements of the acceleration due to gravity illustrate the rate of progress of
precise measurements. They show that relative measurements (- - -) are more accurate than absolute measurements (—) and that each successive method leads to a more rapid rate of progress
(after Torg, 1982).

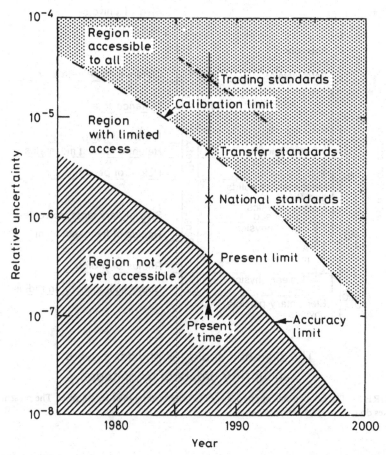

FIGURE 2 The region inaccessible to measurement and the decreased accuracy with dissemination of a physical quantity.

time it is essential to have a coherent length unit throughout. Similar arguments apply to other physical quantities and to transferring measurements from one scientific and technological discipline to another. It is the task of the theoretician as well as of the metrologist to ensure that an adequately rigid and interconnected reference framework has been erected to satisfy the demands of present-day science and technology. In cases where too much accuracy is lost in the transfer process we set up local units, particular examples being the *atomic mass unit* and the *electron volt*. Once set up, the measurement system obeys certain rules; for example, the units are additive and commutative, and have properties which are largely in accordance with those of an Abelian set. Our measurement system is an anthropomorphic one and is changed in response to scientific and technological progress.

The International System (SI) was set up as a globally universal system of units by the Conférences Générales des Poids et Mesures (CGPM) in 1960. The

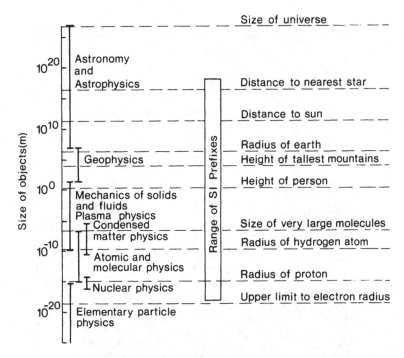

The distance scale for physics covers
about fifty orders of magnitude

FIGURE 3   The subdivisions of physics are characterized by their length scale. The present SI prefixes do not cover the whole range.

International Union of Pure and Applied Physics played a key role in setting up the SI, largely through its Symbols, Units, and Nomenclature Commission (which is today known as the SUN-AMCO Commission). The present definitions of the SI base units are given in Table I. It is evident from Fig. 3 that there cannot be a unique length unit whose magnitude is of a convenient size for all parts of physics. The accuracy of measurement of any dimensioned quantity depends ultimately on the accuracy with which the SI base units are realized; their present accuracies are shown in Fig. 4.

Some Aspects of the SI Base Units

We must always be alert to the possibility that our measurement system may hinder an accurate description of our universe. For example, we sometimes forget that there are two types of physical quantities, namely *extensive* and *intensive* quantities. A simple way of demonstrating these is to consider what happens when we cut an object in half. The extensive properties: length, resistance, mass, ..., would be changed, but the intensive properties: density, permit-

TABLE I.   The present definitions of the SI units.

second (s)
The second is the duration of 9 192 631 770 periods of the radiation corresponding to the transition between two hyperfine levels of the ground state of the cesium-133 atom.

meter (m)
The meter is the length of the path traveled by light in vacuum during a time interval of 1/299 792 458 of a second.

kilogram (kg)
The kilogram is the unit of mass; it is equal to the mass of the international prototype of the kilogram.

ampere (A)
The ampere is that constant current which, if maintained in two straight parallel conductors of infinite length, of negligible cross section, and placed 1 meter apart in vacuum, would produce between those conductors a force equal to 2×10^{-7} newton per meter of length.

kelvin (K)
The kelvin, unit of thermodynamic temperature, is the fraction 1/273.16 of the thermodynamic temperature of the triple point of water.

mole (mol)
The mole is the amount of substance of a system which contains as many elementary entities as there are atoms in 0.012 kilogram of carbon 12.
Note. When the mole is used, the elementary entities must be specified and may be atoms, molecules, ions, electrons, other particles, or specified groups of particles.

candela (cd)
The candela is the luminous intensity, in a given direction, of a source that emits monochromatic radiation of frequency 540×10^{12} hertz and that has a radiant intensity in that direction of (1/683) watt per steradian.

tivity, permeability, pressure, temperature, ..., would not. In order to measure an intensive quantity we usually change the problem into one of measuring one or more extensive physical quantities—almost without further thought. This transfer process relies on there being an exactly known relationship between the intensive property and the extensive property.

Temperature is an intensive quantity and, in order to establish the thermodynamic temperature of an object, we may use the length of a column of mercury, or the volume of a fixed mass of gas, or the resistance of a platinum wire, etc. Such relationships are, at best, linear only over a restricted temperature range. The consequent problems of establishing precise thermodynamic temperatures have necessitated the establishment of a reference scale of fixed points. This reference scale, the International Practical Scale of Temperature, known as IPTS-68, was adopted by the CIPM (Comité International des Poids et Mesures) in 1968. Progress since then has shown that many of the assigned thermodynamic temperatures were slightly in error (approximately as shown in Fig. 5). It is hoped that a revised international temperature scale will be adopted in 1990; this scale is presently termed ITS-90.

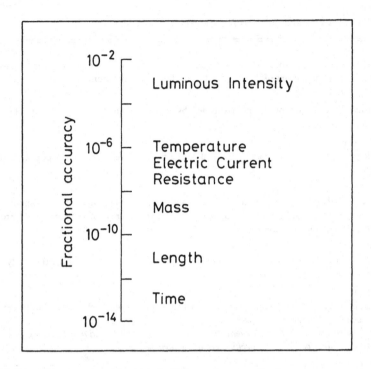

FIGURE 4   The accuracies with which the SI base units are realized at present.

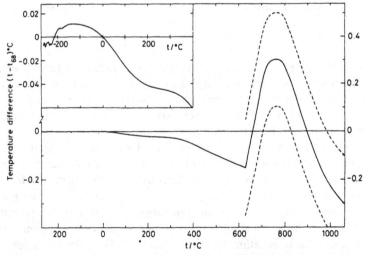

FIGURE 5   The presently measured differences between IPTS-68 and the thermodynamic temperature scale. The inset shows the differences below 400 °C on an expanded scale and indicates their irregular behavior at the lowest temperatures. The dotted lines above 630 °C show the limits of reliable use of platinum–iridium to platinum thermocouples. These are likely to be replaced by extending the resistance thermometer range up to the freezing point of silver (926 °C), and using radiation pyrometry above this temperature (after Rusby, 1987).

Magnetic Flux Density

Another area where nonlinearities between intensive and extensive quantities may be starting to manifest themselves is in the use of magnetic resonance techniques in order to measure a magnetic flux-density. Recently Fletcher *et al.* (1987) found unexpected quadratic effects in ^{85}Rb and ^{87}Rb in the ratio of the nuclear to electronic *g* factors at flux densities between 4.6 and 7.8 T which were consistent with a fractional shift of $-1.68(15) \times 10^{-9}$ T^{-2}. At the other extreme, Leggett (1973) has suggested that there might be a spontaneous breakdown of spin-orbit symmetry in ^{3}He at low temperatures and in very weak flux densities.

It is generally assumed that the NMR spectrum of a rapidly spinning sample is simply the orientationally averaged spectrum—but this may not always be so. The evolution of a quantum-mechanical system involving an adiabatically changing Hamiltonian has been receiving increasing attention of late. One of the manifestations of this is a quantum adiabatic phase, the Berry phase (Berry, 1984), which depends only on certain features of the path and the rate at which it is followed. One consequence of this, and with other effects resulting from an effective magnetic field arising from the rotating-frame transformation, is that when an NMR sample is rotated both line splittings and frequency shifts may occur. This has recently been demonstrated for nuclear quadrupole resonance by Tycko (1987), and is likely to be particularly important for solids.

While on the topic of NMR, we should mention the impressive sensitivity which has now been achieved with NMR imaging, which has enabled Aguayo *et al.* (1986) to image a single biological cell by working at a flux-density of 9.4 T.

The Role of the Fundamental Physical Constants

The fundamental physical constants (e, h, c, m_e, m_p, N_A, k, and G) provide us with invariant quantities which function as natural units in theoretical physics.

Figure 6 shows how the accuracy of the same group of constants has improved since 1965. This further illustrates the pace of measurement technology and shows quite strikingly how our knowledge of the fundamental constants is now limited by our ability to realize the ampere and ohm. For example, an absolute realization of the electrical watt $(K_W)_{LAB}$ when combined with near simultaneous measurements of the quantized Hall resistance $(h/e^2)_{LAB}$, and of the Josephson effect voltage $(2e/h)_{LAB}$, (all in the units maintained by the same laboratory) limits our knowledge of the Planck constant [Petley *et al.* (1987)]. Thus

$$(K_W)_{LAB} = (2e/h)^2_{LAB} (h/e^2)_{LAB} h/4.$$

As a result of these close relationships, the fundamental constants are being increasingly employed to improve our practical measurement system (see Table II). We now know the charge on the electron as accurately as the coulomb, the masses of the electron and proton as accurately as the gram, the

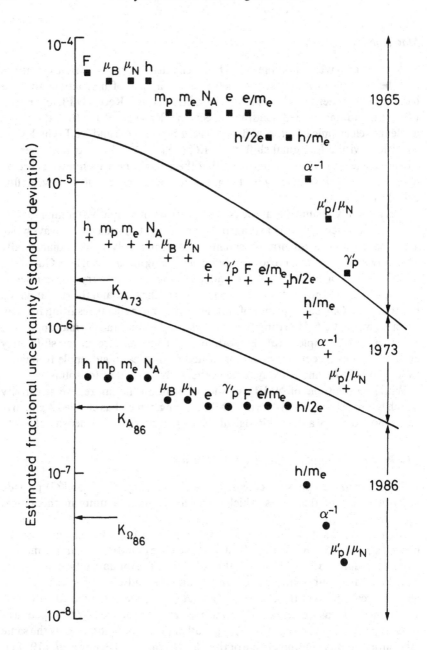

FIGURE 6   Shows how the accuracy of measurements of a particular group of fundamental constants has improved with time. In many cases the accuracy has now caught up with our ability to realize the ampere and ohm. This foreshadows future changes in the definitions of the SI units, much as the velocity of light has been incorporated into the definition of the *meter*.

quantized Hall resistance h/e^2 as accurately as the ohm; and the value of the speed of light is implicitly fixed by the definition of the meter. It seems incredible that this should be so, considering that so many of these constants were barely measured at the beginning of this century. Indeed, what other constants might our successors be able to bring to similar levels of accuracy during the next century?

The Ampere, the Kilogram, and the Meter

Nowhere is the distinction between precision and accuracy of greater importance than with regard to the electrical units, for, despite Josephson-effect voltage measurements being made in terms of the maintained unit with resolutions of the order of 10 nV, recent work has confirmed that the voltage unit disseminated by the various national standards laboratories throughout the world is about 8 microvolts smaller than that implied by the SI definitions of the ampere and watt. The disseminated ohm is similarly incorrect by about 1.5 micro-ohm [Cohen and Taylor (1986), and Giacomo (1987)]. It is expected that all countries will change their disseminated units at about the same time, probably with effect from 1 January 1990.

The Kilogram

The prototype kilogram is kept under three bell-jars in a vault at the International Bureau of Weights and Measures (BIPM) at Sèvres, France, and has long been regarded as being too precious for frequent intercomparisons with secondary standards. In fact this comparison has been effected only three times since the prototype kilogram was cast in 1878—these comparisons (61 to date) involved some 40 weighings prior to 1889, a further 7 weighings in the comparison of 1939 and another 14 in that of 1946. A new series of comparisons is about to be undertaken.

TABLE II. The phenomena involving fundamental physical constants which are already contributing to the SI base and derived units at their full accuracy.

hydrogen ground-state transition	second
the speed of light	meter
the Rydberg constant	meter
the Josephson effect	volt
the gyromagnetic ratio of the proton	tesla
the Faraday constant	coulomb
the quantized Hall resistance	ohm
the Avogadro constant	mole
the gas constant	kelvin
synchrotron radiation	candela

It appears probable that the change from the prototype kilogram to a definition based on an atomic quantity will become necessary (and possible) during the next twenty years or so. It is significant that the masses of the electron and proton are now known with the same fractional accuracy, in terms of the kilogram, as we know the gram. In many respects the problem of replacing the prototype kilogram is allied to that of finding suitable definitions of electrical units. Fundamentally we wish to ensure the identity of the electrical and mechanical units of both force and energy—we ensure this by balancing a mechanical force against an electrical one. At present this means that the definition of the ampere depends on the prior definition of the kilogram, but we might decide otherwise in the future, for example by defining a Josephson-effect volt, or the impedance of free space, or a Quantized Hall Resistance. We might then go on to recast the ampere definition to define the kilogram instead—again balancing an electrical force against a mechanical one. We could also involve the atomic mass unit with the Avogadro constant and take an atomic definition of mass. While it is interesting to speculate on future possibilities, we are unlikely to want to change the definition (of the kg) until we can either detect changes in the absolute mass of the prototype kilogram or until we are sure that realizations of the new definition have the potential to advance to higher levels of metrological accuracy. At present we would merely exchange the imprecision of the electrical units for a diminished accuracy of our unit of mass. Further, although the Josephson effects are now twenty years old, the quantized Hall effects found by Klitzing *et al.* (1980) are still in their first decade, and it is possible that quantum physics may yet produce something even more spectacular relating to this overall metrological problem.

SOME HIGH-ACCURACY MEASUREMENTS

Laboratory Measurements with Ion Traps

Spectacular advances in accurate measurement have become possible through the development of ion traps which are capable of trapping single charged particles for long periods of time. Dehmelt, van Dyck, and Schwinberg at the University of Washington made major contributions to this development, as did others at Boulder and Mainz. A good example of the accuracy attained is that of the comparison of the magnetic moments of single electron and positron by van Dyck *et al.* (1987). They used a Penning trap in a flux density of 5 T and measured the quotient of the *g* values (or magnetic moments in terms of the Bohr magneton) of the electron and positron as:

$$g(e^-)/g(e^+) = 1 - 0.5(2.1) \times 10^{-12}$$

and a value for the electron *g* factor anomaly $(g_e - 2)/2$ of

$$a(e^-) = 1\ 159\ 652\ 188.4(4.3) \times 10^{-12}.$$

The first trapping of an antiparticle was reported by this group in 1979. They also contributed to the trapping of antiprotons in an ion trap at CERN, as a necessary precursor to enabling antihydrogen to be generated and studied at some future date (Gabrielse *et al.*, 1986). In other measurements (i) the ratio of the electron to proton mass was determined to parts in 10^8, (ii) $g - 2$ measurements were made on a single positron for 110 days (Schwinberg *et al.*, 1981), and (iii) relativistic measurements were made on a single trapped electron over a period of eleven months—spectacular work indeed! Their work is setting tough targets for those making the high-order calculations of the anomalous magnetic moments of the electron and positron (e.g., Kinoshita and Lindquist, 1981, 1983, and Samuel, 1986), for the theoretical work already involves some 856 Feynman diagrams to calculate the fourth order term.

Single Atoms and Ions

During the last year, measurements have been made of the transitions of a single atom. These include (i) work with a velocity-selected beam of ^{85}Rb atoms in a high Rydberg state (involving the $5^2S_{1/2}$, $F = 63$, $P_{3/2}$ line) traversing a 21.5 GHz microwave superconducting cavity having a long photon storage time and containing only about 2.5 thermal photons at any one time (Rempe *et al.*, 1987); (ii) the observation of "quantum jumps" in a single ^{198}Hg$^+$ ion (essentially the ion is scattering some 5×10^7, 194 nm photons per second, and this is changed when the ion is pumped to another, 281.5 nm level by two-photon absorption of another dye-laser beam) (Bergquist *et al.*, 1986); and also (iii) the change in transmission of a 3 μW focused beam of 194 nm laser light due to the absorption by a single ^{198}Hg$^+$ ion. This was confined to a $(0.25 \, \mu m)^3$ space in an rf trap (Wineland *et al.*, 1987).

Such measurements help to set the pinnacles of accuracy and achievement—heights towards which others will strive in the future. They also stimulate other technological improvements, for, in order to avoid dephasing gas collisions, it will be necessary to devise methods of attaining (and measuring) a very low pressure, that is below 10^{-10} Pascal (10^{-12} Torr). We can expect similarly spectacular measurements on trapped atoms now that optical resonance radiation has been used to slow and trap atoms for several seconds (Chu *et al.*, 1986).

Hughes–Drever Experiments

Ion traps have enabled tighter limits to be imposed on the Hughes–Drever type of experiment. These involve searching for anisotropies in the observed spin precession frequency as a function of the orientation of the spin axis in space. Prestage *et al.* (1985) reduced the Hughes–Drever limit by a factor of about 300, to about 10^{-4} Hz, by looking for effects on ^{9}Be$^+$ ions in an ion trap at 0.82 *T*. Lamoraux *et al.* (1987) have taken the upper anisotropy limit down to the microhertz level for both cos θ and cos$^2\theta$ terms. They compared optically pumped ^{201}Hg ($I = 3/2$) with ^{199}Hg ($I = 1/2$) in a flux density of 20 mT. This sets constraints, for example, on the fractional size of Lorentz-noninvariant electromagnetic couplings as being $\leqslant 10^{-21}$. Searches for spatial anisotropies by Michelson–Morley experiments have reached the part in 10^{15} level of sensitivity with the work of Brillet and Hall (1979). The above ion trap work also has implications for the isotropy of the speed of light, setting limits on the value of the $[(c_0/c)^2 - 1]$ term at the 10^{-22} and 10^{-18} level, respectively.

GRAVITATION AND FORCE MEASUREMENTS

For the most part, force measurements are not presently made to much better than a tenth part in a million accuracy, and modern gravity meters are approaching the natural limit of meaningful terrestrial measurements at the 10^{-9} *g* accuracy (Marsen and Faller, 1986). We note in this connection that Niebauer *et al.* (1987) have recently compared the gravitational accelerations of copper and uranium masses with an uncertainty of 5 parts in 10^{10}, demonstrating that force comparisons will greatly exceed the absolute limits.

Gravitation provides one of the weakest forces which are amenable to experiment. Technological advances have brought only a factor of 50 in the accuracy of *G* during the last eighty years, and it may be that the best measurements of *G* will be made in space satellite experiments. This slow progress is partly because the gravitational forces between the masses which one can use in a laboratory measurement are not much more than a part in 10^9 of the attraction due to the Earth. Consequently, in order to measure *G* to 0.01%, one must remove the effect due to the Earth at the level of a part in 10^{13}! The suspensions in torsion balances are operated at near breaking stresses in order to achieve the required sensitivity to torsional force.

There is still discussion concerning the exact nature of the law of gravitational force; that is, whether it is an exact inverse-square law. For example, it has been thought that there may be a medium-range non-Newtonian repulsive component of the field. Some work by Stacey and Tuck (1981) seemed to support a short-range force. More recently Fischbach *et al.* (1986), following a reanalysis of some of the earlier Eötvös (1922) torsion balance measurements, suggested that the non-Newtonian term might be due to a hitherto unknown baryon–baryon interaction and that this interaction might also explain some anomalous results found for the $K^0 - \bar{K}^0$ system by Aronson *et al.* (1982).

Although the earlier Eötvös results could have an alternative explanation, it certainly shows that we should not take anything for granted. The recent reanalysis has stimulated a number of experimental investigations. It is too soon for all of these to have been completed, but Thieberger (1987) has already reported a positive result for the sideways movement of a 5 kg evacuated hollow copper sphere, which was almost totally immersed in water, when it was located near the edge of a 161 m high cliff. This movement corresponded to a sideways force of only about 4 nN on the sphere and it may yet prove to have an alternative explanation. Adelberger *et al.* (1987) and Stubbs *et al.* (1987) operated a torsion balance on the edge of a hill and looked for effects between aluminum and beryllium. Niebauer *et al.* (1987) measured the differential acceleration in free-fall between copper and uranium masses and found results consistent with no effect, in accordance with some earlier work. Earlier results are relevant too, and these have been reviewed by Cook (1987) and Stacey *et al.* (1987). Overall, there appears still to be a small window for possible effects in the range between one and ten meters, and the discussion will doubtless continue as a variety of other experiments are completed.

Effects such as gravitational shielding are expected to be very small, for the gravitational permeability of free space has been predicted to be of the order of $16\pi G/c^2$ ($\approx 10^{-26}$ m/kg) and gravitational absorption shown to be less than 6×10^{-16} (see the review by Gillies, 1987).

Donoghue and Holstein (1987) have suggested that, according to the equivalence principle, the fractional decrease in acceleration $\delta g/g$ of a mass m, at a temperature T, is given by

$$\delta g/g = 2\pi\alpha(kT/mc^2)^2 ,$$

that is, heavier and/or cooler objects fall faster than their lighter and/or hotter counterparts, but even for electrons this would only be at around the 10^{-17} level of accuracy—which is well below present detection limits.

Gravitational Effects and SQUIDS

Atomic clocks have, of course, already reached sufficient accuracy to require gravitational corrections in order to allow for the altitude of the clock, and "flying clock" relativity experiments have been performed. For example, the gravity correction between the NBS Boulder laboratories and sea level is about 3×10^{-13}.

Pound and Rebka (1960) observed the gravitational energy increase of a γ-ray photon after falling through a distance of 23 m, and their work has since been extended to verify the gravitational shifts at the 10^{-4} precision level. These shifts are at the part in 10^{16} level of accuracy at optical frequencies and may soon be investigated experimentally by comparing measurements of narrow transitions made in laboratories at different altitudes.

In a different type of experiment Jain, Lukens, and Tsai (1987) have looked for gravitational effects at the 10^{-22} V level between two Josephson junctions which were biased by a 16.8 GHz microwave source and separated by a vertical height of only 72 mm. This separation produced a gravitational red-shift equivalent to a Josephson voltage of 2.35×10^{-21} V. Their voltage detection sensitivity enabled them to verify at the 4% level that the emf round the loop was consistent with the invariance of the gravito-electrochemical potential along the wires. This provides a striking illustration of the sensitivity of present-day voltage measurements. In earlier measurements, Tsai, Jain and Lukens (1983) set material dependence limits on the Josephson effects at the 10^{-15} level of precision.

Their work also reminds us of the energy-detection sensitivity of SQUID magnetometers, which is now less than h per unit bandwidth—that is $\leqslant 6.6 \times 10^{-23}$ J/Hz (Fagley, 1987). Commercially available SQUID magnetometers have flux detection sensitivities of the order of 10^{-4} Φ_0 per hertz$^{1/2}$ (10^{-19} Wb/Hz$^{1/2}$), input current sensitivities of 10 pA/Hz$^{1/2}$, and voltage sensitivities such that one can detect the noise from a 10^{-4} Ω resistor at 4.2 K, that is about 0.2 pV per Hz$^{1/2}$. The new high-temperature superconductors are showing promise as SQUIDS (Zimmerman *et al.*, 1987), and as magnetic shields at liquid nitrogen temperatures, and commercial devices will surely follow.

OVERCOMING SOME PRESENT LIMITS

The highest accuracies achieved so far relate to time and length measurements, and we consider some aspects relating to future improvements in the measurement of these quantities.

The Search for Narrower Linewidths

One of the factors limiting accuracy is the width, or definition, of the quantity measured—whether it be the thickness of a line on a ruler or the frequency width of some quantum transition. Commercially available cw tunable dye lasers have a linewidth of the order of 0.5 MHz, or a Q value ($f/\delta f$) of about 10^9, and the linewidth of a present-day helium–neon laser stabilized to an iodine transition at 633 nm has a similar Q value. Similarly the linewidth in a cesium-beam atomic clock is in the region of 50 Hz in a frequency of 9.2 GHz, or a Q value of 2×10^8. An even narrower transition is the ^{57}Fe γ-ray transition which has been used to study the Mössbauer effect.

The $1S$–$2S$ transition in hydrogen has an intrinsic lifetime of about 0.7 s. It has recently been studied at the 10^{-9} level of accuracy by both pulsed and cw laser techniques by several workers (Barr *et al.*, 1986, Hildum *et al.*, 1986, and Beausoleil *et al.*, 1987). The Rydberg constant is presently being measured in the visible region with an accuracy of 1.6 parts in 10^{10} (Zhao *et al.*, 1987). The accuracy required is stretching that of the present-day realizations of the meter to the limit. Spectroscopic measurements of the Lamb shift are now extending from the Lyman-alpha level in atomic hydrogen, via muonium, through to helium-like U^{90+}—with few enduring differences between theory and experiment.

There may be even narrower transitions than the above which are yet to be explored, with Q values in the region of 10^{18} or more—that is, having lifetimes of the order of hundreds of seconds. Some possible examples are illustrated in Fig. 7. Of course it is debatable whether it will ever prove practicable to study transitions lasting longer than the lifetime of either the metrologist or the apparatus!

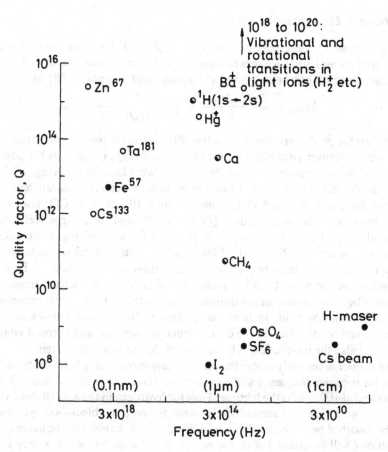

FIGURE 7 Some narrow-linewidth, high "Q" transitions throughout the frequency spectrum: ●—studied, ◑—partly studied, ○—for the future (developed from Letokhov and Chebotaev, 1977).

Suppression of Spontaneous Decay

If the above transitions are not narrow enough, or prove inconvenient to use, the lifetimes of these or other transitions may be extended by increasing the impedance mismatch between the atom and the surrounding medium—for example, by bringing it close to a metal surface. In this way Anderson *et al.* (1987) have reported extending the 1.6-μs lifetime of the 3.49 μm transition of cesium atoms in the $5D_{5/2}$ level by approximately 13 natural lifetimes, that is to 20.8 μs. It must be remembered that, although narrowing the transition, the cavity may pull the center of the resonance. Consequently the narrower line produced by this technique may be useful for the precision required in say, gravity-wave detectors, but be less suitable for measurements where long-term fidelity is required.

The Intrinsic Linewidths of Lasers

Before most of these narrow lines can be studied it will be necessary to develop lasers and other sources to have the necessary monochromaticity. The intrinsic linewidth of a laser is given (Schawlow and Townes, 1958) by

$$\Delta v_{\text{laser}} = \frac{2\pi h v (\Delta v_{1/2})^2}{P_e} \cdot \frac{N_2}{N_2 - (g_2/g_1)N_1},$$

where hv is the photon energy, $\Delta v_{1/2}$ is the full width of the passive resonance, P_e is the emitted power (including mirror losses), and N_2, g_2 (N_1, g_1) are the population and the degeneracy factor of the upper (lower) laser level (Wing, 1983). According to this equation, a 633-nm, 1-mW, helium–neon laser should have an intrinsic linewidth of less than a millihertz and a 10-μm, 3-W, CO_2 laser less than a microhertz, corresponding to Q values of 10^{18} and 10^{20}, respectively, and one could expect even higher intrinsic Q values for a 12-W argon–ion laser. Although the intrinsic linewidth of lasers provides us with an ultimate limit, other noise processes must be greatly reduced if these are to be achieved. Examples would be the noise from the pump source affecting the active atoms, or collisions between atoms, and also spontaneous emission. Other noise processes affect the radiation field, such as fluctuations in the refractive index of the medium, vibrations affecting the cavity structure, acoustic and thermal vibrations, and also the interaction of black-body radiation with the system.

One cannot necessarily ignore the effects of any surrounding black-body radiation, for it causes frequency shifts (which are similar to dynamic Stark effect shifts) and also shortens the lifetime. Thus for hydrogen in the $n = 10$ state, the shift is about 1 kHz, and around 35 000 room temperature black-body photons will be absorbed per second (Farley and Wing, 1981). Evidently, the narrowest transitions will be studied with the atoms, ions, or molecules in a very low temperature environment—particularly those in the infrared region of the spectrum.

The narrowest laser linewidths achieved to date are around fifty mH (Hils *et al.*, 1987), corresponding to a Q value of 5×10^{16}. For these lasers the linewidth is limited by measurement noise processes in the servo-system rather than by the above Schawlow–Townes limit. It is of interest that some commercially available helium–neon lasers (those having internal mirrors) can have short-term linewidths of around 10 kHz, or a Q of 5×10^{10} (Hall *et al.*, 1987).

Overcoming Some Limiting Properties of Light

It is salutary to remember that the lasers used for the elegant present-day spectroscopy have a fractional linewidth that is up to a million times poorer than that of the best radiofrequency sources. We can see therefore that there appears to be some way to go before the ultimate measurement limits are reached.

The linewidth of laser radiation sets a limit to the visibility of laser fringes at long pathlengths. A further property of a laser beam, which prevents us from using very long pathlengths in order to achieve increased measurement accuracy, is that the laser beam spreads out as a consequence of its being a Gaussian beam. That is, a beam of wavelength λ and initial radius w_0 has a radius $w(z)$ at a distance z given by:

$$w(z) = w_0 [1 + (\lambda z / \pi w_0^2)^2]^{1/2} .$$

While a 10-mm-diameter helium–neon laser beam may only have doubled in diameter after traveling about a kilometer, in the laser lunar-ranging experiments (where only a few reflected photons per laser pulse are required at the detector), the ruby-laser beam diameter is several km by the time the laser radiation reaches the moon. This type of beam spreading has been reduced recently in the laboratory by producing a laser beam which has a special intensity profile, essentially comprising several annular rings with the intensity inversely proportional to their radii (Durnin *et al.*, 1987).

The Uncertainty Principle and Squeezed Light

The necessity to improve the sensitivity of apparatus designed to detect gravity waves has led to considerations of how the uncertainty principle might be circumvented and has stimulated the study of squeezed light states. These states have the property that the quantum fluctuations in one phase of the wave motion are less than those of the vacuum state of the electromagnetic field, while those of the quadrature phase are correspondingly greater. This is a very active area at present.

Slusher *et al.* (1985) generated squeezed light by four-wave mixing of two counterpropagating laser beams in a nonlinear medium (a beam of sodium atoms). Either the amplitude noise or the phase noise could be reduced by appropriate movement of the beam-combining mirrors. Wu *et al.* (1987) later reported a reduction of the noise by 63% using parametric down-conversion of photons of frequency $2v$, split to produce correlated pairs of photons at the subharmonic frequency v. Xiao *et al.* (1987) have reported a 3-dB reduction below the shot noise limit. A number of attractive applications are possible once the noise is reduced by a factor of 10 or so, and we discuss next the limitations of shot noise in gravity wave detectors.

Gravity-Wave Detectors

Gravity wave detectors which have strain detection sensitivities in the region of 10^{-18} to 10^{-21} are either in operation or under construction in a number of laboratories. These experiments push the measurement precision to the greatest extremes, although the apparatus is not, of course, required to be dimensionally stable over extended periods.

There are formidable problems which are steadily being overcome; for example, in the 1 km multi-path-length interferometers that are under construction for gravity wave experiments, one must take account of photon shot-noise—even when 200 W lasers are used. This leads to a strain detection limit given by

$$\frac{\Delta l}{l} = \frac{1}{2\pi\tau_s}\sqrt{\frac{\lambda h(1-R)}{c\epsilon I_0\tau_a}},$$

where τ_s = interferometer storage time, τ_a = averaging time, ϵ = photodetector efficiency, and R = mirror reflectivity. For $I_0 = 200$ W, $1-R = 10^{-4}$, $\epsilon = 0.5$, interferometer arm length = 1 km, this gives $\Delta l/l \approx 5 \times 10^{-23}/\sqrt{\tau_a}$ (Weiss, 1987).

SOME ULTIMATE LIMITS

Noise and Resistance Measurements

Fluctuations in the energy, of the order of $\frac{1}{2}kT$ per degree of freedom, affect many types of measurement at some level or other. Thus a 60-mm diameter brass bar of length l fluctuates in length by an rms amount Δl given by

$$\left(\frac{\Delta l}{l}\right)^2 = \frac{4kT\Delta f}{\pi^3 QA\rho c^3}$$

where, for brass, ($c \sim 4.4$ km/s, $\rho \sim 8.4 \times 10^3$ kg/m^3, $Q \sim 10^4$, and $\Delta f \sim 1$ Hz) the fluctuations are about 2 parts in 10^{19} at room temperature. The Nyquist noise, or Johnson noise voltage associated with a resistor R in a bandwidth δf and temperature T, is $(4kTR\delta f)^{1/2}$, and this limits the accuracy with which we may expect to measure resistance. Thus all resistors have both a temperature coefficient and a related power coefficient of resistance which is significant at around the part per million level. Consequently, for accurate resistance measurements the electrical power P_R that may be dissipated in the resistor is limited to between one and ten milliwatt. Taking account of this and the Johnson noise (and taking $T = 1$ K, $\delta f = 0.01$ Hz, $P_R = 10$ mW), the noise-limited fractional measurement precision becomes $(4kTR\delta f/P_R)^{1/2}$, or about 7×10^{-12}. This ballpark limit is independent of the value of the resistance. It does not necessarily apply to the measurement of capacitance or to measurements of other types of 'resistance' such as the quantized Hall resistance.

1/f Noise

The above limit is based on assuming that the $f^{1/2}$ noise process continues out to long averaging times. We know, however, that in solid state devices there is flicker effect, or $1/f$ noise, which is avoided by amplifying radio and other signals at as high a frequency as possible. These $1/f$ processes at present limit the accuracy of atomic clocks (and some lasers too) for longer averaging times (Fig. 8). The $1/f$ noise may not stem from fundamentally limiting physical processes, and may be attributable to inadequate isolation from man-made disturbances—temperature fluctuations, etc. This type of noise is especially worry-

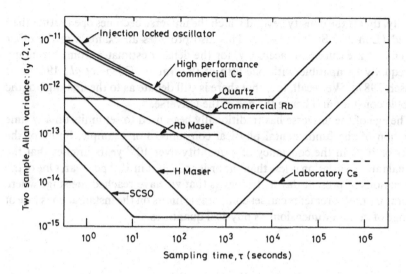

FIGURE 8 The accuracy of atomic clocks as a function of the averaging time, showing that at present $1/f$ noise becomes important in all cases at long averaging times (after Hellwig, Arditi *et al.*).

ing in experiments which look for small effects over a long period, for example in quark detectors, or in gravity-wave detectors.

These $1/f$ processes are being explored in a wide variety of measurement situations, and theories are now being developed (Bak *et al.*, 1987). In addition to lasers and clocks, these include electrical networks (Kai *et al.*, 1987), alpha-particle decay (Prestwick *et al.*, 1986), the interplanetary magnetic field (Matthaeus and Goldstein, 1986), and also in cryogenic devices such as small tunnel junctions and SQUIDs. Measurements of the properties of very thin conductors are revealing $1/f$ effects and also new types of quantum effect consequent on the scattering of the conduction electrons (Benoit *et al.*, 1987, and Skocpol *et al.*, 1987). Such processes may set limits to the miniaturization of integrated circuits and will probably also lead to new devices.

The Age of the Universe and Ultimate Accuracy

Taking account of the Heisenberg uncertainty principle relating the uncertainty of the energy (δE) with the measurement time (δt) and the Planck constant (h) as

$$\delta E \, \delta t \approx h$$

we can see that, from some points of view, the ultimate measurement accuracy might be expected to come from an "experiment" which involved the whole universe and which endured for the life of the universe. The cosmologists have found one such measurement, for apparently the present dimensions of the universe imply that the density must lie within a few parts in 10^{60} of the critical density, that is the density beyond which the universe becomes open rather than closed (Loh and Spillar, 1986). This also provides an example of the other extreme of measurement accuracy, for the density estimates from astronomy are equally compatible with one-fifth of this value (Cowie *et al.*, 1987, and Kaiser, 1987)! We recall, too, that there is still debate as to the first digit of the Hubble constant and hence the age of the universe.

The age of the universe has traditionally been used to set limits on any time variation of the fundamental physical constants. For example, limits of the order of 10% in the constancy of a quantity over 10^{10} years implies that the annual rate of variation is less than the order of a part in 10^{11} per year. One of the fascinations of present-day metrology is that we have reached the stage where laboratory measurements can set comparable limits on the instantaneous rate of change of certain dimensionless physical quantities.

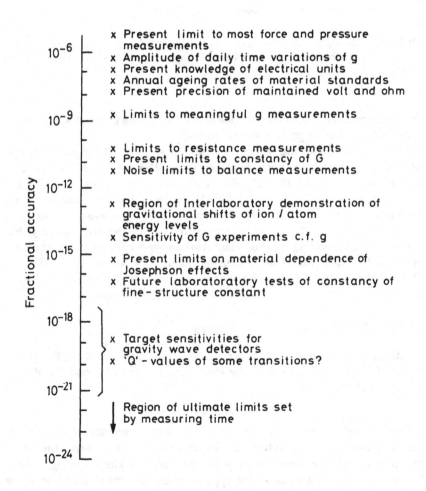

FIGURE 9 Some present accuracies and some ultimate measurement limits.

CONCLUSION

We live in an age where tests of relativity are becoming increasingly common, as too are elegant tests of the subtleties of quantum theory. It is very much a period where small departures from exactness are providing critical tests of theory. It is no longer sufficient for something to exist for the timescale of the universe for it to be considered as infinite; for example, the experimental establishment of a proton lifetime of at least 10^{32} years suddenly became an important test of Grand Unified Theory only a few years ago. Departures from equality of the magnitudes of the charges on the electron and the proton at the level of $10^{-20}\ e$ could have profound effects on fundamental theories, and so it goes on.

We seem to be in a period corresponding to one of the surges forward in both our understanding and our ability to measure which were forecast by Maxwell just over a century ago (see Petley, 1985). It is just this unpredictability of the nth decimal place that motivates us to refine our measurement system, for physicists never know which digit of a measurement may prove the most significant. Thus for the magnetic moment of the proton it was the first digit that provided the surprise, for the g factor of the electron it was the third. For the hydrogen atom the Lamb shifts in the optical region lie in the sixth decimal place and beyond; so, too, do many of the differences between classical physics and relativity. Our journey is far from completed, and in most cases the ultimate frontiers of measurement are sufficiently far away for us to safely leave the beauty of the ultimate nth decimal place as something for our grandchildren to study (Fig. 9)—and, judging by what we have experienced of late, that should be quite something!

REFERENCES

E. G. Adelberger, C. W. Stubbs, W. F. Rogers, F. J. Raab, B. R. Heckel, J. H. Gundlach, H. E. Swanson, and R. Watanabe, Phys. Rev. Lett. **59**, 849 (1987).

J. B. Aguayo, S. J. Blackband, J. Schoeniger, M. A. Mattingley, and M. H. Hintermann, Nature **322**, 190 (1986).

S. H. Aronson, G. J. Bock, H. Y. Cheng, and E. Fischbach, Phys. Rev. Lett. **48**, 1306 (1982).

P. Bak, C. Tang, and K. Wiesenfeld, Phys. Rev. Lett. **59**, 381 (1987).

J. R. M. Barr, J. M. Girkin, J. M. Tolchard, and A. I. Ferguson, Phys. Rev. Lett. **56**, 580 (1986).

R. G. Beausoleil, D. H. McIntyre, C. J. Foot, B. Couillard, and T. W. Hänsch, Phys. Rev. A **35**, 4878 (1987).

A. Benoit, C. P. Umbach, R. B. Laibowitz, and R. A. Webb, Phys. Rev. Lett. **58**, 2343 (1987).

J. C. Bergquist, R. G. Hulet, W. M. Itano, and D. J. Wineland, Phys. Rev. Lett. **57**, 1699 (1987).

M. V. Berry, Proc. R. Soc. Lond. Ser. A **392**, 45 (1984).

A. Brillet and J. A. Hall, Phys. Rev. Lett. **42**, 549 (1979).

S. Chu, J. E. Bjorkolm, and A. Cable, Phys. Rev. Lett. **57**, 314 (1986).

E. R. Cohen and B. N. Taylor, Codata Bull. **63** (1986).

A. H. Cook, Contemp. Phys. **28**, 159 (1987).

L. Cowie, M. Henrickson, and R. Mushotzky, Ap. J. **317**, 593 (1987).

J. F. Donoghue and B. R. Holstein, Eur. J. Phys. **8**, 105 (1987).

J. Durnin, J. J. Miceli, and J. H. Eberly, Phys. Rev. Lett. **58**, 1499 (1987).

R. S. van Dyck, Jr., P. D. Schwinberg, and H. G. Dehmelt, Phys. Rev. Lett. **59**, 26–29 (1987).

R. L. Fagley, Sci. Prog. Oxf. **71**, No. 2, 181–202 (1987).

J. W. Farley and W. H. Wing, Phys. Rev. A **23**, 2397 (1981).

E. Fischbach, D. Sudarsky, A. Szafer, C. Talmadge, and S. H. Aronson, Phys. Rev. Lett. **56**, 3 (1986).

G. D. Fletcher, S. J. Lipson, and D. J. Larson, Phys. Rev. Lett. **58**, 2535 (1987).

G. Gabrielse, X. Kai, K. Helmerson, S. L. Rolston, R. Tjoekler, T. A. Trainer, H. Kalinowsky, and J. Haas, Phys. Rev. Lett. **57**, 2504 (1986).

P. Giacomo, Metrologia **24**, 45 (1987).

G. T. Gillies, Metrologia, Suppl. to Vol. **24** (1987).

J. L. Hall, M. Long-Sheng, and G. Kramer, IEEE J. Quantum Electron. **QE-23**, 427 (1987).

E. A. Hildum, V. Boesl, D. H. McIntyre, R. G. Beausoleil, and T. W. Hänsch, Phys. Rev. Lett. **56**, 576 (1986).

D. Hils, G. Soloman, and J. L. Hall, J. Opt. Soc. Am. B (1987).

A. K. Jain, J. E. Lukens, and J.-S. Tsai, Phys. Rev. Lett. **58**, 1165 (1987).

W. Jhe, A. Anderson, E. A. Hinds, D. Meschede, L. Moi, and S. Haroche, Phys. Rev. Lett. **58**, 666 (1987).

S. Kai, S. Higaki, M. Imasaki, and H. Furukawa, Phys. Rev. A **35**, 374 (1987).

N. Kaiser, Mon. Not. R. Astron. Soc. **227**, 1 (1987).

T. Kinoshita and W. B. Lindquist, Phys. Rev. Lett. **47**, 1573 (1981); Phys. Rev. D **27**, 867, 877, 886 (1983).

A. J. Leggett, J. Phys. C **6**, 3187 (1973).

K. Von Klitzing, G. Dorda, and M. Pepper, Phys. Rev. Lett. **45**, 494 (1980).

S. K. Lamoreaux, J. P. Jacobs, B. R. Heckel, F. J. Raab, and E. N. Fortson, Phys. Rev. Lett. **57**, 3125 (1986).

V. S. Letokhov and V. P. Chebotaev, *Nonlinear Laser Spectroscopy* (Springer, Berlin, 1977).

E. D. Loh and E. J. Spillar, Astrophys. J. Lett. **307**, L1 (1986).

I. Marson and J. E. Faller, J. Phys. E **19**, 22 (1986).

W. H. Matthaeus and M. L. Goldstein, Phys. Rev. Lett. **57**, 495 (1986).

T. M. Niebauer, M. P. McHugh, and J. E. Faller, Phys. Rev. Lett. **58**, 609 (1987).

B. W. Petley, *The Fundamental Physical Constants and the Frontier of Measurement* (Adam Hilger, London, 1985), p. 314.

B. W. Petley, A. Hartland, and B. P. Kibble, Nature **327**, 605 (1987).

R. V. Pound and S. A. Rebka, Phys. Rev. Lett. **4**, 337 (1960).

J. D. Prestage, J. J. Bollinger, W. M. Itano, and D. J. Wineland, Phys. Rev. Lett. **54**, 2387 (1985).

W. V. Prestwick, T. J. Kennett, and G. T. Pepper, Phys. Rev. A **34**, 5132 (1986).

G. Rempe, H. Walther, and N. Klein, Phys. Rev. Lett. **58**, 353 (1987).

R. L. Rusby, Meas. and Control **20**, 9 (1987).

M. Samuel, Phys. Rev. Lett. **57**, 3133 (1986).

A. L. Schawlow and C. H. Townes, Phys. Rev. **112**, 1940 (1958).

P. B. Schwinberg, R. S. Van Dycke, Jr., and H. G. Dehmelt, Phys. Lett. **81A**, 119 (1981).

W. J. Skocpol, P. M. Mankiewich, R. E. Howard, L. D. Jackel, and D. M. Tennant, Phys. Rev. Lett. **58**, 2347 (1987).

R. E. Slusher, L. W. Hollberg, B. Yurke, J. C. Mertz, and J. F. Valley, Phys. Rev. Lett. **55**, 2409 (1985).

F. D. Stacey and G. J. Tuck, Nature **292**, 230 (1981).

F. D. Stacey, G. J. Tuck, G. I. Moore, S. C. Holding, B. J. Goodwin, and R. Zhou, Rev. Mod. Phys. **59**, 157 (1987).

C. W. Stubbs *et al.*, Phys. Rev. Lett. **58**, 1070 (1987).

P. Thieberger, Phys. Rev. Lett. **58**, 1066 (1987).

J.-S. Tsai, A. K. Jain, and J. E. Lukens, Phys. Rev. Lett. **51**, 316 (1983).

R. Tycko, Phys. Rev. Lett. **58**, 2281 (1987).

R. Weiss, private communication (1987).

D. J. Wineland, W. M. Itano, and J. C. Bergquist, Opt. Lett. **12**, 389 (1987).

W. H. Wing, Proc. 1983 Symp. on Precis. Meas. and Gravitation, edited by Wei-Tou Ni (Nat. Tsing Univ., Taiwan, 1983), pp. 325–388.

L.-A. Wu, H. J. Kimble, J. L. Hall, and H. Wu, Phys. Rev. Lett. **57**, 2520 (1986).

M. Xiao, L.-A. Wu, and H. J. Kimble, Phys. Rev. Lett. **59**, 278 (1987).

P. Zhao, W. Lichten, H. Layer, and J. Bergquist, Phys. Rev. Lett. **58**, 1293 (1987).

J. E. Zimmerman, J. A. Ball, M. W. Cromer, and R. H. Ono, Appl. Phys. Lett. **51**, 617 (1987).

High-Temperature Plasma Physics/ HAROLD P. FURTH

1. INTRODUCTION

The basic theoretical principles of plasma physics were developed during the period 1930–1950. The temperatures of experimentally available plasmas, however, were generally not much above the first-ionization energy range. Within this "low-temperature" regime, many practical applications of plasma physics were found in such fields as illumination, electrical switching, and the understanding of the ionosphere. During the 1950s, there emerged a highly ambitious new application: the controlled release of fusion energy from the nuclei of heavy hydrogen. The pursuit of this goal shifted the relevant plasma temperature from the atomic to the nuclear energy scale, and brought about the development of modern plasma physics.

The comprehensive density-temperature diagram for plasma physics[1] is shown in Fig. 1. Hydrogen plasmas are well ionized above a temperature of about 1 eV or $\sim 10^4$ K. The "ideal classical plasma" regime is defined as having a temperature that exceeds both the Coulomb interaction potential and the Fermi energy. The first condition implies the predominance of collective particle interactions through macroscopic electromagnetic fields (or equivalently, the presence of a large number of particles $n\lambda_D{}^3$ within a Debye sphere). The second condition insures that quantum-mechanical effects are generally small.

On this familiar n-T diagram, blank spaces have traditionally been reserved for two projected fusion plasma regimes: (1) magnetically confined plasmas, with densities of 10^{14}–10^{15} cm^{-3} and temperatures of 10–30 keV, approaching the "relativistic plasma" border; (2) inertially confined plasmas with similar temperatures and with densities approaching the border marked "degenerate quantum plasma." During the last several years, real experimental data have begun to appear in these reserved spaces. Two illustrative points are shown in Fig. 1. The corresponding state of progress toward the fusion-energy objective,

HAROLD P. FURTH *has made major contributions to both experimental and theoretical plasma physics, with special reference to controlled fusion. He is Director of the Princeton Plasma Physics Laboratory, and also Professor of Astrophysical Sciences at Princeton University, Princeton, New Jersey.*

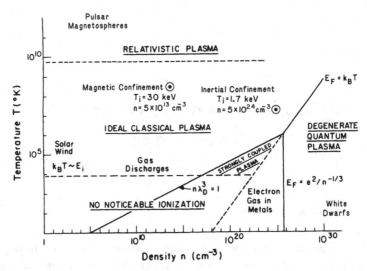

FIGURE 1   Comprehensive density–temperature diagram for plasma physics. Two illustrative experimental data points recently obtained in magnetic and inertial confinement research have been added.

as measured by the product of central ion temperature, central density, and energy confinement time ($T_i n \tau_E$), is indicated in Table I.

Experimental work on magnetically confined high-temperature plasmas began in earnest during the mid-1950s, with inertial confinement experiments following about fifteen years later. These two fields of research have gone through a notably similar pattern: Initial theoretical inspiration has been followed by a phase of experimental vexation and anomaly that forced the growth of complex new physical understanding, yet has ultimately left a clear path to the fusion energy goal.

In the case of magnetic confinement, the *a priori* guidelines consisted mainly of ideal magnetohydrodynamics (MHD) and simple "classical" collisional transport theory. The predictions of MHD instability theory proved to be fairly relevant, but had to be enlarged in the direction of increased pessimism by allowance for finite electrical resistivity. In addition, large new classes of microinstabilities were discovered which were shown to result from nonthermal

TABLE I. Comparison of $T_i n \tau_E$ Values (in keV cm^{-3} sec).

	Magnetic Confinement	Inertial Confinement
Best Experiments	3×10^{14}	3×10^{14}
Break-even	5×10^{14}	
Ignition	5×10^{15}	5×10^{15}
Target Gain = 1		$\sim 10^{16}$*
Target Gain = 100		$\sim 10^{17}$*

*Reached by self-heating, following ignition.

velocity distributions. In a high-temperature plasma, even the mild departures from the Maxwellian that are implied by the presence of spatial gradients in plasma density and temperature proved to be ample for the excitation of significant levels of microturbulence, along with transport rates well in excess of classical expectation.

The seeming failure of the initial theoretical visions in magnetic confinement research was aggravated by various imperfections of the real experimental world, such as geometric asymmetries and residual "atomic effects" due to incomplete ionization. With gradual advances in technology and increases in plasma size and magnetic field strength, these disturbing phenomena have been pushed out toward the edge of the main plasma (where they continue to exert a surprisingly strong effect on the quality of global energy confinement). As a result of better implementation, the performance of present-day devices is actually approaching initial theoretical expectation.

In the case of inertial confinement, the main *a priori* guidelines were provided by ordinary fluid dynamics, including classical collisional transport and radiation-absorption. Somewhat analogous to the magnetic confinement experience, the main experimental difficulties were associated with plasma gradients, edge effects, and imperfections: Intense laser light was found to give rise to a variety of harmful nonlinear plasma-wave interactions in the low-density corona surrounding the fuel pellets. The associated generation of non-Maxwellian energetic-electron components gave rise to "anomalous" heat propagation, preventing efficient pellet compression. Insufficiently uniform illumination of laser targets helped drive Rayleigh–Taylor instabilities to unacceptable levels during the target-implosion phase. The physical understanding of these phenomena was expedited by contributions from theorists who had analyzed similar problems in the context of magnetically confined plasmas.

As in the magnetic confinement case, inertial confinement experimenters soon discovered effective remedies: The use of shorter wave (visible and near-ultraviolet) laser light produced a marked reduction in collective phenomena at the plasma edge. "Indirect illumination" of targets has recently permitted astonishing target-compression factors to be achieved in spite of the Rayleigh–Taylor mode. Again, as in the magnetic confinement case, there is an unfortunate correlation between the direction of most effective experimental parameters (higher magnetic field strength/shorter laser wavelength) and the direction of greater *technological* cost and difficulty.

Ten years ago, when the author of the present paper was asked by the Department of Energy for a short definition of "the scientific feasibility of fusion power," he suggested: "Scientific feasibility is when they stop saying you can't do it and start saying they may not want it." According to this definition, both the magnetic and inertial confinement approaches are reaching the phase of scientific feasibility. There is no longer much serious doubt that ignited fusion plasmas can be achieved in laboratory experiments of affordable magnitude. The central issue has now become the economic viability of fusion power: the projected capital investment per megawatt of generating capacity, the minimum power rating per reactor unit, the estimated fractional availability at full power output, and the degree of environmental attractiveness relative to the two competing "inexhaustible" power sources, fission and solar power.

The pursuit of maximum economic potential tends to drive both magnetic and inertial fusion research into directions that depart somewhat from the proven lines of least experimental resistance. In the magnetic case, the two areas of most intensive research are the enhancement of the β value (the ratio of plasma pressure nT to magnetic pressure $B^2/8\pi$) and the provision of true steady-state plasma confinement. The objectives are to generate higher fusion power density, (proportional to $\beta^2 B^4$) and to avoid the auxiliary costs and mechanical fatigue problems associated with a nonconstant heat source. The most successful experimental approach to date, the "tokamak" toroidal-confinement system, has the potential to satisfy both these objectives by introducing various auxiliary features. There are also a number of alternative magnetic confinement solutions with inherently greater potential for high-β operation or true steady-state operation—but with still unproven capabilities in the area of plasma confinement. The unpredictability of ultimate potentials on the plasma physics side of magnetic fusion research has lately been matched by the emergence of entirely new prospects on the side of magnet technology. Depending on the manufacturability of mechanically reliable magnet coils from the new high-temperature superconducting materials, the magnetic field strength of steady-state fusion reactors may rise significantly. The implied increase of fusion power density could turn out to satisfy the economic requirements even at present-day levels of the plasma β value.

In the inertial confinement case, the main challenges are to minimize the energy input required to ignite a high-gain target, and to develop an efficient target-driver system, using practical high-power technology. Lower energy inputs imply smaller and more frequent target explosions, thus alleviating mechanical problems and minimizing the unit reactor power rating and cost. The currently "front-running" experimental approach, using x rays from a laser-produced plasma for indirect illumination of targets, could be adapted in a number of ways to the economic requirements: One could retain indirect illumination, while seeking to improve its efficiency—for instance, by replacing the laser-light driver system with an energetic-ion driver. Alternatively, if the problem of uniform *direct* target illumination can be overcome by special optical techniques, the input energy requirement for ignition would be lowered, and the use of laser-driver technology could become more attractive for a practical reactor. A convenient improvement for all inertial (and magnetic) D–T fusion systems would be to spin-polarize the nuclei of the fuel, thus enhancing the D–T fusion-reaction rate in the plasma phase by a factor of 3/2. Polarized fuel could have a particularly great benefit in reducing the ignition-energy requirement for inertial fusion.

In addition to the special technological problems that are inherent in their individual approaches, the magnetic and inertial fusion research programs face some common issues in the area of nuclear technology. The 14-MeV neutron and the 3.5-MeV α particle of the D–T reaction represent very "high-grade" energy—but the exploitation of this energy source with present-day fission-reactor technology is problematical. Since α particles are electrically charged, they are useful for self-heating (or "igniting") the fusion fuel, which is important for the achievement of high-energy multiplication, especially in the case of inertial fusion. (For magnetic fusion, there is also a potential for direct conversion of α energy into electrical energy.) The 80% of the *D–T* power output in energetic neutrons can be captured in a meter-thick external blanket and used for nuclear transmutation, for generation of extremely hot fluids—or just for making steam to drive a conventional power plant. About 2/3 of the *D–T* fusion neutron output is needed to regenerate the tritium fuel by transmutation of lithium in the blanket. The balance can be used for other forms of nuclear transmutation, such as disposal of long-lived radioactive wastes or breeding of fission-reactor fuel. Super-hot fluids could have applications in chemical processing or may be exploitable for direct MHD-generation of electrical power. In the present undeveloped state of nuclear fusion technology, the normal practice for conceptual reactor design studies is to seek economic viability simply in terms of a conventional steam plant. While both magnetic and inertial fusion reactor plants are most economical in large scale, the standard design goal has been limited to a maximum unit size of ~ 1000 MW electric output power.

The outstanding fusion-reactor technology problem is the fabrication and maintenance of the "first wall" through which the energetic neutron flux must pass on the way from the burning plasma to the reactor blanket. Present-day structural materials imply seemingly acceptable—but inconveniently short—first-wall replacement times due to radiation damage, resulting in the production of substantial radioactive waste, which fortunately need not be long-lived. Progress in materials research should lead to better performance in both respects. A more radical approach, particularly suitable for inertial confinement, is the use of a first wall made of streaming liquid metal or ceramic pellets. Another possibility is to put the problem back in the area of plasma physics research by looking toward fusion reactions that release most of their energy in charged particles—while unfortunately producing much lower fusion-power densities.

The existence of a wide spectrum of variant solutions constitutes part of the long-term promise of fusion energy. The most direct technological path, however, would seem to be the development of new structural materials that are specialized to the fusion-reactor application. The success of magnetic and inertial burning-plasma experiments during the last years of this century may provide the stimulus for an appropriate materials-development program.

Aside from the obvious historical similarities between magnetic and inertial fusion research, there are also some notable differences. One type of asymmetry results from the relative age in the two fields. Magnetic fusion reactor concepts were born into an uninhabited ecological niche, where scientific feasibility beckoned as the great objective, and economic feasibility was thought to be an easy corollary. The originally envisaged magnetic fusion reactor schemes—including the Sakharov–Tamm proposal for the tokamak[2]—now look so huge and inefficient that one must marvel that they served to mobilize serious support. During subsequent years, the combination of experimental success and more realistic appreciation of economic factors has caused the size of projected magnetic fusion reactors to decrease substantially. Inertial confinement concepts, on the other hand, were born into a highly competitive world—already occupied by magnetic confinement concepts. In this case, the conceptual starting point was minimal reactor size, and the experimental learning process has driven a continual growth in reactor-size projections. By these two different evolutionary paths, the magnetic and inertial confinement efforts appear now to have arrived at similarly realistic and promising reactor plans.

A second important difference relates to classification policy. Magnetic fusion research was a closely kept secret until August 1958, when American–British and Soviet results were unveiled simultaneously in Geneva and found to be roughly identical. During subsequent decades, the international effort in magnetic fusion research has become increasingly cooperative, and planning is now under way for a joint American–European–Japanese–Soviet project to build an International Thermonuclear Experimental Reactor (ITER). Research in inertial confinement, on the other hand, is still largely classified—as evidenced by the curiosity that "indirectly" illuminated targets are frequently mentioned but seldom described. The relationship between the micro-explosions of inertial confinement and the macro-explosions of nuclear testing has been advantageous in developing physical insight—while at the same time depriving inertial confinement researchers of the opportunity to share some of their most significant experimental results.

Recent upward progress in Fig. 1 has been motivated primarily by the fusion application, but there have been a number of important side benefits. First of all, the new field of high-temperature plasma physics has been created, and found to be a scientifically rich and powerful discipline. Laboratory-based plasma physics has provided insight into the ionosphere and near-space plasma surrounding the earth. The emergence of diagnostic satellites has now turned the Earth's neighborhood—and even the whole solar system—into a productive large-scale plasma physics laboratory. The advances in plasma physics have also led to unprecedented success in the understanding of astrophysical objects.

Practical applications of high-temperature plasma physics tend to come in special areas, where the rewards are so great as to justify substantial technological investment. One such area is the extension of lasers into the extreme ultraviolet and soft-x-ray regions (50–300 Å), with prospects of advance to still shorter wavelengths. The techniques of both inertial and magnetic confinement are relevant to this task, and the requirements are sufficiently relaxed compared with those of the fusion energy objective so as to promise much earlier economic returns. Important practical applications of x-ray lasers are expected in many fields, including medicine, biology, and semiconductors.

Even in the pre-1950 period, there was one area of research where "hot plasmas," ranging up towards the relativistic plasma regime, were available in the laboratory: namely, the low-density, hot-electron plasmas of high-frequency

electronics. Over the years, there has been steady progress toward the generation of shorter wavelength, higher-powered microwaves—from magnetrons to
gyratrons and recently to free-electron lasers. Some of the same velocity-space
instabilities that interfere with the maintenance of ordered particle motion in
magnetic fusion reactor schemes underlie the success of the free-electron laser in
extending efficient microwave generation toward the infrared. Among the
promising applications of free-electron lasers, in turn, is the heating of magnetic
fusion plasmas to ignition by electron-cyclotron absorption of high-frequency
microwaves.

The following sections of this paper are designed to provide some technical
material in support of the qualitative discussion offered in the introduction.
Section 2 reviews the elementary physics concepts of magnetic confinement and
describes recent experimental results, with emphasis on tokamak research. Section 3 outlines the basic concepts of inertial confinement and reviews recent
(unclassified) experimental results. Section 4 briefly discusses general applications of high-temperature plasma physics, with emphasis on x-ray lasers.

2. MAGNETIC CONFINEMENT

A. Basic Physical Concepts

A charged particle gyrating in a magnetic field is equivalent to a diamagnetic
dipole with moment $\mu = W_\perp /B$, where $W_\perp$ is the gyration energy. As illustrated
in Fig. 2 for the case of plasma confinement in the earth's magnetosphere,
gyrating particles can be "mirror trapped" by positive gradients $\nabla_\parallel B$ along
magnetic field lines, because μ is an adiabatic invariant, which is strictly conserved when the gyroradius r_g is small compared with the scale height
$L = |B /\nabla B|$. Coulomb scattering, however, can alter μ and permit the escape of
higher-$W_\parallel$ particles along magnetic field lines. From the relative size of the
Coulomb and fusion cross sections, one sees that *unthermalized* fusion reactor
plasmas—such as simple mirror-trapped plasmas depending on $W_\perp > W_\parallel$—

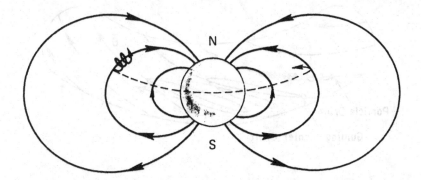

FIGURE 2   A charged particle can be trapped in the Earth's dipole magnetic field, if its velocity is
predominantly perpendicular to **B**. The particle gyro-orbit precesses around the equator.

cannot be expected to produce useful output power, even in the case of D–T fuel at optimal plasma temperature.

One ingenious solution is the tandem mirror machine[3,4] where highly energetic ions are mirror-trapped and in turn serve to generate an electrostatic trap for a well-thermalized reacting plasma. A more radical solution is to provide closed magnetic field lines, as in the toroid of Fig. 3, where both mirror-trapped and untrapped particles are constrained to orbit in the neighborhood of magnetic flux surfaces. In that case, the Coulomb scattering rate ν_c leads only to a slow diffusion process, with transport coefficients scaling as $r_g{}^2\nu_c$. This "classical" transport is somewhat enhanced due to orbital excursions larger than r_g, which occur as the guiding center of the gyromotion oscillates between magnetic mirrors (cf. Fig. 3). In axisymmetric toroids, the conservation of canonical angular momentum implies that these excursions must be smaller than the gyroradius r_{gp} in the *poloidal* magnetic field component (the short way around the torus), and classical transport is therefore limited to $D \lesssim r_{gp}{}^2\nu_c$. The requirement of sufficiently good magnetic confinement to permit the ignition of a D–T plasma then implies a minimum poloidal magnetic flux equal to that of a toroidal current loop carrying several megamperes. By coincidence, this is also the requirement for confining the large orbital excursions of the 3.5-MeV α particles that are expected to produce the ignition phenomenon. Classical confinement is actually better, by one to two orders of magnitude in the relevant transport coefficients, than the minimum needed for purposes of ignited reactor operation.

A magnetically confined plasma is a diamagnetic medium with a magnetization $n\mu$ such that $n\mu B$ equals the plasma pressure $p = nT$. The plasma β value, $8\pi nT/B^2$, is a measure of plasma diamagnetism—and therefore also a measure of the plasma's tendency toward Rayleigh–Taylor-like interchange modes, driven by $\mu\nabla_\perp B$. In ideal MHD theory, where the plasma is approximated as a fluid with null electrical resistivity, the resultant locking of magnetic field lines

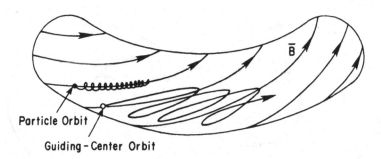

FIGURE 3   In a toroidally symmetric magnetic field configuration, particles of all velocity vectors can be confined. Particles trapped in the weaker-field region on the large-R side of the torus precess in the toroidal direction.

into fluid elements can be used to suppress both the pressure-driven interchange modes and the gross tendency of the plasma current channel toward helical "kinking." Allowance for arbitrarily small but finite resistivity, however, permits field lines to break and rejoin within the fluid, which somewhat aggravates the problem of designing stable configurations with large β values.

Confined plasmas that are stable against MHD modes can experience slower anomalous loss processes. The nonthermal velocity distributions of mirror-trapped plasmas have a strong tendency toward relaxation by high-frequency modes. Even the isotropic plasmas of toroidal confinement necessarily exhibit ordered "drifts" of the particle distribution, corresponding to the diamagnetic magnetization current $(J_m = c|\nabla(P/B)|)$ plus the somewhat slower drift motion of the orbital guiding centers (cf. Fig. 3). These small departures from the Maxwellian can only drive electrostatic microturbulence with a characteristic growth rate that is reduced by the factor r_g/L relative to MHD instabilities, but the associated transport can still be as large as the famous Bohm diffusion rate $D_{Bohm} = (1/16)cT_e/eB$—which extrapolates to prohibitive reactor-plasma size. A principal goal of toroidal confinement research has been to reduce anomalous transport by several orders of magnitude relative to the Bohm value— about halfway, on a logarithmic scale, toward the extremely low transport rates of the classical diffusion model.

A number of toroidal confinement configurations[5] are illustrated in Fig. 4. The "dynamic pinch," which has purely poloidal field and therefore a β value of unity, is predicted by ideal-MHD theory to be highly unstable. This pessimistic prediction has generally proved to be accurate—except, perhaps, in the case of recent pinch experiments using very short (10^{-8}–10^{-7} sec) current pulses through frozen fibers.[6] The "stabilized pinch" increases the rigidity of the current channel (and lowers the β value into the 10% range) by adding a toroidal field component B_t that is comparable in strength to B_p. Residual resistive-MHD kink modes serve to maintain a quasi-stable configuration by means of dynamo processes that lower the magnetic energy while conserving the magnetic helicity $\int dV(\mathbf{A \cdot B})$: The lowest-energy state is the "reversed-field pinch" [7] (RFP), where the toroidal magnetic field outside the plasma is slightly reversed by dynamo currents. The tokamak follows a different strategy of seeking to suppress MHD activity altogether by means of a strong toroidal magnetic field $B_t \gg B_p$. In this case, the toroidal and poloidal periodicity conditions serve to exclude all kink modes or to localize specific residual modes within limited radial zones. If one wishes to go still further and eliminate the destabilizing effect of the toroidal plasma current entirely, one solution is to use a current-carrying ring, levitated inside the plasma, to generate the confining poloidal field. The "levitron" is well suited for plasma physics research, but not for a D–T reactor. Alternatively, a set of external helical multipole windings can serve to create an average poloidal field.[8] The "stellarator" concept would lend itself straightforwardly to the desirable objective of a steady-state reactor; by contrast, the tokamak (and RFP) require some auxiliary means to maintain the toroidal current against resistive decay.

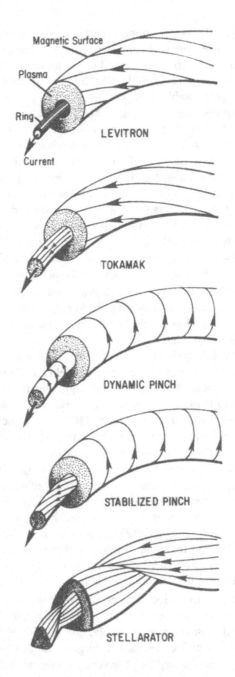

FIGURE 4   A number of toroidal confinement configurations have been studied experimentally. The dynamic pinch has a purely poloidal field, generated by a toroidal plasma current. Other configurations use the combination of a toroidal magnetic field plus a poloidal field component generated by a toroidal current flowing in a solid ring coil (levitron) or within the plasma itself (tokamak and stabilized pinch). The stellarator uses external helical multipole windings to generate an average poloidal field component.

B. Progress Towards the Fusion Reactor Regime

The Lawson diagram of Fig. 5 compares reactor plasma objectives and recent results of magnetic fusion research.[9] Economical reactors must operate close to the "burn" curve in Fig. 5, where the ratio Q of fusion power to auxiliary plasma-heating power goes to infinity as the α particles of the D–T reaction take over the plasma-heating function. The Lawson $Q = 1$ curve refers to break-even in a fully thermalized plasma with $T_i = T_e$. When plasmas are heated by injection and thermalization of an energetic non-Maxwellian ion tail, the break-even condition is shown in Fig. 5 to be relaxed appreciably for an ideal ion-energy distribution. (The Lawson diagram for magnetic fusion usually incorporates hydrogenic bremsstrahlung losses—and therefore has a minimum ignition temperature at 4.4 keV.)

Following the initial years of experimental magnetic fusion research, in the mid-1950s, the product $T_i n \tau_E$ of Table I stood at about 10^8 keV cm^{-3} sec. By the early 60s, it had risen to 10^{10} keV cm^{-3} sec. In 1969, the T-3 tokamak, at the I.V. Kurchatov Institute, advanced the $T_i n \tau_E$ value to 5×10^{11} keV cm^{-3} sec in a device of modest parameters (major radius $R = 150$ cm, plasma current I_p ~ 100 kA). During subsequent years, worldwide progress along the tokamak line has continued steadily (cf. Fig. 5).

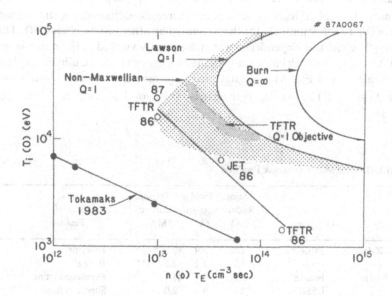

FIGURE 5 The Lawson diagram presents the requirements for equilibrium burn ($Q = \infty$) and for breakeven ($Q = 1$) in thermalized plasma. Non-Maxwellian plasmas with enlarged energetic-ion components can, in principle, achieve break-even under substantially related conditions. The "TFTR $Q = 1$ objective" illustrates an experimental break-even goal.

The key to the advance of tokamak data points in the Lawson diagram has been the improvement of the quality of toroidal confinement. The rigid magnetic field structure of the tokamak eliminated MHD activity as a dominant driver of anomalous transport. At the same time, the simplicity of the tokamak magnet configuration, and the effectiveness of its built-in ohmic-heating mechanism, permitted the production of much larger and denser plasmas in the T-3 device than had been available in contemporary levitron and stellarator experiments. The tokamak edge region, where $\nabla(\log p)$ becomes large, and where atomic processes are an important part of the energy and particle balances, has electrostatic fluctuation levels ($e\delta\phi/T$, $\delta n/n$) that are comparable to those of pre-tokamak plasmas and are accompanied by a Bohm-like level of anomalous transport. The novelty of the tokamak plasma consists in the emergence of a relatively quiescent hot-plasma *core*, where local transport is sufficiently small so that the resultant global energy-confinement time τ_E exceeds the Bohm confinement time by orders of magnitude and comes close to classical expectation.

A dramatic demonstration of this favorable trend was provided by the Alcator tokamak experiments, carried out at MIT, beginning in the mid-1970s. In tokamak devices somewhat smaller than T-3, the product of plasma density times linear size L was raised by using strong magnetic fields ($\gtrsim 10T$) and correspondingly intense ohmic heating. These experiments yielded $T_i n\tau$ values in the range 10^{13}–10^{14} keV cm^{-3} sec. They also led to the empirical "neo-Alcator" scaling law where τ_E does not increase simply in proportion to L^2, as in ordinary classical diffusion processes, but increases additionally with the product Ln (up to a maximum value that is limited by other loss mechanisms). This optimistic cubic size-dependence was subsequently verified in the much larger TFTR experiment at Princeton and in the JET experiment at Culham, England (cf. Table II and Fig. 6). JET has measured record τ_E values, close to one second, and TFTR has reached an $n\tau_E$ value of 1.5×10^{14} cm^{-3} sec (cf. Fig. 5).

TABLE II. Some Major Tokamak Facilities.

Facility	Location	Major Radius (m)	Field Strength (T)	Plasma Current (MA)	Special Feature
ASDEX-U	FRG	1.6	3.9	(2.0)*	Divertor
DIII-D	US	1.7	2.2	2.5 (5)**	High β
Tore Supra	France	2.2	4.5	(1.7)*	Superconducting
T-15	USSR	2.4	4.5	2.0	Superconducting
TFTR	US	2.5	5.2	2.5 (3)**	D–T
JT-60	Japan	3.0	4.5	2.7	Divertor
JET	UK	3.0	3.5	5.0 (7)**	D–T

*Not yet in experimental operation.
**Potential for extended operation.

FIGURE 6   The vacuum chamber of the JET device has approximately the cross-sectional size that will be required for a tokamak reactor—but the major radius will have to be approximately doubled.

The inherent ohmic-heating process of the tokamak raises the plasma temperature into the 1–5 keV range, but for purposes of bringing the plasma to the ignition point the introduction of auxiliary heating appears to be convenient. Tokamak plasmas have been heated (with roughly comparable effectiveness) by the injection of energetic neutral-atom beams and by the absorption of various electromagnetic waves—particularly in the frequency ranges around the ion and electron gyrofrequencies. Auxiliary-heating experiments have proved enlightening in respect to the scaling of anomalous transport: In ohmic heating, the input power P_H can be increased only by raising the plasma current, and therefore also the confining poloidal field. In auxiliary heating, the input power can be raised for fixed magnetic confinement parameters, and the "Goldston" scaling of τ_E was found: An adverse $P_H^{-1/2}$ dependence, together with favorable dependences on size and poloidal field strength B_p. When the P_H dependence is eliminated (using $P_H \propto L^3 nT/\tau_E$), one recognizes that the Goldston scaling fits into the family of Bohm-like scaling laws, but with a much lower magnitude of transport and with the familiar B dependence of D_{Bohm} replaced by a B_p dependence.

The Goldston scaling would be compatible with an ignited tokamak plasma in the 30-MA range of plasma current—which is not unrealistic for a full-scale commercial reactor—but would impose a serious budgetary handicap during the phase of research and development. This problem could be resolved by means of relatively small numerical improvement factors in the Goldston transport coefficient (~ 2), which happily appear to be available through plasma-edge-control techniques.

The ASDEX experiment in Garching, Germany used a magnetic divertor (Fig. 7) to inhibit the reinjection and reionization of outflowing plasma, and found that at sufficiently high heating-power levels the plasma regime became thermally bistable. Starting from the auxiliary-heated ohmic regime, which they named the L-mode, the plasma was able to "jump" into the H-mode, which also exhibits a Goldston-type scaling for τ_E, but with several times longer confinement times. These divertor H-mode experiments have been extended successfully to the 10-MW level of neutral-beam-heating power in the DIII-D tokamak at GA Technologies in San Diego, and also in the JET device, yielding the "JET" data point shown in Fig. 5. While the TFTR is not equipped with a divertor, edge-plasma recycling was effectively suppressed by using a large surface of absorbent graphite tiles as the plasma limiter. In this way τ_E was also improved severalfold, giving the high-T_i TFTR point in Fig. 5 (which has somewhat lower electron temperature $T_e \sim 7$ keV).

The improved TFTR regime has given D–D neutron yields of 10^{16} sec^{-1}, with pulse durations of order 1 sec, which would correspond to a total fusion energy release of about 4 MJ if D–T fuel were substituted for D–D. The energy-multiplication factor Q would then fall into the range 0.20–0.25. A number of machine improvements are currently under way on TFTR, as well as on JET and on the JT-60 device in Ibaraki–ken, Japan, which seem likely to permit the achievement of "equivalent" $Q \sim 1$ conditions in D–D plasmas during 1988. Break-even experiments in D–T plasmas are expected to begin in TFTR during 1990 and in JET during 1992.

Beyond the present generation of experimental tokamaks, international planning is looking toward a major tokamak test reactor facility to produce quasi-steady-state ignited-plasma conditions and to provide an initial test bed for fusion reactor technology. A joint international design for such a machine (the INTOR)[10] was developed during the period 1979–87. Extended experience with cooperative international design and supporting research efforts is now providing the basis for the initiation of a three-year design program (1987–90) aimed at the joint construction of the ITER (International Thermonuclear Experimental Reactor) device during the period 1993–2000. At present, there are also a number of parallel national design projects, aimed at similar or somewhat smaller reactor facilities.

The ITER class of tokamak designs is characterized by superconducting steady-state magnet systems that produce ~ 5-T magnetic fields at the plasma. (Superconducting-coil operation is to be prototyped during the late 1980s by the T-15 tokamak at the Kurchatov Institute, and the Tore Supra at Cadarache, France.) Other characteristic ITER parameters are: major radius $R \sim 5$ m and plasma current $I_p \sim 15$ MA.

The success of the Alcator experiment in achieving relatively high $T_i n \tau_E$ in small machine size has opened up a parallel path to ignition, which is expected

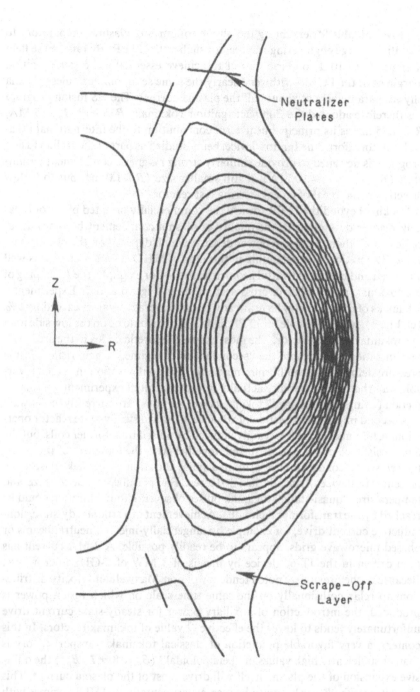

Z

R

Neutralizer
Plates

Scrape—Off
Layer

FIGURE 7 Tokamak confinement properties can be improved by *D* shaping the minor cross section of the magnetic surfaces. The associated magnetic separatrix can be used to divert the plasma outflow into the "scrape-off layer."

to prove valuable for exploring the physics of burning plasmas preparatory to operating a large engineering test reactor such as the ITER. By raising the field strength B to 10 T, one can expect to achieve essentially the same ignition margin as in the ITER (with very nearly the same set of dimensionless plasma physics parameters) at about half the plasma current. The US fusion program has therefore adopted the Compact Ignition Tokamak ($B = 10$ T, $I_p = 9$ MA, $R = 1.75$ m) as its principal near-term contribution to the international toka-mak reactor effort. The Ignitor device, being studied as part of the Italian fusion program, is designed to provide similarly strong magnetic confinement param-eters ($B \sim 13$ T, $I_p \sim 10$ MA) in still smaller size ($R \sim 1.0$ m), but to be less directly prototypical of an engineering test reactor.

Looking beyond the ITER toward the commercially oriented phase of toka-mak reactor development, the quality of plasma confinement becomes a less critical issue than the fusion power density, which depends on the plasma beta value. Theoretical understanding in this area of MHD stability analysis is well advanced and has led to fruitful geometric ideas—for example, the D-shaping of the tokamak minor plasma cross section, as in Figs. 6 and 7. Experimental tokamaks of this geometry have already reached average β values exceeding 6% (cf. Fig. 8), which is ample for ITER and CIT, but would be on the low side for a 5-T tokamak power reactor. The most exciting theoretical idea in the area of β-maximization is entry into the "second stability regime," where rising plasma pressure deforms the equilibrium magnetic field configuration in such a favor-able way that MHD stability actually improves. Initial experiments in this di-rection (using "bean-shaped" tokamak cross sections) are currently under way.

A second important direction of advance is toward steady-state reactor oper-ation. A tokamak current is readily induced by pulsed transformer coils, but the finite resistance of the plasma toroid, together with the finiteness of the trans-former volt-seconds, implies a limited pulse duration. Tokamak plasmas in present-day devices typically last 1–10 sec. For plasmas of reactor size and temperature, "quasi-static" tokamak pulses of several hours duration could be reached by the transformer technique. Achievement of a true steady-state, non-inductive current drive, for example by tangentially-injected neutral beams or phased microwave grids, appears to be readily possible. A 2-MA current has been driven in the JT-60 device by means of 3 MW of 2-GHz microwaves. Because of the previously noted tendency of non-Maxwellian velocity distribu-tions to relax collisionally on the same time scale on which fusion power is produced, the introduction of auxiliary power for steady-state current drive unfortunately tends to lower the effective Q-value of tokamak reactors. In this context, a very favorable prediction of classical tokamak transport theory is that, at sufficiently high values of "beta poloidal" ($\beta_p = 8\pi nT/B_p^2$) the diffu-sive expansion of the plasma itself will drive most of the plasma current. This "bootstrap effect" has begun to be seen experimentally in TFTR regimes with $\beta_p = 2$–3.

In sum, the prospects for advancing the tokamak concept toward ideal reac-tor performance are substantial and promising—but they are not yet so solidly

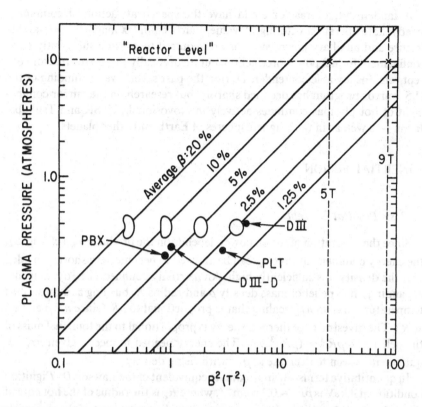

FIGURE 8 The tokamak beta-value (ratio of plasma pressure to magnetic pressure $B^2/8\pi$) is limited by MHD stability conditions that can be relaxed by cross-sectional shaping. (The approximate β values that can theoretically be achieved with various shapes are indicated.) Present experimental points and probable reactor requirements are shown.

founded as to discourage serious competition from other sectors of magnetic confinement research. In regard to the achievement of high β values, the RFP version of the stabilized pinch (Fig. 4) can easily outdistance present-day tokamak performance—maintaining gross stability at the 10–20% level. Multi-MA RFP experiments, aimed at extending plasma confinement parameters into the "tokamak range" of the Lawson diagram (Fig. 5), are currently under construction at Padua, Italy and Los Alamos, New Mexico.

The stellarator (Fig. 4) offers the unique advantage of true steady-state confinement without need for any auxiliary power input. The quality of confinement in stellarators has been found similar to that in tokamaks—though the greater complexity of stellarators has caused them to lag behind tokamaks in affordable plasma size, and therefore also to lag behind in the Lawson diagram. Currently, new stellarators with 200-kA-level equivalent poloidal confining fields are entering operation in Oak Ridge, Tennessee, and Garching, Germany, and a substantially larger stellarator facility is being planned for Nagoya, Japan.

A tandem-mirror reactor could have the special attraction of combining steady-state operation with high β value. Fairly complex magnetic-field architecture and auxiliary power systems are required to provide sufficiently good confinement for net power production, but theoretically the mirror reactor concept remains a viable contender. During the past several years, funding of the U.S. mirror program has declined sharply, but research on the mirror confinement of hot plasmas continues actively in Novosibirsk, USSR and Tsukuba, Japan—as well as in the magnetospheres of Earth and other planets.

3. INERTIAL FUSION

A. Basic Physical Concepts

Since the generation of fusion power depends on the product $n\tau_E$ (cf. Fig. 5), the energy confinement time τ_E can be allowed to become very short, provided that the density n is sufficiently high. An inertially confined (i.e., dynamically expanding) fuel pellet of mass density ρ and radius R, burning at a prescribed temperature, has an $n\tau_E$ scaling that is proportional to ρR (since $n \propto \rho$ and $\tau_E \propto R$). The investment of thermal energy is proportional to the total fuel mass of the pellet $M \propto \rho R^3 = (\rho R)^3 \rho^{-2}$. The energy needed to meet a given $n\tau_E$ requirement is seen to decrease as ρ^{-2} with rising density.

In quantitative terms, an approximate equivalent of the Lawson D–T ignition condition at 4 keV is $\rho R_c \sim 0.3$ g cm^{-2}, where R_c is the radius of the hot core of the pellet. In order that ignition may actually take place, most of the α-particle energy should be deposited within the pellet core, thus imposing an auxiliary minimum-size condition that depends on ρR_c and again happens to be satisfied for $\rho R_c \sim 0.3$ g cm^{-2} (the α-particle range in 10-keV D–T).

The favorable ρ scaling of the energy investment required to reach ignition and high gain suggests a strategy of compressing most of the pellet mass isentropically to the maximum possible density.[11-14] To minimize the specific energy investment required for compression, the fuel temperature must be kept near the Fermi energy, which scales as $E_F \propto n^{1/3}$ (cf. Fig. 1). The "cold" bulk of the pellet can be ignited by careful shaping of the driver pulse, so as to create a small, hot central fuel core.

The proposed experimental approach to the achievement of ignition is illustrated in Fig. 9. A high-powered, short-pulse driver is used to heat the pellet surface, thus forming an expanding atmosphere. As this spherical "rocket blow-off" proceeds, the remainder of the pellet is compressed by the recoil. At maximum compression, the pellet core crosses the ρR_c threshold for ignition. The temperature climbs rapidly into the 10-keV range during an α-particle-driven thermal excursion that spreads outward through the rest of the pellet.

The fusion energy multiplication requirements that are relevant to the case of inertial confinement can be displayed in a Lawson-type diagram (Fig. 10) that

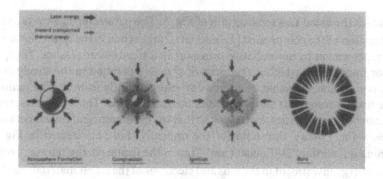

FIGURE 9 In the inertial confinement approach, high-powered laser beams vaporize the surface of a pellet, which is compressed by the recoil, reaching high central temperatures and densities. The compressed *D–T* plasma ignites and burns.

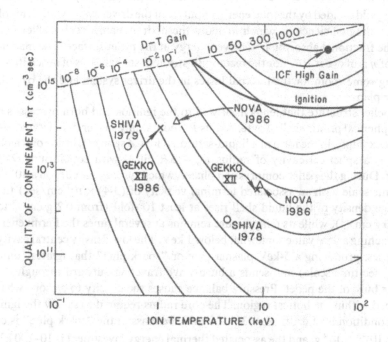

FIGURE 10 An inertial-confinement diagram analogous to Lawson's shows a similar ignition condition. The practical reactor objective of 100-fold target gain calls for substantial advance beyond the ignition curve.

resembles the usual Lawson diagram of Fig. 5. The parameter τ_E now refers to the duration of the compressed plasma state, rather than to the time constant of heat replacement by nonradiative transport in a steady-state plasma. The ignition or burn curve, which is the locus of $Q = \infty$ operation in the steady-state case, is now the threshold condition for energy multiplication. The ignition curve itself is somewhat more lenient in the inertial case: Thermal bremsstrahlung can be trapped in dense plasmas, so that the minimum ignition temperature is below the 4.4-keV threshold for the optically thin case treated in Fig. 5. Defining an inertial "D–T pellet gain" G_{DT} as the fusion energy release divided by the energy investment in the ions and electrons of the target material, one has $G_{DT} \sim (3 \text{ MeV}/T)\phi$, where ϕ is the burn-up fraction, which is approximately proportional to $n\tau_E$ or ρR, when ϕ is small. The D–T gain indicated in Fig. 10 for a plasma core that ignites with $\rho R_c \sim 0.3 \text{ g cm}^{-2}$ and heats up into the 10-keV range is of order 10–20, corresponding to a burn-up fraction of 3–6%. If the ρR of the pellet as a whole is larger by an order of magnitude than that of the ignited core, the final burn-up fraction increases and the gain is also improved, because the surrounding fuel is initially colder. The reactor goal indicated in Fig. 10 is $G_{DT} \sim 10^3$. It is useful to define the "target gain" $G_T = \eta G_{DT}$ (the fusion yield divided by the total energy input from the driver system) in terms of the coupling efficiency η, which includes the hydrodynamic rocket efficiency and the fractional absorption of driver energy at the pellet surface. The magnitude of η is of order 10^{-1} for the reactor illustration, so that G_T is of order 10^2—leaving some room for the internal losses in the driver system and the overall power plant.

A pellet structure that lends itself well to the ignition and burn process is a thin spherical plastic shell of outer radius R_o, containing a layer ΔR of liquid D–T (for example, in the form of a liquid-saturated low-Z sponge), inside of which there is a spherical cavity of radius $R_o - \Delta R$, filled with equilibrium D–T vapor. During the pellet-compression phase (which typically occurs on a 10^{-8}-sec time scale with an associated terminal velocity of $(3-4) \times 10^7$ cm/sec) the average density of the liquid shell rises at least 10^3-fold (from 0.2 g cm^{-3} to ~ 200 g cm^{-3}), while its temperature remains at several times the Fermi energy, reaching a final value somewhat below 1 keV. The low-density central cavity collapses, producing a 5-keV plasma core or "spark plug" that ignites (on a 10^{-11} sec time scale) and sends a burn-wave traveling outward through the colder bulk of the pellet. Pressure balance causes the density to be somewhat reduced within the hottest region. The core radius required to satisfy the ignition condition is of order $R_c \sim 30$–$100 \ \mu m$, the mass of the "spark plug" is of order 10^{-5}–10^{-4} g, and the associated thermal energy investment is 10–100 kJ. To achieve the desired target-energy multiplication by virtue of a propagating burn, the surrounding pellet material should have a radius of order defined by $\rho R \gtrsim 3 \text{ g cm}^{-2}$, with a mass of several times 10^{-3} g, corresponding to an additional energy investment of several hundred kilojoules. For such target parameters, the fuel burn-up fraction ϕ reaches 30–40%.

The successful formation of a small, hot, central ignition region (with R_c/R_o typically a few times 10^{-2}) is perhaps the most critical physics challenge of inertial confinement research. Initial asymmetries associated with either the pellet or the illumination tend to be amplified exponentially by a Rayleigh–Taylor-type instability mechanism. When a light fluid (ρ_l) pushes a heavy one (ρ_h) to generate an acceleration $\pm g$, interchange modes are unstable with a growth rate $(Akg)^{1/2}$, where k is the wave number and A is the Atwood number ($\rho_h - \rho_l)/(\rho_h + \rho_l)$. This type of unstable situation arises both during the initial acceleration process and during the final deceleration, where the converging bulk-pellet material compresses the small central cavity of hot plasma. Asymmetries that have exponentiated during the initial phase can lead to turbulent mixing of hot and cold fuel during the final phase, thus preventing ignition. The most dangerous instability wave numbers k are of order $(\Delta R)^{-1}$, where ΔR is the (strongly reduced) shell thickness following initial acceleration. The associated number of e-foldings during compression is proportional to the square root of the "in-flight aspect ratio," i.e., $(R/\Delta R)^{1/2}$. The use of high-aspect-ratio shells is constrained by a maximum permissible number of about eight e-foldings, set by the limitations of target-fabrication technology. Mitigating factors result from the finiteness of the density gradient and from the nature of the rocket blow-off mechanism itself: As the ablation front eats into the pellet material, it tends also to carry away the vorticity associated with Rayleigh–Taylor interchanges. If the velocity characterizing the ablation front is comparable to $(g/k)^{1/2}$, a significant stabilizing effect can result. In this context, careful shaping of the driver-power pulse as a function of time is found to be advantageous.

The motivation for the use of pellets with large $R/\Delta R$ arises from the severity of the requirements on the driver system. The energetic efficiency of the pellet compression would be highest if the implosion velocity were to exceed the rocket blow-off velocity somewhat. The pellet, being denser than the surrounding plasma atmosphere, tends, however, to implode more slowly—resulting in a hydrodynamic "rocket efficiency" that can easily fall below 10%, unless $R/\Delta R$ is large. This effect becomes more severe with rising laser intensity, thus giving an additional advantage to pellets that are more hollow, with relatively larger initial radius R_0 and lower surface-power density requirements. Aside from the aspect ratio $R_0/\Delta R$, the depth of penetration of the pellet illumination has an important effect on the compression process: Absorption of the driver energy in a deeper, denser layer of the plasma corona surrounding the pellet tends to blow off more material at lower rocket velocity—two features that are valuable, respectively, for inhibiting Rayleigh–Taylor modes and improving hydrodynamic efficiency.

For an illustrative laser-driver energy output of 10^6 J, delivered in 10^{-8} sec onto a pellet of radius $R_0 = 3$ mm, the required power density would be of order 10^{14} W cm^{-2}. As the laser light propagates into the edge of the pellet, it encounters a local plasma frequency ω_p that increases proportionally to $n^{1/2}$—until propagation cuts off at the critical density n_c where ω_p equals the light frequency. (For 1-micron light, n_c is about 10^{21} cm^{-3}, so that cutoff occurs very near the pellet surface.) The "classical" process for absorbing the laser light is inverse bremsstrahlung, where the input energy is thermalized collisionally. This process is favored by using shorter laser wavelengths, which give deeper penetration and allow the energy to be deposited into denser, colder pellet material. On the other hand, in the hot, lower-density plasma corona surrounding the pellet, illumination at very high power densities lends itself to the excitation of a variety of plasma waves (cf. Fig. 11), which have largely undesirable effects on the pellet-compression process. The most serious consequences are: (1) back-scattering of incident light, which reduces the absorption efficiency; (2) channeling of light into local density depressions, which spoils the illumination symmetry and aggravates phenomena associated with high laser intensity; and (3) plasma-wave acceleration of energetic electrons, which causes premature propagation of heat into the bulk-pellet material, and interferes with isentropic compression. The streaming of nonuniformly produced energetic electrons is one of several mechanisms that can give rise to the generation of very strong magnetic fields $(B \sim 10^2 \, T)$—which are generally unhelpful in the ICF context, since reduced surface heat conductivity tends to interfere with efficient, uniform absorption of the driver energy.

A promising alternative to direct laser-pellet illumination is to heat up the interior surface of a hohlraum, with the resultant radiation "bath" acting to illuminate and compress an interior pellet. Since the thermal radiation in the hohlraum is in the soft x-ray range, the energy deposition onto the pellet surface is relatively penetrating, which improves the rocket efficiency. The primary process of laser-heating at the hohlraum surface will, of course, tend to excite the usual plasma instabilities, but this problem is mitigated by the use of surface materials with high atomic number, which enhances the collisionality of the plasma. Indirect illumination also lends itself conveniently to various alterna-

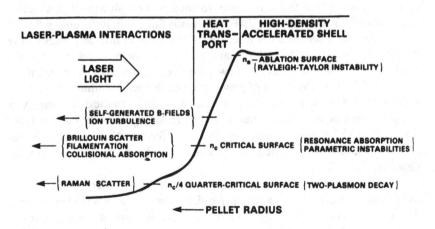

FIGURE 11 Intense illumination by laser light can give rise to a variety of collective plasma interactions.

tive drivers, such as energetic ion beams: a hohlraum has the obvious potential for improving the uniformity of pellet illumination. The drawback of "indirect drive" is that it adds one more energy-conversion stage, thus tending to lower the ultimate efficiency of the coupling process.

B. Progress Toward the Fusion-Reactor Regime

The somewhat idealized theoretical concepts of inertial confinement that were current during the late 1960s envisaged compressed-pellet densities of order 2×10^3 g cm^{-3}—some 10^4 times ordinary liquid density—and estimated that, for peak powers of order 10^{14} W, driver energies of order 10 kJ (or even less) might suffice to reach ignition. Delivering an energy pulse of this magnitude on a sub-nanosecond time scale seemed to be within the reach of the emerging solid-state laser technology. By the mid-70s, neodymium glass laser facilities in the 10–100 J, 10^{11}–10^{12} W range became available and provided the initial body of experimental information on laser-pellet interactions at a wavelength of 1 μm.

As in the case of magnetic fusion research, the earliest experiments were oriented toward fusion-neutron production. Thin-walled *D–T*-gas-filled glass "microballoons" were imploded, and measurable neutron yields were obtained. These experiments departed fundamentally from the theoretical strategy for reaching ignition (cf. Sec. 3.A), in that high temperatures were produced at relatively low densities, by exploding the outer glass shell and generating a convergent spherical shock wave to heat the central *D–T* fuel. While these "exploding pusher" experiments were not on the direct path to energy production and did not address the critical physics issues of ablation and fluid stability, they have provided valuable experience with laser-irradiation physics and diagnostics for the study of the compression process.

By the end of the 1970s, neutron yields in the range 10^{10} per shot were being produced in the SHIVA experiments at the Lawrence Livermore National Laboratory (LLNL), with driver energies in the range 1–10 kJ. By 1985, the OMEGA laser (cf. Table III) at the University of Rochester, New York, reached over 10^{11} neutrons per pulse, the GEKKO program at Osaka, Japan then reached 10^{12}, and both GEKKO XII and the NOVA facility at Livermore (Fig. 12) have meanwhile exceeded 10^{13}. In the latest experiments,[9] central temperatures of order 10 keV were reached, along with $n\tau_E$ values of order 10^{13} cm^{-3} sec (cf. the high-temperature points in Fig. 10). The associated target gains, however, were only of order 10^{-3}, and it would seem difficult to reach the break-even range (equivalent to $Q = 1$ for magnetic fusion) by further development along this line.

Alongside the microballoon experiments, the reactor-relevant compression technique described in Sec. 3.A has been pursued with increasing realism and success. Laboratory experiments to date have avoided the complexities of actually using liquid-*D–T*-filled pellets, but have otherwise addressed the physics issues involved in the ablative compression of Rayleigh–Taylor-unstable shells.

TABLE III. Some Major Inertial-Confinement Facilities.

Facility	Location	Driver Type	Energy (kJ)	Power (TW)
VULCAN	UK	Nd-glass**	1	2
PHAROS III	US	Nd-glass**	1.5	1
OMEGA	US	Nd-glass**	4	12
DELPHIN	USSR	Nd-glass**	2	1
GEKKO XII	Japan	Nd-glass**	20	40
PHEBUS	France	Nd-glass**	24	20
NOVA	US	Nd-glass**	120	100
AURORA	US	KrF	(5)*	(1)*
PBFA-II	US	Light ion	(2000)*	(100)*

*Not yet in experimental operation.
**Output parameters are for 1 μm and should be roughly halved for 2nd and 3rd harmonics.

Densification factors as high as 130 (relative to liquid D–T) have been achieved in the NOVA facility, by using indirect drive for the ablative compression of thick gas-filled shells. Experiments of this type have provided the high-$n\tau_E$ points in Fig. 10. The corresponding peak plasma pressures (Fig. 13) are of order 3×10^{10} atm—far beyond all other laboratory experience, and well on the way toward the inertial reactor goal of 10^{11}–10^{12} atm.

These NOVA experiments have successfully addressed another figure of

FIGURE 12   The 4.4-m diameter target chamber of the NOVA facility is designed for ten-beam laser-light input.

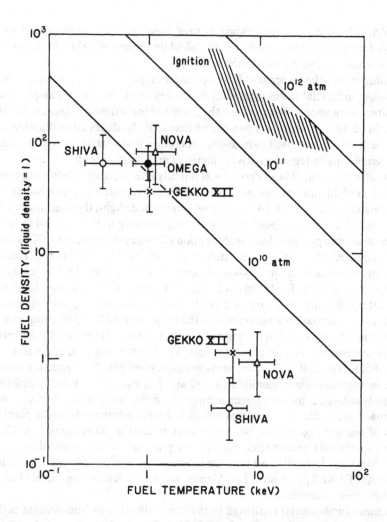

FIGURE 13 Plasma pressures in the 10^{11}–10^{12} atm range will be needed for an inertial confinement reactor. The 10^{10} atm range has already been reached by the experiments.

merit that is of fundamental importance for achieving ignition: the convergence ratio R_o/R_c, relating the initial pellet radius to the final radius of the hot pellet core. Computer-modeling studies have shown that for purposes of substantial energy multiplication, the convergence ratio should be greater than 40. In recent NOVA experiments, convergence factors exceeding 35 (i.e., a fuel volume reduction of 4×10^4) have been reported. The somewhat unexpected achievement of this result, with less-than-optimal driver-pulse shaping, would seem to support the concept of Rayleigh–Taylor-mode inhibition and has favorable implications for high-gain experiments. The indirect-drive results represent a major improvement relative to those reported for direct-drive experiments in

NOVA, where convergence ratios greater than 10 were accompanied by a marked deterioration of actual versus calculated neutron yield—indicative of unstable mixing and cooling in the hot core.

Hohlraums (like tokamaks) have opened a major path to experimental success—but in inertial (as in magnetic) fusion research, there is still ample room for alternative approaches to claim the potential for ultimate economic advantage. In particular, there is a persuasive case that the direct laser-illumination approach would lend itself to high-gain operation with reactor systems that are less demanding in regard to driver efficiency and energy output.[9]

The development of laser drivers was initially hampered by a defect of experimental implementation: the excitation of excessive plasma turbulence at the illuminated surface. With 1-micron neodymium-laser light, the coupling to various plasma waves was found to be so strong, relative to the classical inverse-bremsstrahlung process, that the absorption efficiency dropped to the 30% level for the desired illumination intensities in the 10^{14}–10^{15} W cm^{-2} range. Thanks to improvements in frequency-multiplication technique at the University of Rochester and at the École Polytechnique in France, the neodymium light was converted efficiently to its 3rd and 4th harmonics, respectively, with an attendant rise of laser-target absorption into the range above 70%. More important, there was a marked accompanying drop in the fraction of laser energy delivered to the undesirable energetic-electron tail. The outstanding results achieved in the GEKKO and NOVA experiments were obtained with the frequency-multiplied neodymium-glass laser light (at 1/2 and 1/3 microns)—which was found to be advantageous for both direct and indirect drive. By contrast, the CO_2-laser approach uses a driver system that would be highly attractive from the point of view of reactor engineering, but unfortunately lases at 10 microns. The CO_2-laser experiments encountered such severe plasma-interaction problems with both direct and indirect drive that this particular alternative (formerly pursued in the ANTARES program at Los Alamos and the LEKKO program at Osaka) has now been discontinued.

A more fundamental challenge to the use of direct laser illumination is the requirement for a high degree of uniformity. During the past several years, a number of promising "optical-smoothing" schemes have been proposed to counteract imperfections of the laser-beam source. The basic strategy is to split the laser light into a large set of randomly phased independent "beamlets," each of which produces almost perfectly uniform illumination over a small area. Thousands of such beamlets could be focused onto the pellet surface, the interference between neighbors could be kept small, and the desired quality of overall illumination could theoretically be achieved. These ideas are being tested in the PHAROS III laser facility at the Naval Research Laboratory in Washington D.C., the GEKKO laser at Osaka, and the OMEGA laser at Rochester as well as at the Shanghai Institute of Optics in China. In the NRL approach, random *time delays* have been introduced among beamlets, with a view to combining spatial and temporal incoherence so as to avoid a stationary interference pattern.

The reactor concept envisaged for the direct-laser-drive approach is compatible with a moderately efficient short-wavelength source, such as the krypton-fluoride laser system (source efficiency ~7%, wavelength 1/4 micron), which is currently under development at Los Alamos and elsewhere. The improvement in target gain that is expected for the direct-drive approach would give this reactor system an overall energy multiplication comparable to that achieved by the indirect approach when using a 20%-efficient source, such as an energetic ion-beam driver (or possibly a 10%-efficient source, with a compensating improvement in x-ray energy-conversion efficiency.) In either case, one envisages a reactor with about 1-GW electric output power. For the currently expected minimum driver-input energies into the pellet, which are in the 3-10 MJ range, the fusion microexplosions would have yields of order 0.3-1 GJ and corresponding repetition rates of 3-10 Hz.

If the direction of future inertial-confinement research is toward indirect ion-beam drivers, there are two main possibilities. Light ions (such as lithium) have been accelerated into the low MeV range in the PBFA experiments at the Sandia National Laboratory in New Mexico. The PBFA II facility is designed to yield 1-2 MJ of 20-30 MeV ions at a peak power of 10^{13}-10^{14} W. Another promising alternative is to use 10-GeV-range ions of somewhat greater mass ($A \sim 200$), with the high nuclear charge helping to reduce their stopping distance. Appropriate accelerator technology is being developed at the Lawrence Berkeley National Laboratory, but no large-scale prototype driver has as yet been undertaken.

4. GENERAL APPLICATIONS

The rapid progress in high-temperature plasma physics that has taken place thanks to the stimulus of the world's long-term energy problem has created major new scientific and technological resources for other areas of practical application. One important illustration is the development of x-ray lasers—which has drawn on the expertise of both inertial and magnetic fusion research.

For such purposes as x-ray microscopy, holography, and surface etching, lasers capable of operating at wavelengths of ~40 Å (energies of ~300 eV) would be of great value, and even the soft x-ray range around 100-200 Å has interesting applications. The plasma-temperature requirements are much less demanding than for fusion: Depending on the nature of the population-inversion scheme, electron temperatures T_e in the low keV range lend themselves to the requirements for a 40-Å laser, and temperatures as small as a few hundred eV can suffice for the longer-wavelength range. (The ion temperatures T_i are desired to be somewhat lower than T_e, so as to avoid excessive Doppler broadening, which reduces the gain of the medium.)

The optimal plasma densities for the x-ray laser application are intermediate between those of magnetic and inertial fusion: typically, $n = 10^{19}$-10^{22} cm^{-3}, with the higher densities corresponding to shorter x-ray wavelengths. The plasma density is limited on the high side by the need to avoid collisional depopulation of excited energy levels, but the choice of n is driven upward—especially in the case of short wavelengths λ—by the desire to maximize the gain G (which is proportional to $n\lambda^\alpha$, where $2 \lesssim \alpha \lesssim 4$).

The burdensome condition on the energy confinement time τ_E that preoccupies fusion researchers has no precise counterpart in the x-ray laser application. If the x-ray laser scheme is based on the "electron-pump" mechanism,[15] where population inversion is achieved by collisional filling of upper energy levels and radiative depletion of lower ones, there is no evident advantage in prolonging the heat-up time τ_H. When a laser scheme based on three-body ion-recombination is used,[16] the best strategy is to cool the plasma electrons as rapidly as possible, following the initial heating and ionization phase, while seeking to keep the plasma density from falling (i.e., requiring the *ratio* of energy to particle confinement time, τ_E/τ_p, to be as small as possible). A scaling that is at least somewhat reminiscent of Lawson's $n\tau_E$ condition can, however, be elicited by requiring the gain length GL to be sufficiently large (of order 20), so as to saturate the lasing medium. Defining $\tau_L = L/c$ as the transit time of the laser pulse over the path length L through the medium, the GL requirement translates into an $n\tau_L$ requirement. This type of $n\tau$ criterion could make sense, for example, when the gain length is limited by the number of passes that can be made between a pair of xuv-mirrors before the amplifying medium is lost; even for x-ray laser parameters as large as $nL = 10^{23}$ cm^{-2}, however, the $n\tau_L$ requirement amounts to a relatively lenient 3×10^{12} cm^{-3} sec. (For *single*-pass amplification, there is no requirement that the lifetime of the plasma should exceed τ_L, since the driver-power input can be timed to move along L, keeping in phase with the x-ray pulse.)

The x-ray laser experiments carried out at LLNL[15] using the NOVETTE laser driver in 1984, and more recently, using two of the ten NOVA beams, have led the way toward the achievement of high gain lengths. As illustrated in Fig. 14, the targets consist of thin foils ($\sim 0.1\,\mu$m). The heated foil material provides very uniform transient ($< 10^{-9}$ sec) plasmas up to 5 cm in length. Initial success came by using plasmas of $T_e \lesssim 1$ keV, $n \sim 5 \times 10^{20}$ cm^{-3} for electron-pumping of neon-like selenium, thus producing gain lengths of order eight at wavelengths of 206 Å and 209 Å. The peak gain length of these lines has now been raised to 15–20, corresponding to a remarkable 10^6-fold amplification relative to the spontaneous-emission level. Three other lines with smaller gains have also been observed—including the 182-Å line which had initially been expected to dominate. Other neon-like systems have been used to produce wavelengths down to 106 Å, and nickel-like systems have gone down to 50 Å—but the gain lengths in these short-wavelength cases are in the range $\lesssim 4$.

A quite different approach to x-ray lasers, inspired by the magnetic confinement experience, has been followed at PPPL[16] (cf. Fig. 15). The basic idea has been to strip light ions, such as carbon, and use the recombination mechanism to produce laser amplification with relatively predictable hydrogen-like systems.

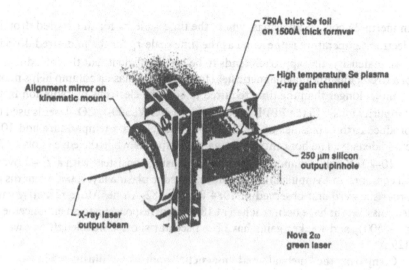

FIGURE 14 By means of a cylindrical lens, the NOVA laser beam can be used to produce a thin column of hot plasma from a metal-foil target. For suitable choice of ion species and plasma parameters, a longitudinal x-ray laser beam is generated within this plasma column.

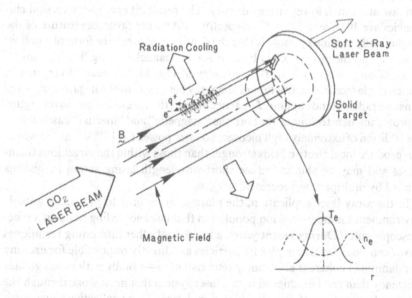

FIGURE 15 A strong magnetic field can be used to confine a thin column of hot plasma evolving from a laser-illuminated solid target. A tubular x-ray laser beam is generated at the radiation-cooled surface of the plasma column.

In inertially confined plasmas, where the time scale τ_E for the desired drop in electron temperature is the same as the time scale τ_p for the undesired drop in plasma density, this approach tends to be very inefficient, but the introduction of a strong longitudinal magnetic field to confine the plasma column helps make τ_p much longer than the time required to cool the electrons by radiation from "impurity ions." In the PPPL experiments, a 3-GW, 300-J, CO_2-laser is used to produce carbon plasmas of 200–300 eV initial electron temperature and 10^{19} cm^{-3} density. The hot central plasma then expands while the edge cools to T_e = 10–20 eV, while maintaining its initial density, consistent with a $\beta \sim 1$ overall equilibrium. Amplification occurs in the edge plasma layer: Gain lengths of order six were first observed in 1984 for the 182-Å line. More recently, gain lengths of eight have been reached at 182 Å (corresponding to an enhancement of ~ 500), and weaker gains have been seen at shorter wavelengths down to 129 Å.

Comparing the "inertial" and "magnetic" approaches illustrated in Figs. 14 and 15, one finds that the former approach currently has the advantage of producing much denser and hotter plasmas, which lend themselves to the achievement of high gains and short wavelengths at high instantaneous power, while the latter approach has some intrinsic merits that may become important for the practical application of x-ray lasers. The highest output energies to date are comparable (1–3 mJ for "magnetic" and ~ 0.5–1.0 mJ for "inertial"), but the input energy is much less in the former case—precisely because the plasma can lase at a much lower energy density. The resultant energy-conversion efficiencies are 10^{-5} and 2×10^{-7}, respectively. Another favorable feature of the magnetically confined plasma is that the density profile has the form of a hollow cylinder, and the interior x-ray beam is stably channeled along **B**, with a divergence of only ~ 5 mrad. In the geometry of Fig. 14, the beam divergence is somewhat larger, and refraction of the x-ray beam due to the negative outward density gradient tends to limit L, particularly in the case of longer wavelengths. A promising new technique for both the "magnetic" and "inertial" cases will be the addition of extremely high picosecond laser power ($\sim 10^{18}$ W cm^{-2}) which can produce local electric fields stronger than those within the target ions themselves and may be able to induce short-wavelength lasing action in systems excited by multiphoton processes.

In the x-ray laser application, the plasma serves only to provide a suitable environment for an excited ion population that lases according to its own spectroscopic rules. During recent years, a number of other interesting techniques have been invented where plasma particles are directly responsible for exciting (or damping) coherent electromagnetic radiation—usually with much greater efficiency than can be achieved with a laser system that must work through the intermediary agency of an excited ion population. These collective wave-particle interactions are of a type long studied in magnetic fusion schemes that depend on strongly ordered motion—for example, mirror machines.

The free-electron laser[17,18] (FEL) imparts a transverse wiggle to a high-quality energetic electron beam, by means of a periodic magnet structure. The FEL electron population is unneutralized, but when seen in its rest frame it looks just like an ordinary plasma. The periodically accelerated electrons radiate collectively, transferring their kinetic energy into coherent electromagnetic waves. With a magnet system that is "tapered" so as to keep in phase with the slowing-down electron beam, efficiencies as high as 30–40% have been achieved. Both short-pulse and steady-state FEL's have been demonstrated. The emitted frequencies can range up from the microwave domain into the visible (or higher), depending on the energy of the wiggled electron beam.

Among the many promising applications of the FEL is the high-powered heating of magnetic fusion plasmas at the electron gyrofrequency (or its harmonics). Another possibility is to invert the FEL scheme in the form of an electron-beam accelerator (driven by FEL's). Still another ingenious concept uses the interference of two input laser beams to excite high-electric-field heat waves at the plasma frequency of an ordinary neutralized plasma, thus providing a powerful electron acceleration structure. The particular inventions mentioned here are part of a larger continuum of new ideas that holds the promise of producing a plasma technological revolution during the coming decades.

ACKNOWLEDGMENTS

I am indebted to the inertial confinement research community—particularly to Drs. S. E. Bodner, E. M. Campbell, D. A. Goerz, and R. L. McCrory—for their help in preparing this paper. Work supported by U.S. Department of Energy Contract No. DE-AC02-76CH03073.

REFERENCES

1. *Physics Through the 1990s. Plasmas and Fluids* (National Academy Press, Washington, DC, 1986).
2. I. E. Tamm and A. D. Sakharov, *Plasma Physics and the Problem of Controlled Thermonuclear Reactions*, edited by M. A. Leontovich (Pergamon, Oxford, 1961), Vol. I, p. 1.
3. G. I. Dimov *et al.*, Sov. J. Plasma Phys. **2**, 326 (1976).
4. T. K. Fowler and B. G. Logan, Comments, Plasma Phys. Controlled Fusion Res. **2**, 167 (1977).
5. *Fusion*, edited by E. Teller (Academic, New York, 1981), Vol. I, Part A.
6. J. D. Sethian *et al.*, Phys. Rev. Lett. **59**, 892 (1987).
7. J. B. Taylor, Phys. Rev. Lett. **33**, 1139 (1974).
8. L. Spitzer, Jr., Phys. Fluids **1**, 253 (1958).
9. Plasma Physics and Controlled Nuclear Fusion Research, Proc. 11th Int. Conf. (Kyoto, Japan, 1986); 3 Volumes, IAEA (Vienna, 1987).
10. International Tokamak Reactor, Phase Two A, Part II, International Atomic Energy Agency (Vienna, 1986).
11. J. Nuckolls, Phys. Today, **35**, 24 (1982).
12. R. L. McCrory and J. M. Soures, Sci. Am. **225**, 68–79 (August 1986).
13. N. G. Basov *et al.*, Sov. Phys. JETP **19**, 123 (1964).
14. J. Nuckolls *et al.*, Nature **239**, 139 (1972).
15. D. L. Matthews *et al.*, Phys. Rev. Lett. **54**, 110 (1985).
16. S. Suckewer *et al.*, Phys. Rev. Lett. **55**, 1753 (1985).
17. D. A. G. Deacon *et al.*, Phys. Rev. Lett. **38**, 892 (1977).
18. T. J. Orzechowski *et al.*, Phys. Rev. Lett. **57**, 2172 (1986).

Frontiers of Atomic
Physics/ DANIEL KLEPPNER

INTRODUCTION

Considering its status among the oldest fields of modern physics, atomic physics today is so vigorous that its behavior borders on the unruly. Advances are rapid-fire and any summary is likely to be obsolete almost as soon as it gets into print. To illustrate this, let me call your attention to the most recent overview of the field. This is the volume *Atomic, Molecular, and Optical Physics* in the recent physics survey by the National Academy of Sciences, *Physics Through the 1990s*. The physics survey was published in 1986, but the scientific reporting was essentially finished in 1985. That is not so long ago. Nevertheless, hardly any of the work I will describe here is described in the physics survey. With respect to atomic physics, the survey is already out of date.

Such rapid progress presents an obvious advantage for this sort of review: there is an abundance of "plums" from which to select one's topics. It also presents an obvious disadvantage: the selection is necessarily arbitrary and incomplete. So let me apologize at the outset to my friends and colleagues whose work is overlooked here. It would also be in order to apologize to my friends and colleagues whose work is *not* overlooked here for giving rather short shrift to their accomplishments.

Notwithstanding Arthur Schawlow's notorious dictum "a diatomic molecule is a molecule with one atom too many," atomic and molecular physics have always been close allies. Now, however, there is a new partner in the alliance. The boundary that separated atomic and molecular physics from physical optics has disappeared: the collective subject "Atomic, Molecular, and Optical Physics" has come into being. The title lacks grace but at least its acronym— AMO—is friendly.

DANIEL KLEPPNER *is Lester Wolfe Professor of Physics, and also Associate Director of the Research Laboratory of Electronics, at the Massachusetts Institute of Technology, Cambridge, Massachusetts, where he works on various problems in experimental atomic physics. Previously, he was for a number of years a member of the faculty at Harvard University, where he worked with Norman Ramsey in the development of the hydrogen maser.*

Let me start by describing some recent advances in quantum optics. Among the most dramatic is the production of squeezed light by Slusher and Yurke at Bell Labs, by Kimble at the University of Texas, Austin, and by others. The creation of light with new statistical properties, and the implications of "squeezing" for measurement beyond the normal quantum limit, have attracted wide attention. However, this topic is taken up elsewhere in these Proceedings by Paul Fleury, and we shall turn to a related topic, *cavity quantum electrodynamics*.

CAVITY QUANTUM ELECTRODYNAMICS

It is difficult to think of a subject more exhaustively studied and beautifully developed than radiation theory. Nevertheless, a growing body of radiative phenomena is being recognized which, while being essentially completely incorporated within the theory, has never been seriously discussed because it was too remote from experiment to warrant much thought. Among these is the dynamics of individual atoms radiating in cavities.

The "single atom maser" provides a good introduction to this line of research. Several years ago Walther[1] showed that it was possible to sustain maser action with a stream of excited atoms passing through a superconducting cavity. The arrangement was similar in many respects to the original ammonia maser, except that the average number of radiators in the cavity was not the typical value of 10^7 or more, but was less than one! This was possible because the atom was in a Rydberg (highly excited) state whose dipole transition matrix element is more than 5×10^3 larger than a typical dipole moment, and because the cavity had such a high Q that the radiation emitted by one atom would linger until another atom happened by. "Maser" is an acronym for microwave amplification by stimulated emission of radiation: in the one-atom maser the atom is stimulated by the ghost of another atom that has long since departed.

Masers use cavities because cavities enhance radiation. However, cavities can also suppress radiation. If the cavity is seriously mistuned, spontaneous emission can be essentially turned off.[2] This was demonstrated a few years ago with Rydberg atoms,[3] and now the effect has been observed at infrared and optical wavelengths. In an experiment by Haroche *et al.*[4] cesium atoms were excited to a 5D state that radiates spontaneously at a wavelength of 3.5 μm. The excited atoms passed through a channel 8 mm long but only 1.1 μm wide. The lifetime was measured by monitoring the transmission of atoms in the 5D state. In one orientation the radiation was inhibited and the atoms were transmitted; at other orientations the radiation was allowed and the atoms were lost.

Feld and his colleagues[5] have quantitatively demonstrated suppressed and enhanced spontaneous emission at optical wavelengths by observing directly the radiation from a beam of ytterbium atoms in a confocal resonator. Figure 1 shows the signal for radiation along the axis of the resonator as the cavity length is varied. In resonance, the radiation rate into the cavity mode is enhanced

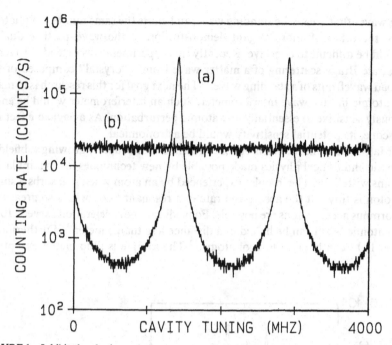

FIGURE 1   Inhibited and enhanced spontaneous emission. The signals are a measure of the spontaneous radiation rate into a confocal cavity. Curve *b* is the rate with the cavity blocked: this is the rate in free space. Curve *a* shows the effect of enhanced and inhibited emission as the cavity is tuned through resonance. (From Ref. 5.)

compared to free space by a factor of about thirty. When the system is antiresonant, the rate is suppressed by roughly the same factor.

Accompanying the change in an atom's radiation rate caused by a cavity, there are changes in its energy. "Cavity pulling" effects occur,[6] the Lamb shift is altered[7] and the atom-cavity van der Waals interaction comes into play.[4] Although the effects are small, they can be important in precision measurements such as the determination of the electron *g* factor.[9] In addition, cavity quantum electrodynamics has opened the way to creating new coherence states of light and small systems of atoms.

MANIPULATING ATOMS WITH LIGHT

Diffraction of a light wave by a periodic grating of matter is commonplace in optics and spectroscopy. Recently, the complementary process—diffraction of a matter wave by a periodic grating made of light—has been realized. Figure 2 illustrates an experiment by Pritchard and his colleagues[10] in which an atomic beam of sodium is diffracted by a standing wave of light. The interference pattern can be calculated from the deBroglie wavelength of the atom and the period of the standing wave nodes; it can equally well be calculated from the momentum recoil of the atom as it absorbs one or more photons from one of the traveling waves that create the standing wave, and emits the same number of photons into the other. A more elegant demonstration of the wave–particle duality would be difficult to conceive. Recently this experiment was extended to demonstrate Bragg scattering of a matter wave from a "crystal" composed of the broad wavefronts of a standing wave.[10] The next goal for this research is to make an atomic matter-wave interferometer. Such an interferometer would be enormously sensitive to essentially any atomic perturbation. As a Sagnac-effect gyroscope, its potential sensitivity would be astronomical.

Matter-wave diffraction is but one illustration of a rapidly growing subfield of atomic and optical physics made possible by new techniques for manipulating atoms with light. The impulse experienced by an atom when it absorbs a single photon is tiny but the absorption rate in a resonant laser field is so great that enormous accelerations are possible. For sodium, the acceleration exceeds 10^5 g! An atomic beam can be halted in a distance less than a meter.[11] Furthermore, laser light can be used to cool atoms.[12] The sample is illuminated by optical

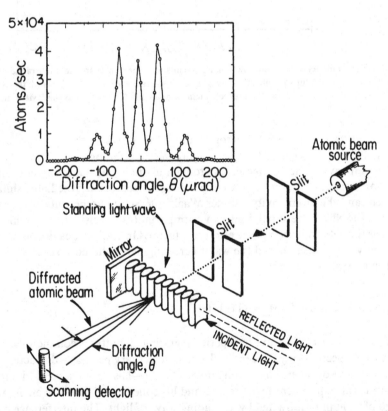

FIGURE 2   Diffraction of matter by light. The schematic diagram illustrates how a carefully prepared matter wave can be diffracted by the periodic structure of a standing light wave. The data in the inset reveals a peak due to the undiffracted beam, and two diffraction maxima on either side. (Courtesy of D. E. Pritchard.)

standing waves—or perhaps one should say oppositely directed running waves—from a laser tuned slightly below the principal transition. A moving atom experiences a Doppler shift that allows it to absorb radiation from the running wave which opposes its motion. The resulting recoil causes the atom to experience an enormous viscous force. The gas forms what is aptly called "optical molasses". Its temperature rapidly drops to the sub-millikelvin regime. The very first experiment by Chu *et al.* achieved a temperature of 240 microkelvin and it appears that much lower temperatures should be achievable.

Trapped atoms in the microkelvin regime constitute an extraordinary state of matter. Because collisions can occur only by *s*-wave scattering, transport phenomena are dominated by the atomic statistics. The atoms can form a degenerate Fermi gas or a Bose condensate. Atom-surface interactions are expected to become vanishingly small. The nature of free-bound molecular transitions is fundamentally altered because the continuum becomes so sharp that it behaves like a bound state. The time scales for radiation and collisional interaction in excited state collisions are reversed microkelvin temperatures. In addition, trapped ultra-cold atoms constitute a superb target for ultra-high resolution laser spectroscopy and possibly the creation of optical frequency standards.

CHAOS AND QUANTUM MECHANICS

The appearance of James Gleick's book *Chaos* on the *New York Times* best-sellers list provides cheerful news for anyone who is worried by the estrangement of science from the public. Its popularity, however, belies the discomfort of scientists who cherish orderly systems and don't quite know what to make of chaos. In particular, quantum mechanics appears to have little use for chaos: the Schrödinger equation is linear, and linear equations have orderly solutions.

The connection—if any—between chaos and quantum mechanics is a small but important problem in the field of nonlinear dynamics. The subject is more or less loosely known as "quantum chaos." Two atomic systems turn out to provide ideal testing grounds for studying quantum chaos. The first is a hydrogen atom in a strong microwave field.

Planetary motion in the presence of a uniform but oscillating gravitational field constitutes an interesting but fundamentally academic problem in classical nonlinear dynamics. However, if the problem is expressed in terms of the motion of an electron around a proton while the system is subjected to a microwave electric field, the problem becomes experimentally accessible. Furthermore, the system provides a most interesting testing ground for examining quantum chaos.

If the classical equations of motion for the hydrogen atom in a microwave field are integrated numerically, it is found that as the field amplitude is increased the orbits evolve from slightly perturbed Kepler orbits to elaborately distorted curves that suddenly become chaotic. Experimentally, for a given binding energy the atoms start to ionize at the precise field at which classical chaos first appears. Furthermore, the onset of chaos, or ionization, is by no means monotonic with the principal quantum number n. There is a great deal of structure in both the experimental data and the results of the classical numerical calculations. What is surprising is that the experimental and numerical data agree in great detail. It is not obvious why quantum behavior should mirror the details of classical behavior in regions of chaos. Quantum mechanically, one normally portrays ionization in terms of tunneling, multiphoton absorption, and photoionization. None of these processes are suggestive of classical processes.

Figure 3 shows experimental data taken by Koch and his colleagues, and the results of calculations by Jensen.[13] The data are plotted in scaled coordinates: the unit of field is the static field at which ionization is energetically allowed for a given value of n, and the unit of frequency is the Kepler frequency for the same level. The peaks in the data occur at low order rational values of the scaled frequency: 1, 2/3, 1/2, 2/5, ... Classically, for these subharmonic frequencies

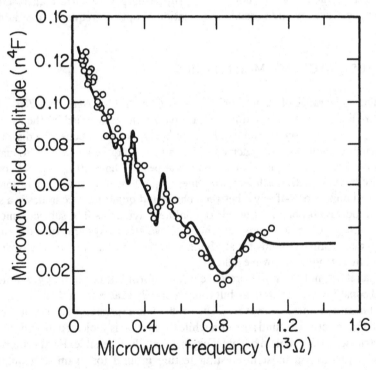

FIGURE 3   Chaos in atomic physics: microwave ionization of hydrogen. The measured intensity of microwave field F needed to ionize a highly excited state of hydrogen is plotted against the frequency of the field, using scaled coordinates that eliminate global variations due to the changing energy scale. The unit of frequency is the classical orbital frequency. The solid line shows the onset of chaos for classical orbits corresponding to the same experimental conditions. The peaks are related to islands of stability in the phase plane at subharmonics of the driving frequency. (Courtesy of R. V. Jensen and P. E. Koch.)

islands of stability appear in the phase plane. Quantum mechanically, one might expect that ionization would be more probable at frequencies harmonically related to resonance excitation and thus require relatively weak microwave field amplitudes. On the contrary: at these frequencies the field required to ionize the atom is *stronger* than for nearby frequencies.

Quantum-mechanical solutions have been carried out by a number of authors.[13] The onset of chaos turns out to be mirrored by a change in the dynamical nature of the wave function. When the system is stable the wave function remains localized among a small group of adjacent eigenstates. The onset of chaos is reflected by a transition to diffusive motion in n space: the wave function becomes delocalized. At high scaled frequency there appears to be a discrepancy between the classical and quantum behavior. This is unexpected since this is the region where one might expect the Correspondence Principle to be valid. Evidently this system continues to offer surprises.

The second atomic system for studying quantum chaos is the hydrogen atom in a strong magnetic field. Chaos occurs in the regime of high quantum numbers, and for highly excited hydrogen the principal magnetic interaction is diamagnetism. The diamagnetic interaction is normally feeble—hardly big enough to be detectable—but it can be shown that the ratio of the diamagnetic interaction to the electrostatic interaction scales with principal quantum number as n^6. In a field of 6 T, for instance, one enters the regime where the atom is essentially bound magnetically at approximately $n = 50$. In spite of the simplicity of the system, general quantum-mechanical solutions are still lacking. The problem remains as perhaps the last unsolved problem in elementary quantum mechanics. Nevertheless, progress in the past few years has been substantial.

Classical studies of the diamagnetic hydrogen problem reveal a clear transition to chaos.[15-17] Figure 4, from calculations by Delande and Gay,[16] shows Poincaré plots in the transition region. These are plots of the intersections of the orbit on successive passes through a plane of phase space. Although a precise definition of chaos can be elusive, there is no mistaking it here.

Accurate calculations of the energy-level structure in the regime of chaos have been carried out by a number of researchers.[16-18] The statistics of energy-level distribution are generally believed to contain the signature of chaos in quantum mechanics. The distribution of spacings between adjacent levels shows the characteristic evolution from Poisson-like in the regime of orderly motion to Wigner-like in the regime of chaos. The most conspicuous change is the elimination of spacings at small distances, that is, the suppression of degeneracies. Physically, this reflects the breakdown of symmetry and the loss of an approximate constant of motion.

Welge and his colleagues[19] have carried out a series of elegant experimental and theoretical studies of diamagnetic hydrogen in the regime of chaos. By Fourier-analyzing their spectrum they have found regularities in the motion over short periods, roughly the times for single classical orbits. They have discovered new families of orbits in this regime, and have observed the period doubling behavior that often characterizes the onset of chaos. These orbits ap-

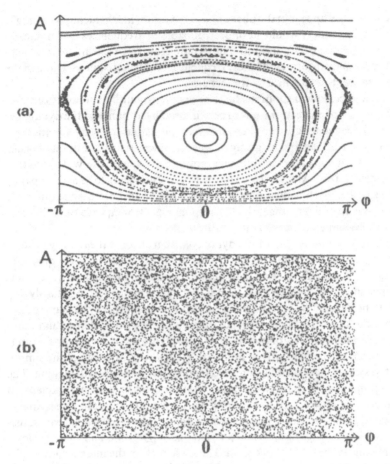

FIGURE 4 Chaos in atomic physics: hydrogen in a strong magnetic field. These plots of the intersections of the classical orbit with a plane in phase space (Poincaré plots) show orderly behavior (upper plot) and the random behavior of chaotic motion (lower plot). (From Delande and Gay, Ref. 16.)

pear to be related to islands of stability in the phase plane, islands which are too small to find in the sea of chaos unless you know exactly where to look. The existence of other islands, and the lifetimes of these quasi-stationary states, remain open questions.

The diamagnetic hydrogen problem is close to ideal for studying quantum chaos because of the simplicity of the Hamiltonian, and because the magnetic field provides an experimental "handle" for turning chaos on and off.

ADVANCES WITH TRAPPED IONS

Ion traps are widely used for mass analysis, spectroscopy and collisional studies. In the past few years, however, new opportunities have been opened by techniques for studying small numbers of ions at ultra-low temperatures. And at the other end of the energy scale, a route to the study of stable antimatter has been opened by techniques for catching antiprotons in flight. Here is a brief rundown of some of the advances.

Single Ions and Shelving

A single trapped ion can be an ideal subject for ultra-high resolution spectroscopy. Trapped in space by the gentlest of perturbations, the ion can be observed under conditions of essentially complete isolation. By observing forbidden transitions to metastable levels, optical linewidths in the subhertz regime become accessible.

The ultimate subject for ultra-precise spectroscopy could well be a single trapped ion in the ground state of the trap, confined in a distance small compared to the optical wavelength. The small confinement volume assures that the ion is observed under conditions of "Dicke narrowing" where Doppler broadening is absent. Placing the ion to the ground state assures that the energy shift due to the second-order Doppler shift—time dilation—is small and precisely known. The Dicke narrowing regime has been achieved,[20] and one can look forward to the isolation of an ion in the ground state.

There is, however, a serious obstacle to subhertz spectroscopy of single ions: the data acquisition rate is intrinsically low. Detecting a signal of less than one photon per second presents a formidable challenge to locating a resonance whose position may not be known to better than 10^5 Hz. To overcome this problem, Dehmelt[21] suggested a strategy called "shelving". Three levels are involved: a ground state G, a metastable state M, and some level A that has an allowed optical transition to G. The allowed transition $G \rightarrow A$ is driven by a laser and intense fluorescence is radiated, perhaps 10^7 photons per second. A second laser explores the narrow transition $G \rightarrow M$. When it hits resonance the atom is elevated to M where it resides "shelved" for perhaps a second or more. As soon as this occurs the fluorescence from A abruptly ceases. Thus the signal-to-noise ratio for detecting the occurrence of the transition $G \rightarrow M$ is enhanced by the ratio of the lifetimes, roughly a factor of 10^7.

Shelving has been demonstrated independently by three groups.[22-24] Figure 5, by Itano *et al.*,[24] shows the clear signature of shelving with one, two, and three ions in the trap. The ion is Hg^+. In this experiment the metastable level is not populated from the ground state by laser light, but by spontaneous emission from the excited state. Furthermore, two metastable levels play a role. The lifetimes of these levels and their branching ratios were found by analyzing the "on–off" statistics of the fluorescence.

Aside from its promise as a tool for high resolution spectroscopy, "shelving" has attracted attention because it dramatically illustrates how a state function evolves, and the reality—if one can use that word—of "quantum jumps" between eigenstates.

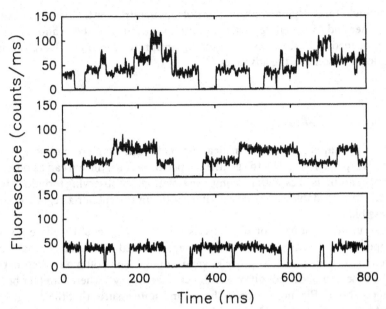

FIGURE 5 Shelving. Signal from small numbers of laser-excited ions shows the turn-off of fluorescence when one of the ions decays to a metastable state. Upper curve: three ions. Middle curve: two ions. Lower curve: single ion. (From Itano *et al.*, Ref. 24.)

A Few Trapped Ions—Pseudomolecules and Microcrystals

The search for single-ion techniques has produced fascinating discoveries about the motions of small numbers of ions. Briefly, a single trapped ion will execute characteristic harmonic motions in the trap potential. The motion manifests itself by well-defined sidebands on its absorption spectrum. With two ions in the trap, the center of mass continues to execute the same fundamental motion. However, the interaction between the ions creates additional modes. The two ions form a pseudomolecule with internal vibrational degrees of freedom much like a conventional molecule, except that the binding is due to the trap rather than the usual molecular interactions. Wineland and his colleagues have observed the vibrational structure of a diatomic pseudomolecular ion and, in what can perhaps best be described as a physicist's foray into molecular physics, have calculated the spectrum from the trap parameters.[25]

When a few ions are added to the trap and slowly cooled they abruptly solidify into a regular structure.[25,26] Figure 6 is a photograph of the fluorescence from such a cluster. Using laser heating and cooling, the microcrystal can be repeatedly melted and refrozen in a cycle that displays large but highly repeatable hysteresis.

Trapped Ions and Strongly Coupled Plasmas

An ion cloud is cold when its mean thermal energy falls below its mean electrostatic energy. In the parlance of plasma physics, such an ion cloud is called a strongly coupled non-neutral plasma. In contrast to ordinary plasmas, strongly coupled plasmas display liquid- and solid-like properties. A considerable experimental literature exists for strongly coupled plasmas, but experiments have been scarce. Gilbert, Bollinger, and Wineland[27] recently trapped a cloud containing up to 15 000 ions at a temperature of 10 mK. The ratio of electrostatic to thermal energy was as large as 200. One might expect such an exotic system to display correspondingly exotic behavior, and it did: the ions spontaneously formed a crystal of apparently liquid shells. Within each shell the ions moved freely in a liquid-like state, but the shells themselves were solid-like. Figure 7 shows a section of the structure of concentric cylindrical shells.

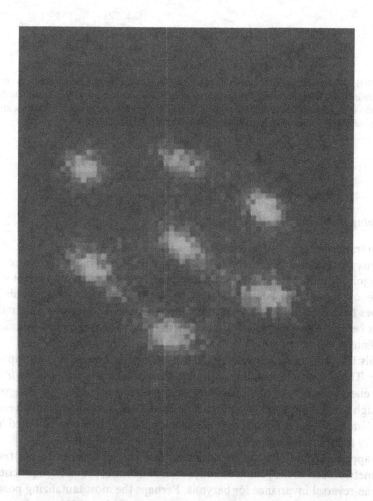

FIGURE 6   Ionic microcrystal. Fluorescence from seven Mg$^+$ ions in the millikelvin temperature regime held in a rigid lattice by their electrostatic repulsion and the confining field of an ion trap. The mean distance between ions is 23 μm. (From Ref. 27.)

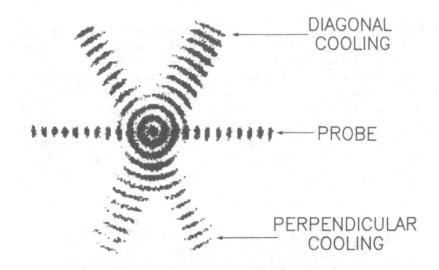

DIAGONAL
COOLING

PROBE

PERPENDICULAR
COOLING

FIGURE 7   Strongly coupled plasma. The photograph shows rings of fluorescence from shells of ions confined in a trap, illuminated by three laser beams. The ions move freely within the shells, but the shells themselves have a crystalline rigidity. The diameter of the pattern is 500 microns, and the number of ions is approximately 10 000. (Courtesy D. E. Wineland, Ref. 27.)

Trapping the Anti-Proton

Ion traps are typically loaded with ions formed at a few electron-volts. Anti-protons are formed at several billion electron volts. Considering the nine orders of magnitude difference in energy, attempting to trap an antiproton might strike one as overambitious. Nevertheless, it has been done. Gabrielse and his colleagues have captured antiprotons and kept them around for one hundred seconds before releasing them.[28] It appears that they can be trapped practically indefinitely.

This feat was made possible by LEAR—CERN's Low Energy Antiproton Ring. This incredible machine captures, slows and cools antiprotons from a high energy source. 21-MeV particles were ejected in short bursts and passed through a beryllium window; approximately 10^{-4} of them emerged with energy less than 3 keV. Using some fancy electronics, a 3-keV trap was switched on so rapidly that the slow antiprotons were grabbed in mid-flight.

Trapped anti-protons can be cooled to low energy and put to use testing symmetries such as the proton to antiproton mass ratio, providing a precise test of time-reversal invariance for baryons. Perhaps the most tantalizing possibility, however, is making low-energy antihydrogen. This would open the way to the study of stable anti-matter, and possibly give rise to a new branch of AMO physics: anti-atomic physics.

SOME OMISSIONS

The topics sketched above are only a sample of the frontier activities in atomic physics. Conspicuous among the omissions are the numerous studies of basic symmetries and fundamental laws—the isotropy of space, charge conservation, quantum electrodynamics, neutral current effects, the discrete symmetries, and the fundamental constants. An excellent review article by Vernon Hughes can be found in the proceedings of the last international conference on atomic physics.[29] Other topics omitted include advances in theory, studies of highly stripped ions, spectroscopy with synchrotron sources, multi-photon research, precision x-ray spectroscopy, the role of atomic physics in astrophysics and astronomy, and the broad advances in atomic and molecular collisions. These stories will have to wait for another occasion.

REFERENCES

1. D. Meschede, H. Walther, and G. Müller, Phys. Rev. Lett. **54**, 551 (1985).
2. D. Kleppner, Phys. Rev. Lett. **47**, 233 (1981).
3. R. G. Hulet, E. S. Hilfer, and D. Kleppner, Phys. Rev. Lett. **55**, 2137 (1985).
4. W. Jhe, A. Anderson, E. A. Hinds, D. Meschede, L. Moi, and S. Haroche, Phys. Rev. Lett. **58**, 666 (1987).
5. D. J. Heinzen, J. J. Childs, J. E. Thomas, and M. S. Feld, Phys. Rev. Lett. **58**, 1320 (1987).
6. D. J. Heinzen and M. S. Feld, Phys. Rev. Lett. **59**, 2622 (1987).
7. P. Dobiasch and H. Walther, Ann. Phys. (Paris) **10**, 825 (1985).
8. L. S. Brown and G. Gabrielse, Rev. Mod. Phys. **58**, 233 (1986).
9. P. L. Gould, G. A. Ruff, and D. E. Pritchard, Phys. Rev. Lett. **56**, 823 (1986).
10. P. J. Martin, B. G. Oldaker, A. H. Miklich, and D. E. Pritchard, Phys. Rev. Lett. **60**, 515 (1988).
11. A. L. Migdall, J. V. Prodan, W. D. Phillips, J. H. Bergman, and H. J. Metcalf, Phys. Rev. Lett. **54**, 2586 (1985).
12. References to the original proposals and a review of work through early 1986 can be found in the article by S. Chu, J. E. Bjorkholm, A. Ashkin, and A. Cable, *Atomic Physics 10*, edited by H. Narumi and I. Shimamura (North-Holland, Amsterdam, 1987), p. 377.
13. S. Chu, J. E. Bjorkholm, A. Cable, and A. Ashkin, Phys. Rev. Lett. **55**, 48 (1985).
14. P. M. Koch, K. A. H. van Leuwen, O. Rath, D. Richards, and R. V. Jensen, *The Physics of Phase Space*, edited by Y. S. Kim and W. W. Zachary (Springer–Verlag, Berlin, 1987), Lecture Notes in Physics, Vol. **278**, p. 106 and references therein.
15. J. B. Delos, S. K. Knudson, and D. W. Noid, Phys. Rev. A **30**, 1208 (1984).
16. D. Delande and J. C. Gay, Phys. Rev. Lett. **57**, 2006, 1986; also, J. C. Gay, *Atoms in Unusual Situations*, edited by J. P. Briand (Plenum, New York, 1986), p. 107.
17. G. Wunner, U. Woelk, I. Zech, G. Zeller, T. Ertl, F. Geyer, W. Schweitzer, and H. Ruder, Phys. Rev. Lett. **57**, 3261 (1986).
18. D. Wintgen and H. Friedrich, Phys. Rev. Lett. **57**, 571 (1986).
19. J. Main, G. Wiesbusch, A. Holle, and K. H. Welge, Phys. Rev. Lett. **57**, 2789 (1986); and K. H. Welge *et al.* (to be published).
20. W. M. Itano, J. C. Bergquist, and Randall G. Hulet, Phys. Rev. A **36**, 2220, 1987.
21. H. G. Dehmelt, Bull. Am. Phys. Soc. **20**, 60 (1975).
22. W. Nagourney, J. Sandberg, and H. Dehmelt, Phys. Rev. Lett. **56**, 2797 (1986).

23. Th. Sauter, W. Neuhauser, R. Blatt, and P. E. Toschek, Phys. Rev. Lett. **57**, 1696 (1986).

24. J. C. Bergquist, R. G. Hulet, W. M. Itano, and D. J. Wineland, Phys. Rev. Lett. **57**, 1699 (1986).

25. D. J. Wineland, J. C. Bergquist, W. M. Itano, J. J. Bollinger, and C. H. Manney, Phys. Rev. Lett. **59**, 2935 (1987).

26. F. Diedrich, E. Peik, J. M. Chen, W. Quint, and H. Walther, Phys. Rev. Lett. **59**, 2931 (1987).

27. S. L. Gilbert, J. J. Bolinger, and D. J. Wineland, Phys. Rev. Lett. **60**, 2022 (1988).

28. G. Gabrielse, X. Fei, K. Helmerson, S. L. Rolston, R. Tjoelker, and T. A. Trainor, Phys. Rev. Lett. **57**, 2504 (1986).

29. V. W. Hughes, *Atomic Physics 10*, edited by H. Narumi and I. Shimamura (North-Holland, City, 1987), p. 1.

Quarks and Gluons in Nuclear and Particle Physics/ LEON VAN HOVE

1. QUARKS AND GLUONS IN THE STANDARD MODEL OF PARTICLE PHYSICS

The last twenty-five years have completely changed our understanding of the *strong interactions* (which in nuclear physics are often called the nuclear forces) and of the particles undergoing them, for which one uses the generic name of *hadrons* (they consist of *baryons* such as the proton, neutron, hyperons, and their resonances, their antiparticles the antibaryons, and *mesons* such as the pion, kaon, and their resonances). The key to the new understanding is the representation of all hadrons as composites of basic fermions called *quarks* (q) and *antiquarks* ($\bar{q}$), with *quantum chromodynamics* (QCD) as the nonabelian gauge field theory describing the strong interaction between q's and $\bar{q}$'s, the *gluons* being the quanta of the QCD gauge fields. Experimentally, one knows five types ("*flavors*") of quarks, labelled u and d (up and down) for the constituents of nucleons, s for strange, c for charmed, and b for bottom quarks. The constraints of the theory require the existence of a sixth flavor (t for top quark), not found as yet. The generic name of *partons* is commonly used for gluons and all types of (anti)quarks.

In addition to the partons, matter as we know it today also consists of *leptons*. All leptons are fermions. The electrically charged ones are the electron e^-, its antiparticle the positron e^+, the positive and negative muons and tau leptons, while the neutral ones are neutrinos associated one to one with the charged leptons. The leptons do not undergo the strong interaction, i.e., are not directly coupled to the gluons. All basic fermions (leptons and quarks) undergo with various couplings the *electroweak interactions*, described by another nonabelian gauge field theory called *electroweak theory* (EWT) with the *photons* and the recently discovered *weak bosons* W^+, W^- and Z^0 as quanta of the gauge fields. The photons, very familiar to all physicists, of course mediate the well-known electromagnetic interactions, whereas the weak ones are mediated by the very

LEON VAN HOVE *was born in Belgium. His professional affiliations have been with the Princeton Institute for Advanced Study, the Institute for Theoretical Physics in Utrecht, the Max Planck Institute in Munich, and, most recently, CERN, Geneva, where he is Research Director in theoretical physics.*

massive W and Z bosons. The description of matter in terms of quarks and leptons, QCD and EWT with their gauge field quanta, is usually called the *Standard Model* of particle physics.[1]

Nonabelian gauge field theories, first invented in 1954 by C. N. Yang and R. L. Mills, are generalizations of quantum electrodynamics (QED), itself the relativistic, quantum version of Maxwell's theory of electromagnetism and now an integrated part of EWT. Both QCD and EWT are nonabelian in the sense that, contrary to QED, their gauge groups are nonabelian. These groups are $SU(1) \times SU(2)$ for EWT and $SU(3)$ for QCD, and they are commuting subgroups of the overall symmetry group of the Standard Model. The EWT gauge symmetry is spontaneously broken, in the sense that the ground state of the theory, i.e., the *vacuum state* where no matter is present, is not invariant for EWT gauge transformations. EWT postulates a symmetry-breaking scheme called the Higgs mechanism, which can be regarded as a relativistic analogue of the Landau–Ginsburg phenomenological theory of the superconducting phase transition in metals. Its main effect is to generate masses for three of the EWT gauge quanta, the weak bosons W and Z, without spoiling the renormalizability of the theory. It could also be responsible for generating lepton and quark masses. The vacuum is characterized by a nonvanishing value of a gauge-dependent scalar field and is therefore not gauge invariant. The Higgs mechanism implies also the existence of new scalar particles called Higgs bosons, as yet undiscovered. It constitutes the most uncertain and *ad hoc* part of EWT (see T. Applequist's paper in this volume).

One must also postulate spontaneous symmetry-breaking in QCD, but this time the broken symmetry is not the gauge invariance, it is the scale invariance and chiral invariance properties exhibited by the field equations in the absence of quark mass terms. The *scale invariance* is broken by the renormalization prescription which specifies a scale for all lengths, masses, momenta, and energies. This is achieved with a single scale parameter Λ_{QCD}. [Remember that all the above quantities are related in terms of $\hbar$ and c; we adopt the familiar convention of choosing units such that $\hbar = c = 1$, which gives for example $(200 \text{ MeV})^{-1} = 1 \text{ fm} = 10^{-13}$ cm.] Comparison with experiment gives $\Lambda_{QCD} \sim 100$ MeV. *Chiral invariance* is broken by postulating that the products $\bar{\psi}_q \psi_q$ of quark spinor fields have nonvanishing expectation values in the vacuum state (the so-called quark vacuum condensates).

As every physicist knows, the extraordinary thing about quarks and gluons is that they are never seen as free particles. This is another, much more fundamental difference of QCD with EWT, and for that matter with any field theory constructed previously. All charge operators expressing the electroweak couplings to photons and weak bosons are directly measurable. In QCD, on the contrary, the so-called *color charge operators* which express the couplings between quarks and gluons (including the direct couplings between gluons imposed by the nonabelian nature of the gauge group) are assumed to be completely hidden. This property, called *confinement* and so far not derived mathematically from the field equations, implies that all observable particles with strong interactions, i.e., all hadrons and nuclei, have zero color charges. They must be "colorless" composites of the basic color-charge carriers, i.e., the quarks and gluons, the strong interactions between them being analogous to the van der Waals forces between the electrically neutral atoms and molecules of ordinary physics. Indeed, all known hadrons are composites of this type: qqq for baryons, $\bar{q}\bar{q}\bar{q}$ for antibaryons and $q\bar{q}$ for mesons. Colorless composites involving gluons are also possible, the simplest ones being gluon–gluon "glueballs", but they have never been seen experimentally.

Despite its mathematical complexity and its many free parameters, the Standard Model has been extraordinarily successful in describing the very extensive and diversified experimental data now available in particle physics. So far all attempts to prove it wrong, or to perfect it by further unification of interactions, have failed or have remained inconclusive through lack of experimental support. In particular, our present understanding of particle physics rests at the most fundamental level on the assumed existence of quarks and gluons. As we shall explain qualitatively in Sec. 3, the fact that these partons are never produced as free particles (the confinement property), and are therefore never directly observed, does not prevent them from manifesting themselves individually in high energy reactions, where they produce *hadronic jets*, i.e., jets composed of several energetic hadrons flying off within narrow angular cones. In view of the overwhelming evidence, one can say that all modern advances of particle physics would collapse without quarks and gluons.

2. NUCLEAR PHYSICS—WHERE ARE THE QUARKS AND GLUONS?

Paradoxically, the situation is exactly opposite in nuclear physics, where there is an amazing lack of any sizable manifestation for a composite structure of nucleons (protons and neutrons). In fact, it has become a major challenge for theory to explain the almost perfect hiding of any quark structure in nuclear spectroscopy and nuclear reactions up to energies of a few GeV. From simplest QCD estimates, the radius of the nucleon as a composite of three quarks is believed to be about 1 fm, which is far from negligible compared to the nearest-neighbor distance between nucleons in nuclei. Why then can one explain so many nuclear phenomena in terms of nucleons undergoing exchanges of virtual mesons with occasionally the excitation of the delta, the lightest nucleon resonance?

It is not surprising that a variety of theoretical approaches has been developed in recent years to cope with this problem. One has studied various bag models (MIT bag, followed by little bag and chiral bag), and more recently skyrmion models (a skyrmion is a localized solution of certain pre-QCD field theories, discovered in 1962 by T.H.R. Skyrme). Perhaps the most curious aspect of this work, linked with the chiral bag model, is embodied in the "Cheshire cat principle" (a name coined by H.B. Nielsen), which asserts that the physics of nuclei should not depend on where one draws the boundary between subspaces described by meson effects and subspaces described by quark effects (this principle can be shown to hold exactly in a one-dimensional space). This work has reached a level of great mathematical sophistication without yet achieving conclusive results and it is in a state of rapid evolution, so that the reader must be referred to very recent technical review papers.[2]

For the purpose of the present discussion, a more relevant question is how high an energy will be needed to observe the appearance of nuclear effects *patently* linked with the quark structure of hadrons. Section 4 will be devoted to very high energy processes involving nuclei (beam energies of order 100 GeV and higher) which obviously require a quark–gluon description. There the problem is to detect and understand how the QCD properties of nucleons are modified by the nuclear environment. There must be some intermediate energy region above which any Cheshire-cat principle could only be maintained at the cost of highly artificial mathematical complications, and the quark–gluon description therefore imposes itself on grounds of simplicity. It would be wonderful if this intermediate region were low enough to become accessible with the new 5 Gev electron accelerator CEBAF under construction in the USA at Newport News, Virginia.

3. QUARKS AND GLUONS IN HIGH-ENERGY COLLISIONS

While quarks are basic to the understanding of hadron spectroscopy, for which they were invented in 1964 by M. Gell-Mann and G. Zweig, it is in the field of high-energy collisions that they manifest most directly their great explanatory and predictive power. We briefly review the various types of reactions that have been studied systematically so far.

A. Electron–Positron Annihilation

The cleanest case is electron–positron annihilation into hadrons at center-of-mass (CM) energies E_{CM} above a few GeV. The most striking feature here is that, although the e^+e^- annihilate first in a virtual photon γ^* (see Feynman diagram in Fig. 1), a state of maximum spin 1, most of the final states consist of two narrow hadron jets J, J' oriented back to back in the CM frame, with only a few low-energy particles moving in directions transverse to these jets. The initial and final CM configurations are depicted in the drawings on the right of Fig. 1, the particle composition of the jets being illustrated at the bottom of the figure. This jet structure is naturally explained by QCD. The partons that carry electric charge are quarks (q) and antiquarks ($\bar{q}$), the gluons (g) being electrically neutral. Hence the virtual photon materializes into a q and a $\bar{q}$ which fly apart with great energy in opposite directions. Since they carry color charges, a QCD field (more precisely a color-electric field) builds up between them. In a first phase, this field transforms into many $q, \bar{q}$ and g by formation of a *parton shower* (the basic branching processes of QCD are $q \to qg$, $\bar{q} \to \bar{q}g$, $g \to gg$ and ggg, $g \to q\bar{q}$). Then comes the *hadronization* phase in which partons of compensating color charges form the outgoing hadrons. In Fig. 1 the parton shower and the

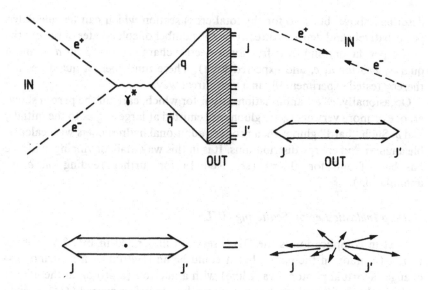

FIGURE 1   Electron-positron annihilation.

hadronization are "contained" in the shaded box, the outgoing hadrons being represented by pairs of (anti)quark lines emerging from the box (these pairs represent mesons $q\bar{q}$, triplets of lines would represent baryons qqq or antibaryons $\overline{qqq}$, see Figs. 2 and 3).

The low spin of the virtual photon (0 or 1) implies that the angle θ between the jets J, J' and the incident e^+e^- (right-hand side of Fig. 1) has a broad distribution, close to $1 + \cos^2\theta$. This is brilliantly confirmed by experiment and contrasts with the narrow jet structure of single events. The latter is characterized by the fact that the outgoing hadrons have low values, of order 300 MeV/c, for their transverse momenta (i.e., the components of their momenta perpendicular to the average jet direction). That at large E_{CM} the initially created q and $\bar{q}$ give rise to such narrow jets is linked with the relative weakness of QCD couplings for *hard processes*, i.e., for processes involving large momentum transfers. This is the celebrated property of "*asymptotic freedom*" of QCD, which makes *renormalized perturbation calculus* applicable to the parton shower formation. In contrast, the hadronization cannot yet be calculated systematically because the momentum transfers involved are low (of order 1 GeV/c or smaller) and for these so-called *soft processes* the QCD couplings are too large for perturbative methods to apply. For the moment one describes them with plausible *nonperturbative algorithms* which were developed over the years in a close interplay of theory and experiment.

Fortunately, many features of e^+e^- annihilation at high energy do not depend on the details of the soft processes. This is true not only for the jet structure described above, but also for the total cross section which can be calculated perturbatively and depends directly on the number of color states of the quarks (three per flavor) and their fractional electric charges ($-$ 1/3 for d, s, and b quarks, 2/3 for u, c, and expected for t). These fundamental quantities are thereby tested experimentally in a very direct way.

Occasionally, e^+e^- annihilations occur for which, early in the parton shower, one or more very energetic gluons are emitted at large angles to the initial q and $\bar{q}$. Such "hard" gluons then produce additional hadronic jets with calculable angular and energy distributions. It is in this way that convincing evidence has been found for gluons (see Ref. 1a for further reading on e^+e^- annihilation).

B. Deep Inelastic Lepton Scattering (DILS)

Next in complication is the DILS reaction illustrated in Fig. 2A where a lepton (a muon in the figure, but it could be an electron or a neutrino) exchanges a virtual photon (wavy line) with a nucleon (a proton in the figure) under such conditions that the four-momentum transfer squared (Q^2) is much larger than 1 GeV^2 (hence the name deep inelastic scattering). The quantity Q^2 is defined by

$$Q^2 = |\mathbf{P} - \mathbf{P'}|^2 - |P_0 - P_0'|^2$$

where $\mathbf{P}$, $\mathbf{P'}$ are the momentum vectors of the lepton before and after the scattering, and P_0, P_0' the corresponding energies (note that Q^2 is always positive from energy–momentum conservation). The virtual photon is absorbed by a q or $\bar{q}$ in the nucleon and deflects it violently (hard process!), while the remaining partons in the nucleon tend to fly on, leading to the formation of two hadronic jets J, J' for reasons similar to those explained for e^+e^- annihilation (Fig. 1). In the present case, the jets are back to back in the rest frame of the total hadronic system created by the photon exchange (see Fig. 2A, shaded box in the Feynman diagram and drawings on the right).

In Fig. 2A the photon is absorbed by one of the three "valence" quarks which compose the nucleon (line marked q), the remaining two flying on ("diquark" marked qq). It also frequently happens that the photon is absorbed by one of the additional "sea" q and $\bar{q}$, the sea being the cloud of virtual partons which are present in the nucleon wave function as a consequence of their coupling with the valence quarks (the nucleon sea is analogous to the cloud of virtual photons and e^+e^- pairs attached to an electron in QED, but it is quantitatively more important because the QCD couplings are stronger). Also in this case of DILS on a sea parton one has formation of two hadronic jets by color separation.

The great interest of DILS is that it probes directly the internal quark structure of the nucleon. One measures the cross section as a function of Q^2 and the variable

$$x = Q^2/2m_p (P_0 - P_0')_{\text{lab}} \; ,$$

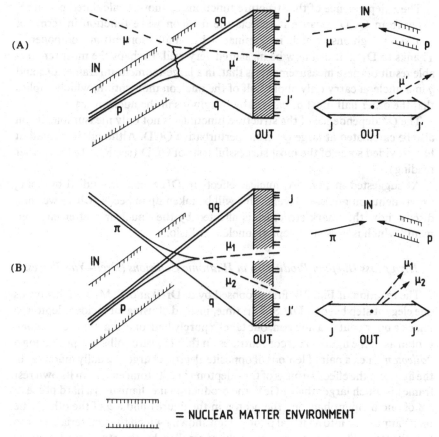

= NUCLEAR MATTER ENVIRONMENT

FIGURE 2   (A) Deep inelastic lepton scattering (DILS).
(B) Drell-Yan effect.

where m_p is the proton mass and $(\cdots)_{\text{lab}}$ refers to energies in the rest frame of the initial nucleon. Energy–momentum conservation implies that x is restricted to values between 0 and 1. The very simple QED formulae for single photon exchange show that DILS is controlled by *structure functions* $f(x,Q^2)$ which describe the distribution in x of (anti)quarks in the nucleon at a space-time resolution of order Q^{-1}, which is much smaller than $(1 \text{ GeV})^{-1} = 0.2 \text{ fm}$. The variable x has a direct physical interpretation if one considers the nucleon in a frame where it has ultra-relativistic velocity ("large-momentum frame"); x is then the fraction of the nucleon momentum carried by the (anti)quark which absorbs the virtual photon. There are several structure functions corresponding to various spin configurations, but the dominant one, usually called $f_2(x,Q^2)$, simply expresses the probability distribution in x of all q and $\bar{q}$ in the nucleon (valence and sea).

The x dependence of the structure functions cannot be calculated perturbatively from QCD because it reflects the nucleon wave function in terms of quarks and gluons, which is dominated by small momentum components. Thanks to DILS it can now be measured very well. Perhaps the most remarkable result of these measurements is that, in a large-momentum frame, all q and $\bar{q}$ in a nucleon carry only about half of the nucleon momentum, which implies that the other half must be carried by the gluons in the nucleon sea.

The Q^2 dependence of the structure functions is not only measurable, it can also be calculated at large Q^2 from perturbative QCD. Although it is small, it has provided some of the most successful tests of QCD (see Ref. 1a for further reading).

As suggested in Fig. 2A, nuclear effects in DILS can be studied by doing experiments on nuclear targets. This will be taken up in Sec. 4, where we shall discuss how the quark structure is affected by the "nuclear matter environment" which is present in lepton–nucleus collisions.

C. High Mass Dilepton Production in Hadronic Collisions (Drell–Yan Process)

The reaction of Fig. 2B, first proposed by S. Drell and T. M. Yan, brings us one elegant step beyond DILS. This time, instead of using an incident lepton to knock a quark out of a nucleon, one takes a purely hadronic reaction (e.g., pion-proton as in the figure) and concentrates on the very rare collisions producing a "dilepton", i.e., a pair of leptons of opposite electric charges, usually muons as in the figure. If the effective mass of the dilepton (i.e., its total energy in its own rest frame) is much larger than 1 GeV, the production mechanism is a hard process; a q of one incident hadron (the proton in the figure) and a $\bar{q}$ of the other (the pion) annihilate into a virtual photon (not shown) which then materializes into the dilepton. The process is obviously controlled by the structure functions (SF) of the two incident hadrons. Since the nucleon SF is known from DILS, the πp Drell–Yan process leads to an experimental determination of the pion SF. The validity of the argument can be tested by doing the same experiment for pp collisions where both SF are known (see Ref. 1a for further reading).

While the results agree qualitatively with expectations, discrepancies by about a factor of 2 (called the K factor by the specialists) appear in the comparison with perturbative QCD calculations. This shows that further effects are at work, which is not surprising since the incident hadrons can briefly interact before the $q\bar{q}$ annihilation, and the momentum transfers involved are not as large as in the best DILS experiments. The study of dilepton production has therefore led to a fruitful exploration of what are called *semihard effects*, i.e., QCD effects which are at the borderline of applicability of perturbation calculus and require the use of *ad hoc* approximations.

The Drell–Yan process is again such that nuclear environment effects can be studied by using nuclear targets (Fig. 2B). First results are mentioned in Sec. 4.

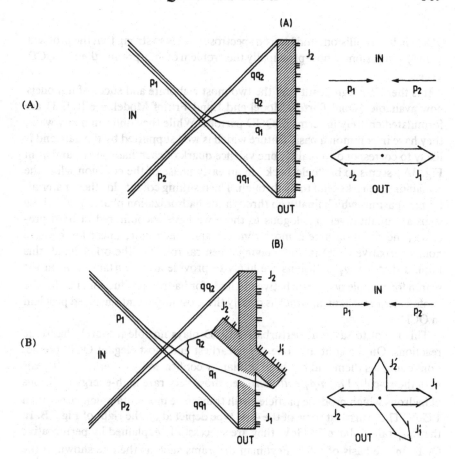

FIGURE 3　Hadronic collisions, low p_T (A) and high p_T (B).

D. Purely Hadronic Reactions

We now discuss hadron–hadron collisions at high E_{CM}. The most common type is sketched in Fig. 3A for two incident protons. It produces two hadronic jets J_1, J_2 flying out in directions very close to those of the incident hadrons, all the outgoing particles having low p_T, where p_T is defined as the momentum component perpendicular to the incident direction.

Curiously, these "*low-p_T collisions*", which have by far the largest cross sections and for which the largest amount of experimental information is available, turn out to be the most difficult ones to interpret. Theoretical work here is much more descriptive and less deductive than for hard processes. It extracts empirical regularities from the data and constructs *plausible dynamical models* able to reproduce them. A lack of consensus among model builders makes the situation rather confusing.[3] In my opinion, however, all the *other* evidence in favor of

QCD (in hard collisions and hadron spectroscopy) is so strong that the problem of low-p_T collisions is now principally the problem of interpreting them in QCD terms.

It is therefore significant that the two most elaborate and successful models now available (Dual Parton Model and Lund Fritiof Model, see Ref. 3) are formulated entirely in terms of QCD partons. While they differ in many ways, they have in common a basic feature which is well supported by the data and is likely to correspond to reality; one valence quark of each hadron (q_1 and q_2 in Fig. 3A) seems to be "held back" in an early phase of the collision while the remaining diquarks tend to fly through. The resulting color fields then materialize into partons which finally go through the hadronization phase. While these steps are qualitatively analogous to those we have encountered in hard processes, the situation here is more complex and much more uncertain because nonperturbative QCD is now playing a central role. On the other hand, this implies that low-p_T collisions are likely to provide an important input in the search for reliable nonperturbative methods for hadron production, i.e., for the confinement mechanism which is clearly the most important unsolved problem in QCD.

This is not to say that perturbative QCD plays no role in purely hadronic reactions. On the contrary, one of the earliest and most elegant QCD predictions concerns them and has been brilliantly confirmed by experiment. It deals with the so-called *high-p_T collisions*, i.e., those very rare high-energy collisions of hadrons which produce particles with transverse momenta much larger than 1 GeV. They turn out to be of the 4-jet type depicted on the right of Fig. 3B. In the E_{CM} range above 50 GeV, they are successfully explained by perturbative QCD on the basis of a few Feynman diagrams such as the one shown in the figure. In this example one valence quark q_1 of proton p_1 makes a high-momentum-transfer, large-angle collision (hard process) with one valence quark q_2 of proton p_2, the diagram shown corresponding to single gluon exchange. In addition to the jets J_1, J_2 generated by the remaining diquarks (which fly through and are similar to those of Fig. 3A), the QCD fields created sideways by q_1 and q_2 give rise to two further jets J'_1, J'_2 which fly off with large p_T. The distribution of J'_1 and J'_2 in angle and energy can be calculated perturbatively from the Feynman diagrams by making use of the structure functions of the incident hadrons as measured in DILS (see Sec. 3 B). The high-p_T collisions observed at the CERN proton–antiproton collider ($E_{CM} = 630$ GeV) give impressive agreement with theory.[4] The calculations show that at these high E_{CM} the hard process occurs not only between valence quarks as illustrated in Fig. 3B, but more frequently between sea partons, dominantly gluons. A high-energy hadron collider is therefore an excellent producer of jets initiated by gluons.

Finally, it should be noted that the separation between hard and soft collisions of hadrons is not sharp. There is a very important intermediate regime of semi-hard processes, still poorly known. As one reaches CM energies of several hundred GeV, more and more collisions show "minijets", i.e., hadronic jets of a few GeV emitted in transverse directions, the study of which will undoubtedly be useful for unraveling semihard QCD physics.

4. ULTRA-RELATIVISTIC NUCLEAR PHYSICS

We now get back to nuclear physics to discuss those very-high-energy processes where quarks *must* play an important role, simply because they are known to be essential for our understanding of the same processes on free nucleons. The problem here will be: is the quark behavior modified by the nuclear matter environment, or is it the same as in the case of free nucleons?

A. Deep Inelastic Lepton Scattering on Nuclei—The EMC Effect

The first modification of the quark behavior of nucleons in nuclei was found in 1982 by the European Muon Collaboration (EMC) using the CERN muon beam at laboratory energies 120 to 280 GeV, and comparing iron with deuterium targets.[5] Contrary to previous expectations, the ratio

$$R(x) = f_{Fe}(x)/f_D(x)$$

of the structure functions f_2 for the two nuclei (normalized to a single nucleon) was found to be substantially different from unity. Similar results were found soon afterwards with electrons at SLAC. This so-called *EMC effect* immediately attracted much theoretical attention, but it became apparent that different mechanisms are at work in various ranges of the variable x. Little is known so far on the Q^2 dependence, which we left out in the definition of R; the trends reported below are averaged over Q^2.

In Fig. 4 the band covering the x regions I, II, and III gives the overall trend of the data available in Summer 1987 (report by K. Rith at Quark Matter 87 Conference, Ref. 6). The dashed curve in region IV, where no measurements are yet available, is essentially kinematical. It gives the calculated effect of the Fermi motion of the nucleons in the target (see Fig. 1 of Ref. 5 for various theoretical estimates). One can easily understand its growth for x approaching unity. Because of the Fermi motion, a nucleon in a nucleus of baryon number A can carry more than a fraction $1/A$ of the momentum of the nucleus, so that x can exceed the value 1 which is its maximum for a free nucleon, and this effect is stronger for a heavy nucleus than for deuterium.

The interesting dynamical effects are concentrated in regions I, II, III. The decrease of R in region I (the most recent results show values below 1 at very small x) is interpreted as a shadowing effect. This effect is well known at low Q^2, and the interesting new point is that it seems to persist at Q^2 values well above 1 GeV^2. Regions II and III show the strongest deviations of R from 1. It is here that the number of proposed explanations is largest. An incomplete list of possibilities is:

(i) Nuclei contain (virtual) pions; their quarks, which have small x, contribute positively to R in region II.

(ii) Nucleons in nuclei have an effective mass smaller than the free mass m_p. This tends to reduce R, especially in region III.

(iii) Nucleons in nuclei may be "swollen" in size, with perhaps color conductivity effects (quark leakage between neighboring nucleons). This may increase R in II and decrease it in III.

(iv) The sharing of momentum between quarks and gluons in a nucleon may be modified by the nuclear-matter environment. This can affect R in either direction.

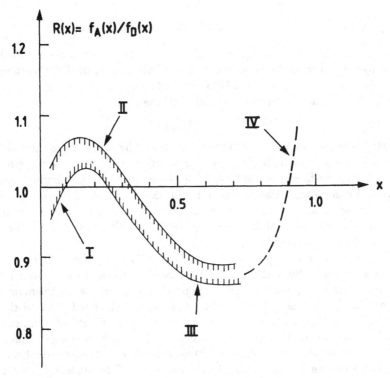

FIGURE 4   EMC effect.

With present data, there is little hope of reaching definite conclusions on the dynamical explanation of the complex behavior of $R(x)$, but further clues can be expected from high-precision measurements to be performed at CERN by the "New Muon Collaboration" over the ranges $x = 0.005 - 0.75$, $Q^2 = 1 - 200 \text{ GeV}^2$ and $A = 1 - 119$. This ambitious program may go a long way toward unraveling how the quark structure of nucleons is distorted inside nuclear matter.

B. Drell–Yan Process in Nuclei

Comparing Figs. 2A and 2B, one sees immediately that two nuclear effects must be expected in the Drell–Yan process on nuclei. Firstly, the structure function of the nucleon in the nuclear target is modified as in the EMC effect. Secondly, the incident pion must penetrate nuclear matter before its antiquark annihilates a quark of the target to produce the virtual photon which then materializes in the muon pair. Both effects have been detected recently by the NA10 Collaboration in a negative pion beam at CERN (laboratory energy of the pions 140 and 286 GeV), in a comparison of deuterium and tungsten targets.[7] The tungsten structure function is found to be modified similarly to the EMC effect. One observes in addition a broadening of the transverse momentum distribution of the muon pair, which is attributed to soft interactions of the incident pion in the target nucleus before the dimuon is produced. As in DILS, although the effects are small, systematic studies are possible by varying the target and the kinematical variables of the dimuon.

C. Ultra-Relativistic Nuclear Collisions

Our last topic in this section is high-energy hadron–nucleus and nucleus–nucleus collisions, i.e., the modifications of the processes considered in Sec. 3 D when one or both of the colliding objects is a nucleus. Since the 1960s, a number of hadron–nucleus experiments have been performed. For low-p_T collisions, the observed increase in the multiplicity of outgoing hadrons can be readily understood in terms of Glauber's multiple collision theory in nuclei, based on the eikonal approximation. The most interesting feature resulting from this analysis is that there is *very little cascading*, i.e., that hadrons produced in a collision on one nucleon in the nucleus very rarely create additional particles by colliding with a second nucleon. The only exception is a small amount of cascading for hadrons produced with energies which are low in the rest frame of the nucleus.

The near absence of cascading is easily understood in terms of relativistic time dilation. Soft-hadron production takes a time of order 1 fm/c in the rest frame of the produced particle (corresponding to momentum transfers of order 200 MeV/c). When the produced particle is relativistic with respect to the nucleus rest frame, time dilation implies that it forms mostly outside the nucleus, and cascading is then not possible. It would be of obvious interest to make this argument more quantitative. Very little has been done so far in this direction, partly because of the incompleteness of the experimental information, but also because of the lack of knowledge concerning the elementary collisions themselves.[8] Now that good models for low-p_T hadron–hadron collisions have become available (Dual Parton Model, Lund Fritiof Model, as mentioned in Sec. 3 D), more refined work should be possible if very accurate and complete data could be provided. An interesting contribution in this direction is found in Ref. 8 (it uses an earlier version of the Lund model).

Since the early 1980s, however, the interest has massively shifted toward the study of ultra-relativistic nucleus–nucleus collisions, mainly in the hope of creating very dense and hot blobs of hadronic matter. It is conceivable that, in such blobs, the QCD partons (q, $\bar{q}$ and gluons) could exist for times up to 10–20 fm/c as a disordered fluid, i.e., a *quark–gluon plasma* (QGP, see also Sec. 5), with all their color and spin degrees of freedom liberated, before the expansion and cooling of the blob forces the partons back into confinement as constituents of final-state hadrons. There are also theoretical reasons to believe that the quark vacuum condensates mentioned in Sec. 1 disappear in the QGP state.

After several years of purely theoretical speculations,[9] contact with reality began late in 1986 with the first heavy-ion experiments at Brookhaven (^{16}O and ^{28}Si beams, laboratory energy up to 15 GeV per nucleon) and at CERN (^{16}O and ^{32}S beams, up to 200 GeV per nucleon). First results, often still preliminary, were presented at the Quark Matter 87 Conference.[6] Not surprisingly, they indicate that common collisions, despite multiplicity and transverse energy distributions extending to spectacularly large values,[10] are essentially explainable by combining Glauber theory with good low-p_T nucleon–nucleon models (the transverse energy is the sum of $[m_i^2 + p_{Ti}{}^2]^{1/2}$ for all observed particles). In fact, small discrepancies will be taken care of by improvements in the nucleon–nucleon models, which still contain considerable flexibility (see Sec. 3 D). Conversely, the model calculations for normal collisions will, one hopes, provide guidance on how to select special events likely to have produced particularly hot and dense blobs of hadronic matter. This may give a more realistic basis for rediscussing the difficult problem of defining reliable signals for QGP formation, a question which up to now has had to be treated on highly uncertain theoretical grounds.

For hard processes, the most interesting result so far concerns the production of muon pairs (dimuons) in the CERN oxygen beam (200 GeV per nucleon) on a uranium target. The NA38 experiment[11] measured the ratio between dimuons from the decay of the J/psi meson (mass 3.1 GeV, quark composition $c\bar{c}$ with c the charmed quark) and those in the dimuon mass continuum around the J/psi. They find that this ratio decreases substantially when the transverse energy E_T of the underlying event is large (E_T is measured on neutrals only for instrumental reasons). One finds a reduction of the ratio by about a factor 2/3 when comparing $E_T > 50$ GeV to $E_T < 28$ GeV. In addition, the "J/psi suppression" is strongest at low transverse momentum p_T of the dimuon.

It is remarkable that this behavior had been predicted under the assumption that the J/psi is produced in a blob of QGP.[12] The main argument underlying the prediction is that the color force between the c and the $\bar{c}$ quarks is screened in the plasma (the partons in the QGP carry color charges and therefore produce the QCD analogue of Debye screening in an ordinary "electric" plasma). This weakens and may suppress the $c\bar{c}$ binding needed for J/psi formation, especially in high-E_T collisions which are more likely to contain a short-lived blob of QGP, and preferably at low p_T since the $c\bar{c}$ then stays longer in the blob.

It is nevertheless too early to conclude that the quark–gluon plasma has been seen, because J/psi suppression may be due to other causes. What is already clear, however, is that this type of experiment reveals remarkable effects of considerable size, and much more detailed work will undoubtedly be done in the years to come.

5. THE QUARK–GLUON PLASMA (QGP)

While the problems around the possible production and detection of blobs of QGP in nuclear collisions are still mostly unsolved, years of work in "lattice QCD" (i.e., the Monte Carlo computer treatment of QCD in a lattice approximation of space and imaginary time) have accumulated impressive theoretical evidence for the prediction that, at sufficiently high temperatures ($T \gtrsim 200$ MeV $= 2.3 \times 10^{12}$ K), hadronic matter in bulk melts into an equilibrium plasma of "deconfined" quarks and gluons, with the simultaneous disappearance of the quark vacuum condensates.[13]

No way is known to create macroscopic quantities of QGP in the laboratory, but the hot big bang theory of the expanding universe implies that at early times all hadronic matter must have been in the form of QGP. The problem of how this plasma of colored QCD partons made its transition to confined hadronic matter (eventually only nucleons) has recently attracted much attention among cosmologists; also the possibility of having some, perhaps abnormal, forms of cold QGP inside neutron stars has occupied astrophysicists. These interesting considerations, and more generally the indisputable fact that very hot and dense hadronic matter in bulk plays an important role in astrophysics and cosmology, provide additional reasons for a vigorous experimental program on ultra-relativistic nuclear collisions.

While there is little doubt that the early universe, at an age of perhaps 10–100 μsec after the big bang, went through the transition of QGP to confined hadronic matter, many scenarios are possible for the detailed course of events in this transition. As an example of current speculations in this field, we mention the work of J. H. Applegate et al.[14] who discuss the possibility that the hadronic phase transition may have created inhomogeneities in the space distribution of nucleons, with possible consequences up to the time of early nucleosynthesis (T around 10^9 K, age of the universe around one minute). The inhomogeneities would have appeared during the coexistence period of regions of plasma and of confined hadrons. While they would then have disappeared rapidly for the neutrons by diffusion, they would have survived much longer for the protons which, due to their electric charge, would have had very small mobility in the electric plasma prevailing at those times. It turns out that such effects may have important consequences for the dark matter problem which has become so central in modern astrophysics.

6. OUTLOOK

The future of particle physics is the theme of T. Applequist's lecture at this meeting. I shall therefore concentrate on the outlook for the further development of high energy nuclear physics, more precisely nuclear physics at the ultra-relativistic energies discussed in Sec. 4. This is where the dynamics of quarks and gluons dominate the reactions on single nucleons and where the quark and gluon effects in nuclei can therefore be studied best. With the experimental discovery of the EMC effect and the early theoretical work on the quark–gluon plasma in 1982–83, the interest in *ultra-relativistic nuclear physics* started to grow at a surprisingly fast rate, attracting numerous experimentalists and theorists from more sedate areas of nuclear and particle physics. The new field offers a very complex but also very rich domain of exploration. The instrumental and experimental challenges are great, requiring completeness and precision of measurement in an environment of very large multiplicities. For the theorists, the problems range from intricate phenomenological work on the new data to totally novel questions such as nonequilibrium processes in QCD at positive temperature. And for all physicists, much interest is added by the relevance to astrophysics and cosmology.

Experimentally, the future outlook is bright. At BNL (Brookhaven), after construction of the booster, the Alternating Gradient Synchrotron (AGS) will provide beams of very heavy nuclei with energies around 10 GeV per nucleon. If the Relativistic Heavy Ion Collider (RHIC) is approved, these beams will be accelerated further and collide against each other at energies up to 100 GeV per nucleon.

At CERN, a proposal is under consideration to extend the present ion program up to lead with beam energy 177 GeV per nucleon. The construction of the Large Electron–Positron Collider (LEP) is nearing completion. Its tunnel is so designed that it has room for the future addition of the LEP Hadron Collider (LHC). If approved, LHC will accelerate and collide against each other proton beams of up to about 8000 GeV = 8 TeV and lead beams of up to about 3.5 TeV per nucleon. In addition to collisions between the two proton and/or ion beams, LHC would also offer the possibility of colliding one of them with the LEP electron or positron beam (energy up to 100 GeV), making it a unique facility for probing the quark–gluon structure of protons and nuclei to extreme space–time resolution.

REFERENCES

1. For further reading on the Standard Model, see for example (a) F. Halzen and A. D. Martin, *Quarks and Leptons* (Wiley, New York, 1984), and (b) at a more theoretical level, St. Pokorski, *Gauge Field Theories* (Cambridge University Press, Cambridge, U.K., 1987).

2. M. Rho, Summary talk at Workshop on Skyrmions and Anomalies, Krakow, Poland, 20–24 February 1987 (to be published by World Scientific, Singapore); U. G. Meissner, "Quantum Chromodynamics at Nuclear Length Scales," Invited talk at Workshop on Low Energy Effective Theory of QCD (Nagoya, Japan, 13–14 April 1987).

3. For a very recent review, see Proc. Shandong Workshop on Multiparticle Production, edited by R. Hwa, Jinan, China, 28 June–6 July 1987 (to be published by World Scientific, Singapore).

4. See, e.g., A. K. Nandi in *Proc. of XVII Int. Symp. on Multiparticle Dynamics*, edited by M. Markytan *et al.*, Seewinkel, Austria, 16–20 June 1986 (World Scientific, Singapore, 1987), p. 709.

5. J. J. Aubert *et al.* (EMC Collaboration), Phys. Lett. **123B** (1983) 275.

6. Quark Matter 87 Conf., edited by H. Satz *et al.*, Nordkirchen, F.R. Germany, 24–27 August 1987 Z. Phys. **38** (1981) 1.

7. P. Bordalo *et al.* (NA10 Collaboration), Phys. Lett. B **193** (1987) 368 and 373.

8. A. Białas and M. Gyulassy, Nucl. Phys. B **291** (1987) 793.

9. See, e.g., Proc. Quark Matter 86 Conf., edited by L. S. Schroeder and M. Gyulassy, Asilomar, California 13–17 April 1986, Nucl. Phys. A **461**, Nos. 1 and 2 (1987).

10. See A. Bamberger *et al.* (NA35 Collaboration), Phys. Lett. B**184** (1987) 271, for an event with 267 charged particles and 80 GeV transverse energy in O + Pb, the oxygen beam having a laboratory energy of 200 GeV per nucleon.

11. M. C. Abreu *et al.* (NA38 Collaboration), presented by A. Bussière at Quark Matter 87 Conference, Ref.6.

12. T. Matsui and H. Satz, Phys. Lett. B**178** (1987) 416; F. Karsch and R. Petronzio, Phys. Lett. B **193** (1987) 105.

13. See, e.g., M. Fukugita, in *Lattice Gauge Theory Using Parallel Processors*, edited by X. Li *et al.* (Gordon and Breach, New York, 1987), p. 195, and for latest status, F. Karsch in Ref. 6.

14. J. H. Applegate and C. J. Hogan, Phys. Rev. D **31** (1985) 3037; J. H. Applegate *et al.*, Phys. Rev. D **35** (1987) 1151.

Particle Physics Beyond
1 TeV/ THOMAS APPELQUIST

During the 1970s, the experimental and theoretical work of many decades culminated in the creation of what has come to be known as the standard model of elementary particle interactions. The standard model is really two theories, one an unbroken gauge field theory of strong interactions (QCD), the other a spontaneously broken gauge field theory of the electroweak interactions. By providing a simple framework accommodating all the known elementary constituents of matter and describing their interactions, the creation of the standard model signals the end of an era. Quantitative tests have sometimes proved difficult, but a large body of experimental data is well described by the standard model and nothing stands in contradiction to it.

In this paper, I will try to describe the new era of deeper questioning and speculation that has now opened up. These questions lead us naturally to energy scales of 1 TeV and beyond. To appreciate the questions that are being asked, it is important to keep in mind what the standard model is—and what it is not. The mathematical equations of the standard model describe exactly the nature of the forces by which the elementary particles—the quarks, leptons, and gauge bosons—interact. The way in which the particles respond to these forces, however, depends also on the overall strength of the forces and on the masses of the particles. These ingredients in the standard model are described by a set of parameters, the coupling constants and masses. They are in no way specified by the standard model itself; instead they must simply be adjusted to fit the experimentally measured values.

Perhaps nothing shapes current research in particle theory more than the effort to understand why these parameters have their particular values. This is a search for new theories, deeper and more unified than the standard model. These theories often predict new phenomena taking place at tinier distances than we have probed so far, phenomena that might be revealed by the higher energies available at the next generation of large accelerators. Even if the direct predictions of a new theory involve distance scales not yet reached, if it can give some understanding of the structure of the standard model and the measured

THOMAS APPELQUIST *is a theoretical particle physicist. After holding a faculty appointment at Harvard University for a number of years, he took his current position as Professor of Theoretical Physics at Yale University, New Haven, Connecticut.*

values of its masses and coupling constants, it will have to be taken very seriously.

The list of standard-model parameters is full of curiosities. Among them is the near equality of the strange quark mass and the QCD confinement scale Λ. The former determines the mass of the K meson and the latter the mass of the proton. An explanation of why these masses are of the same order will almost surely require a unified treatment of the electroweak theory and QCD. An even older mystery is the ratio of the electron mass to the confinement scale Λ and therefore to the proton mass. Why is the electron about two thousand times lighter than the proton? This fact is not at all understood and yet the structure of all the atoms and molecules of matter depends on it.

The first step toward a deeper theory and the computation of the masses and coupling constants is the completion of the standard model itself. Although the electroweak part of the standard model has met every experimental challenge, including the 1982 discovery of the $W^{\pm}$ and Z^0 bosons, an important part of the theory remains unknown. That the W and Z bosons have masses at all means that the underlying local gauge symmetry that they share with the massless photon is partly broken. Why and how this happens, leaving the photon massless and making its W and Z partners very massive, is the outstanding unanswered question of the standard model.

A natural suggestion is that the W and Z masses, like the rest of the parameters of the standard model, must simply be introduced by hand and that their explanation must await the creation of a deeper theory. It can be shown using unitarity arguments, however, that this is impossible. The physical mechanism responsible for the W and Z masses must take place at energy scales not much different from the W and Z rest-mass energies themselves. In this sense, it is natural to view the mechanism as a part of the standard model. Whatever the mechanism is, it must reveal itself at energies no more than about ten times the rest-mass energies of the W and Z, about 1000 GeV. It is therefore accessible to direct experimental study, either with existing accelerators or with the new colliders now being built or planned.

While experimentalists have continued to search for some sign of this physics, theorists have explored a variety of possibilities. The classic idea, tracing its roots to the middle 1960s, is the introduction of a new set of elementary, spinless particles called Higgs bosons into the theory. The interactions of these particles can be arranged to trigger a spontaneous breakdown of the local gauge invariance of the underlying theory. Some of the bosons stay massless as they must, according to the Goldstone theorem, and they are then "eaten" by the gauge bosons, giving them the large mass observed experimentally. It was 't Hooft's study of this idea, beginning in 1971, and his demonstration that it would lead to a theory without consistency problems, that ushered in the modern era of gauge field theories in particle physics. Theories of this sort contain at least one new massive, physical particle, the Higgs boson, whose rest-mass energy is unknown except that it must be less than about 1000 GeV. The search for this particle has been under way for nearly fifteen years now, but it has so far not been seen. Since it couples very weakly to other particles, and since in the simplest version of the theory the particle carries no electric charge, it can be difficult to detect experimentally.

More elaborate theories with elementary Higgs particles have also been proposed. They can contain extra symmetries such as the Peccei–Quinn U(1) symmetry employed to explain the absence of *CP* violation in the strong interactions. The observation that this leads to a new light, neutral particle, the axion, has made such theories the center of much attention in recent years. The axion, too, has been searched for without success. The simplest extension of the standard model to include supersymmetric partners (so far unseen) leads naturally to this extra symmetry. Even larger symmetries and even more Higgs bosons have also been contemplated.

Even if elementary Higgs particles do not exist, spontaneous breaking of local gauge invariance can still be responsible for the masses of the $W^\pm$ and Z^0. The alternative is that the breaking can be driven by some deeper dynamics, in the same way that QCD brings about the spontaneous breaking of an approximate chiral $SU(2)_L \times SU(2)_R$ symmetry associated with the near masslessness of the u and d quarks. This breaking produces the effective masses of about 300 MeV that the quarks exhibit inside hadrons such as the proton. If the underlying symmetry were exact, there would have to be massless particles produced, according to the Goldstone theorem. Because of the small quark masses, the theory instead produces very light particles, the π mesons. Rather than being elementary, however, they are composites—quark–antiquark bound states.

The possibility that a similar mechanism could be responsible for the formation of the Higgs particles absorbed by the $W^\pm$ and Z has been considered throughout the development of the electroweak theory. Much of this work has grown out of a theoretical problem with the elementary Higgs idea, along with the fact that no Higgs particles have been seen experimentally. Theories with elementary Higgs particles have a suspicious character. If a given mass is initially assigned to the particle, the quantum-mechanical interactions tend to raise that mass far beyond its initial value. Although the massless Higgs bosons will not be affected by this problem, their partner, the physical Higgs boson, will. A rest-mass energy substantially less than 1000 GeV can be achieved by a mysterious renormalization, but this mathematical operation is regarded as unnatural, not likely to be allowed by nature. Beyond this theoretical hierarchy problem, the elementary Higgs boson theory seems to offer no hint of the deeper theory we are looking for. Since each fermion mass is simply put into the theory by adjusting the coupling of the Higgs boson to that fermion, no understanding of these masses is provided.

A simple idea for dynamical symmetry-breaking that has attracted a great deal of attention has come to be known as Technicolor. The Higgs bosons are imagined to be bound states of a new fermion and its antifermion interacting through a new asymptotically free gauge field. The theory has the same structure as QCD except that the confinement scale is approximately 1000 times greater—about 250 GeV. The Higgs bosons are then like the pions of QCD.

Here, however, they are absorbed by the gauge fields, and because of the large confinement scale they can produce the observed large gauge-boson masses. The QCD pions could not have done this. The natural mass scale of QCD—on the order of 1 GeV—is far too low to explain the $W^{\pm}$ and Z^0 masses.

Technicolor theories make the exciting prediction that a new kind of matter, similar in some ways to the known hadrons, should exist with a rest-mass energy of order 1000 GeV. This is a striking alternative to the elementary Higgs idea. The new physics demanded by the incompleteness of the standard model is there, but the new matter would be as heavy as the general arguments allow, and it could have wonderfully rich structure. If the idea is correct, the new colliders designed to open up this mass range will have a great deal to discover.

The major problem on the horizon for technicolor theories is the generation of fermion mass. In the elementary Higgs theory, this is incorporated through the direct coupling of the Higgs particles to the fermions. This must somehow be mimicked in technicolor theories by a coupling of the technifermions to the ordinary fermions. Most of the recent work on technicolor theories has been an effort to do this in a way that seems natural, that gives reasonable fermion masses, and that does not lead to processes ruled out by experiment. It has so far not been possible to do all these things. In particular, the new fermion-mass-generating interactions seem to bring with them the notorious flavor-changing neutral currents that plagued weak-interaction theories before the development of the GIM (Glashow, Iliopoulos, and Maiani) mechanism. The incorporation of a simple GIM mechanism into technicolor theories remains a major challenge for theorists.

Beyond the standard model and its completion lie a host of unanswered questions. Are there more quarks and leptons bringing with them more masses to understand? Are there more gauge bosons? Do the neutrinos have nonzero masses, too small to have been detected so far? Why is parity maximally violated in the electroweak theory? What is the origin of *CP* violation? Are the quarks and leptons composed of more fundamental objects? Does supersymmetry play a role at accessible energies, leading to a new set of observable partners for each of the known particles? Can the strong and electroweak theories be unified into a single grand unified theory?

While the energy scales associated with these phenomena are unknown, they could be low enough to be directly probed by a multi-TeV collider. Let me conclude my paper by describing just a few of the above possibilities:

(1) Are there more quarks and leptons?

If the top quark is discovered, three families of quarks and leptons will populate the standard model. There may be more, perhaps arranged in similar families, but so far no real insight into this question exists. What is needed is a "theory of flavor," a dynamical framework accounting for the number of quarks and leptons, perhaps predicting new ones, and very likely including transitions between the generations—flavor-changing neutral currents. The experimental search for transitions among the known fermions—processes such as $K^0 \to \mu^+ e^-$ and $\mu \to e\gamma$—continues at laboratories around the world. Until higher energies become available, these experiments may be our best hope for the new information we may need to construct a theory of flavor.

(2) Are the quarks and leptons composed of more fundamental objects?

The standard model describes the quarks and leptons as pointlike structure-less particles. We know this to be an accurate description down to at least 10^{-16} cm, but a composite structure could be revealed at smaller distances. Theoretical speculation along these lines is motivated partly by the seeming complexity of the list of standard-model parameters. There are already so many quarks and leptons that some theorists find it difficult to believe that they can be truly fundamental. The many families hint at the kind of level structure that has appeared before, in systems that we now know to be composite—atoms, nuclei and hadrons. There is, however, an important difference between these systems and the quarks and leptons. The energy required to break up the atoms, nuclei, or hadrons is either much less than or, in the case of the hadrons, on the order of the relativistic rest energy of the system. If the quarks and leptons are composite, they are far more deeply bound. Because they appear pointlike down to at least 10^{-16} cm, the rules of quantum mechanics tell us that an energy of at least 100 GeV is required to be sensitive to the composite structure. This is several orders of magnitude more than the rest energies of these particles! If the quarks and leptons are composite, they are quite different from any composite structures we have seen before.

(3) Does supersymmetry play a role at accessible energies?

In supersymmetric theories, bosons and fermions enter as partners and share certain properties. Even though there is not one shred of experimental evidence for supersymmetry, the theories have such a compelling beauty that nearly everyone believes they will play a role at some level. Supersymmetry is now almost universally employed in the effort to unify gravity with gauge field theories. There, the symmetry would be completely manifest at the Planck length $\simeq 10^{-33}$ cm, a distance scale that could only be probed with energies on the order of 10^{19} GeV.

While some breaking of this supersymmetry is expected at these large energies, it is attractive to imagine that some of it survives at much lower energies. It has not been seen at 100 GeV, the highest energy probed so far, but it could well appear in the soon-to-be-explored 1000 GeV range. Perhaps the simplest possibility is that a supersymmetric version of the standard model will manifest itself at or below these energies. Each known particle would then be associated with a superpartner—a spin $\frac{1}{2}$ photino for the photon, a spin 0 squark for the quark, etc. The supersymmetry would be broken, but only on the scale of a few hundred GeV, not 10^{19} GeV.

If the experimental study of the multi-TeV regime reveals evidence for any of these phenomena, it will be very exciting. Apart from these possibilities, however, I have tried to emphasize that one thing is guaranteed to be revealed by this kind of experimental study. The mechanism of electroweak symmetry breaking (the Higgs mechanism) must manifest itself at or below energies of a few TeV. The experiments will tell us whether an elementary Higgs boson or some deeper dynamical structure is responsible for this symmetry breaking, giving mass to the $W^{\pm}$ and Z^0. There is also very good reason to believe that multi-TeV experiments will give some direct insight into the mechanism of fermion mass generation. One can hope that this insight will play a role in the construction of a theory of flavor and the computation of fermion masses.

Acknowledgments

The American Physical Society, the American Institute of Physics, and the National Academy of Sciences wish to express their appreciation to the following for their significant contributions in support of the 1987 joint meeting of the XIX General Assembly of the International Union of Pure and Applied Physics and the annual meeting of the AIP Corporate Associates.

Federal Agency Grants

The National Science Foundation
The Department of Energy

Voluntary Contributors from Corporate Associates

Arthur D. Little, Inc.
AT&T Bell Laboratories
Ball Technical Products Group
Bell-Northern Research, Ltd.
Boeing Company
Calspan Corporation
Celanese Research Company
COMSAT Laboratories
Control Data Corporation
Eastman Kodak Company
Elsevier North-Holland, Inc.
Exxon Res. & Eng'g. Co.
Ford Motor Company
General Electric Company
General Motors Research Lab.
Hitachi, Ltd.
Honeywell, Inc.
Hughes Aircraft Company
IBM Corporation
Imperial Chemical Industries

Knowles Electronics, Inc.
LeCroy Research Systems Corp.
Lockheed Corporation
Marathon Oil Company
Monsanto Company
Northrup Corporation
Olivetti & C,s.p.A.
Philips Laboratories
Phillips Petroleum Company
The Rand Corporation
David Sarnoff Res. Ctr.
Rockwell International
Schlumberger-Doll Research
Scientific American
Sony Corporation of America
Springer-Verlag New York
Standard Oil Company
Westinghouse Electric Company
Xerox Company

NATIONAL ORGANIZING COMMITTEE FOR THE 1987 IUPAP GENERAL ASSEMBLY

D. Allan Bromley (Chairman)
Wright Nuclear Structure Lab
Yale University

Mildred S. Dresselhaus
Massachusetts Institute of Technology

Kenneth W. Ford
American Institute of Physics

Anthony P. French
Massachusetts Institute of Technology

Melvin Gottlieb
Princeton University Plasma Physics Lab

William H. Havens, Jr.
The American Physical Society

Robert Hofstadter
Department of Physics
Stanford University

John R. Klauder
AT&T Bell Laboratories

H. William Koch
Director Emeritus
American Institute of Physics

Robert Park
The American Physical Society
Office of Public Affairs

NRC STAFF

Donald C. Shapero
Staff Director

Helene E. Patterson
Admin. Specialist
Board on Physics and Astronomy
National Research Council

Scientific Program

General Assembly, International Union
of Pure and Applied Physics

Annual Meeting, Corporate Associates
American Institute of Physics

30 September–2 October 1987
National Academy of Sciences, Washington, DC

SESSION I: 9:00, Wednesday, 30 September

Presiding: Larkin Kerwin
Université Laval, Canada

Welcome and Overview

A. Welcome and Introductory Remarks:

Frank Press	President, National Academy of Sciences
Val Fitch	President, American Physical Society
Hans Frauenfelder	Chairman of the Board, American Institute of Physics

B. The State of Physics, 1987: A Tour d'Horizon

D. Allan Bromley President, International Union of Pure and Applied Physics

SESSION II: 14:00, Wednesday, 30 September

Presiding: Praveen Chaudhari
IBM Thomas J. Watson Research Center

The Roots of High Technology

Pierre Aigrain	The Roles of Government
H.B.G. Casimir	Industrial Technology, Knowledge and Skills
	Break
Hans Mark	The International Exploration of Space
Joseph P. Allen	Physics at the Edge of the Earth
Joseph Edward Demuth	The Scanning Tunneling Microscope

SESSION III: 9:00, Thursday, 1 October

Presiding: John Rowell
Bell Communications Research

Understanding and Tailoring Matter

Alfred Cho	Artificially Structured Materials
Michael Fisher	Phases and Phase Transitions in Less than Three Dimensions
	Break
Horst L. Stormer	The Quantized Hall Effect
C.W. Chu	Superconductors above 90 Kelvin
Yu. A. Ossipyan	Superconductivity and its Applications (Modern and Traditional Approach)

14:00, Thursday, 1 October

Government Laboratory Tours for AIP Corporate Associates Representatives

SESSION IV: 9:00, Friday, 2 October

Presiding: Denys Wilkinson
University of Sussex, UK

Frontier Applications of Physics

Hans Frauenfelder	Physics and Biology
Paul Lauterbur	Medical Imaging
	Break
Paul Fleury	Physics and the Information Age
Brian W. Petley	Towards the Limits of Precision and Accuracy in Measurement

SESSION V: 14:00, Friday, 2 October

Presiding: K. C. Chou
Academia Sinica

Frontiers of Modern Physics

Harold Furth	High-Temperature Plasma Physics
Daniel Kleppner	Atomic Physics Today
	Break
Leon Van Hove	Quarks and Gluons in Nuclear and Particle Physics
Thomas Appelquist	Particle Physics Beyond 1 TeV

Printed in the United States
By Bookmasters